Methods in Molecular Biology

Series Editor
John M. Walker
School of Life and Medical Sciences
University of Hertfordshire
Hatfield, Hertfordshire, AL10 9AB, UK

For further volumes:
http://www.springer.com/series/7651

Imaging Flow Cytometry

Methods and Protocols

Edited by

Natasha S. Barteneva

Cellular and Molecular Medicine Program, Boston Children's Hospital, Boston, MA, USA; Department of Pediatrics, Harvard Medical School, Boston, MA, USA

Ivan A. Vorobjev

Department of Cell Biology and Histology, M.V. Lomonosov Moscow State University, Moscow, Russia; A.N. Belozersky Institute of Physico-Chemical Biology, M.V. Lomonosov Moscow State University, Moscow, Russia

Humana Press

Editors
Natasha S. Barteneva
Cellular and Molecular Medicine Program
Boston Children's Hospital
Boston, MA, USA

Department of Pediatrics
Harvard Medical School
Boston, MA, USA

Ivan A. Vorobjev
Department of Cell Biology and Histology
M.V. Lomonosov Moscow State University
Moscow, Russia

A.N. Belozersky Institute of Physico-Chemical Biology
M.V. Lomonosov Moscow State University
Moscow, Russia

ISSN 1064-3745 ISSN 1940-6029 (electronic)
Methods in Molecular Biology
ISBN 978-1-4939-3300-6 ISBN 978-1-4939-3302-0 (eBook)
DOI 10.1007/978-1-4939-3302-0

Library of Congress Control Number: 2015953670

Springer New York Heidelberg Dordrecht London

Printed on acid-free paper

Humana Press is a brand of Springer
Springer Science+Business Media LLC New York is part of Springer Science+Business Media (www.springer.com)

Dedication

To the memory of my grandmothers and to my dad whose continuous optimism through their suffering and success of their lives inspired me in doing science.

Ivan A. Vorobjev

To my daughter, Sasha, for her enduring love and courage.

Natasha S. Barteneva

Preface

This will be the first time the cutting-edge technology of imaging flow cytometry appears in the *Methods in Molecular Biology* series. Recent development of the quantitative imaging flow cytometry (IFC) has revolutionized our ability to analyze cells, cellular clusters, and populations in a remarkable fashion. The aim of this book is to present to the scientific community the compendium of IFC techniques and protocols that have been recently developed.

This volume is organized such that the initial part (*Part I: General*) provides a basic introduction to imaging flow cytometry and its different variants. The first chapter is a general overview of fluorescent techniques used to study cells and the comparison of IFC with its most similar methods, i.e., fluorescent microscopy and conventional flow cytometry. In the second chapter, the principles of the Imagestream imaging cytometer that is now responsible for a lot of studies using IFC are described by its inventor, David A. Basiji. However, to give a maverick guide on IFC, Chapters 3 through 4 describe other IFC instrumentation.

The following part of the volume contains a variety of protocols that we grouped into the following sections:

Part II (Chapters 5–9) explains protocols for studies on the cell nucleus, nucleic acids, and FISH techniques using an IFC instrument.

Part III (Chapters 10–14) is devoted to the immune response analysis and drug screening including detailed analysis of apoptosis.

Part IV (Chapters 15–16) deals with IFC protocols for apoptosis and cell death analysis.

Part V (Chapters 17–20) deals with morphological analysis and the identification of rare cells.

This volume follows the popular format of the *Methods in Molecular Biology* series by providing step-by-step instructions that can be either directly used or readily adapted by the researchers to the design of their own experiments. The majority of IFC protocols included in the current volume also describe the very complicated IDEAS software, and we think it will be especially useful to the neophytes to see personal emphasis on the issue to start designing their own experiments.

This volume brings together the expertise and dedication of the accomplished international panel of scientists that made this book possible by sharing details of their favorite techniques with the scientific community. The authors of individual chapters are well-known scientists who have made major contributions toward the continued development of this exciting technology, and all of them have provided a detailed description of their protocols and shared important tips and tricks necessary for reproducible initiation of their laboratory protocols by other research teams. We thank all of the contributors and the MiMB Series Editor, John Walker, for their time and effort. It is our hope that this book will be a critical source for all laboratories seeking to implement IFC for their research studies.

Boston, MA, USA — *Natasha S. Barteneva*
Moscow, Russia — *Ivan A. Vorobjev*

Contents

Contributors

RAFFAELA ABBRIANO • *Marine Biology Research Division, Scripps Institution of Oceanography, University of California, San Diego, CA, USA*

BENJAMIN ALDERETE • *Amnis part of EMD Millipore, Seattle, WA, USA*

CATARINA R. ALMEIDA • *Institute for Research and Innovation in Health, Instituto de Engenharia Biomédica (INEB), University of Porto, Porto, Portugal*

YELDAR BAIKEN • *Nazarbayev University, Astana, Kazakhstan; Cellular and Molecular Medicine Program, Boston Children's Hospital, Boston, MA, USA*

DANIEL R. BARREDA • *Department of Biological Sciences, University of Alberta, Edmonton, AB, Canada; Department of Agricultural, Food and Nutritional Science, University of Alberta, Edmonton, AB, Canada*

NATASHA S. BARTENEVA • *Cellular and Molecular Medicine Program, Boston Children's Hospital, Boston, MA, USA; Department of Pediatrics, Harvard Medical School, Boston, MA, USA*

DAVID A. BASIJI • *Department of Bioengineering, University of Washington, Seattle, WA, USA*

LINDSAY A. BEATON-GREEN • *Radiobiology, Consumer and Clinical Radiation Protection Bureau, Health Canada, Ottawa, ON, Canada*

EDUARD J. VAN BEERS • *Department of Clinical Hematology, Academic Medical Centre, Amsterdam, The Netherlands*

ORNA COOK • *Marine Biology Research Division, Scripps Institution of Oceanography, University of California, San Diego, CA, USA*

AUBREY DAVIS • *Marine Biology Research Division, Scripps Institution of Oceanography, University of California, San Diego, CA, USA*

WILLIAM DAY • *FACS Laboratory, London Research Institute, Sir Francis Crick Institute, Holborn, UK*

ANA M. DIAS • *Institute for Research and Innovation in Health, Institute of Molecular Pathology and Immunology, University of Porto (IPATIMUP), Porto, Portugal; Institute of Biomedical Sciences of Abel Salazar (ICBAS), University of Porto, Porto, Portugal*

ELIZAVETA FASLER-KAN • *Department of Biomedicine, University of Basel and University Hospital Basel, Basel, Switzerland; Department of Pediatric Surgery and Department of Clinical Research, Inselspital, University Hospital, University of Bern, Bern, Switzerland*

KLEBER Y. FERTRIN • *Department of Clinical Pathology, School of Medical Sciences, University of Campinas, Campinas, Brazil; Hematology and Hemotherapy Center—Hemocentro Campinas, University of Campinas, Campinas, Brazil*

ANDREW FILBY • *Flow Cytometry Core Facility, Newcastle Biomedicine, Newcastle University, Newcastle-upon-Tyne, UK*

BRIAN E. HALL • *EMD Millipore, Seattle, WA, USA*

ANDREA L. HENNING • *Applied Physiology Laboratory, University of North Texas, Denton, TX, USA; Department of Biological Sciences, University of North Texas, Denton, TX, USA*

MARK HILDEBRAND • *Marine Biology Research Division, Scripps Institution of Oceanography, University of California, San Diego, CA, USA*

GREGORY J. KATO • *Division of Hematology/Oncology, Vascular Medicine Institute, University of Pittsburgh, Pittsburgh, PA, USA*

ANDY K.S. LAU • *Department of Electrical and Electronic Engineering, Faculty of Engineering, The University of Hong Kong, Pokfulam, Hong Kong*

YAN QUAN LEE • *Department of Microbiology, National University of Singapore, Singapore, Singapore; NUS Graduate School for Integrative Sciences and Engineering, National University of Singapore, Singapore, Singapore*

ORLA MAGUIRE • *Department of Flow and Image Cytometry, Roswell Park Cancer Institute, Buffalo, NY, USA*

KALPANA MANANDHAR-SHRESTHA • *Marine Biology Research Division, Scripps Institution of Oceanography, University of California, San Diego, CA, USA*

NURIA MARTINEZ-MARTIN • *Lymphocyte Interaction Laboratory, London Research Institute, Sir Francis Crick Institute, Holborn, London, UK*

J. PHILIP MCCOY JR. • *NHLBI Flow Cytometry Core Facility, NHLBI, NIH, Bethesda, MD, USA*

BRIAN K. MCFARLIN • *Applied Physiology Laboratory, University of North Texas, Denton, TX, USA; Department of Biological Sciences, University of North Texas, Denton, TX, USA*

KATHLEEN E. MCGRATH • *Department of Pediatrics, Center for Pediatric Biomedical Research, University of Rochester Medical Center, Rochester, NY, USA*

LAUREL MENDELSOHN • *Hematology Branch, NHLBI, NIH, Bethesda, MD, USA*

HANS MINDERMAN • *Department of Flow and Image Cytometry, Roswell Park Cancer Institute, Buffalo, NY, USA*

LISA M. NISWANDER • *Department of Pediatrics, Center for Pediatric Biomedical Research, University of Rochester Medical Center, Rochester, NY, USA; Department of Pathology and Laboratory Medicine, University of Rochester Medical Center, Rochester, NY, USA*

JAMES PALIS • *Department of Pediatrics, Center for Pediatric Biomedical Research, University of Rochester Medical Center, Rochester, NY, USA; Department of Pathology and Laboratory Medicine, University of Rochester Medical Center, Rochester, NY, USA*

SALOMÉ S. PINHO • *Institute for Research and Innovation in Health, Institute of Molecular Pathology and Immunology, University of Porto (IPATIMUP), Porto, Portugal; Institute of Biomedical Sciences of Abel Salazar (ICBAS), University of Porto, Porto, Portugal*

NICOLE J. POULTON • *Bigelow Laboratory for Ocean Sciences, East Boothbay, ME, USA*

ERIC A. PRADO • *Applied Physiology Laboratory, University of North Texas, Denton, TX, USA; Department of Biological Sciences, University of North Texas, Denton, TX, USA*

HALEY R. PUGSLEY • *Amnis part of EMD Millipore, Seattle, WA, USA*

SUKHVEER PUREWAL • *FACS Laboratory, London Research Institute, Sir Francis Crick Institute, Holborn, UK*

CELSO A. REIS • *Institute for Research and Innovation in Health, Institute of Molecular Pathology and Immunology, University of Porto (IPATIMUP), Porto, Portugal; Institute of Biomedical Sciences of Abel Salazar (ICBAS), University of Porto, Porto, Portugal; Medical Faculty, University of Porto, Porto, Portugal*

AJA M. RIEGER • *Department of Biological Sciences, University of Alberta, Edmonton, AB, Canada*

LEIGH SAMSEL • *NHLBI Flow Cytometry Core Facility, NHLBI, NIH, Bethesda, MD, USA*

YVONNE SAMSTAG • *Institute of Immunology/Section Molecular Immunology, University of Heidelberg, Heidelberg, Germany*

EVA L. SÁNCHEZ-ALVAREZ • *Marine Biology Research Division, Scripps Institution of Oceanography, University of California, San Diego, CA, USA*

ROSHAN P. SHRESTHA • *Marine Biology Research Division, Scripps Institution of Oceanography, University of California, San Diego, CA, USA*

HO CHEUNG SHUM • *Department of Mechanical Engineering, Faculty of Engineering, The University of Hong Kong, Pokfulam, Hong Kong*

SARAH R. SMITH • *Integrative Oceanography Division, Scripps Institution of Oceanography, University of California, San Diego, CA, USA*

KEVIN S.W. TAN • *Department of Microbiology, National University of Singapore, Singapore, Singapore; NUS Graduate School for Integrative Sciences and Engineering, National University of Singapore, Singapore, Singapore*

JESSE C. TRALLER • *Marine Biology Research Division, Scripps Institution of Oceanography, University of California, San Diego, CA, USA*

KEVIN K. TSIA • *Department of Electrical and Electronic Engineering, Faculty of Engineering, The University of Hong Kong, Pokfulam, Hong Kong*

ADAM S. VENABLE • *Applied Physiology Laboratory, University of North Texas, Denton, TX, USA; Department of Biological Sciences, University of North Texas, Denton, TX, USA*

IVAN A. VOROBJEV • *Department of Cell Biology and Histology, M.V. Lomonosov Moscow State University, Moscow, Russia; A.N. Belozersky Institute of Physico-Chemical Biology, M.V. Lomonosov Moscow State University, Moscow, Russia*

GUIDO H. WABNITZ • *Institute of Immunology/Section Molecular Immunology, University of Heidelberg, Heidelberg, Germany*

PAUL K WALLACE • *Department of Flow and Image Cytometry, Roswell Park Cancer Institute, Buffalo, NY, USA*

RUTH C. WILKINS • *Radiobiology, Consumer and Clinical Radiation Protection Bureau, Health Canada, Ottawa, ON, Canada*

RANDALL R. WILLIAMS • *Applied Physiology Laboratory, University of North Texas, Denton, TX, USA; Department of Biological Sciences, University of North Texas, Denton, TX, USA*

KENNETH K.Y. WONG • *Department of Electrical and Electronic Engineering, Faculty of Engineering, The University of Hong Kong, Pokfulam, Hong Kong*

TERENCE T.W. WONG • *Department of Electrical and Electronic Engineering, Faculty of Engineering, The University of Hong Kong, Pokfulam, Hong Kong; Department of Biomedical Engineering, Washington University in St. Louis, St. Louis, MO, USA*

Part I

General

Chapter 1

Quantitative Functional Morphology by Imaging Flow Cytometry

Ivan A. Vorobjev and Natasha S. Barteneva

Abstract

This chapter describes advantages and limitations of imaging flow cytometry (IFC) based on Imagestream instrumentation using a hybrid approach of morphometric measurement and quantitation of multiparametric fluorescent intensities' distribution in cells and particles. Brief comparison is given of IFC with conventional flow cytometry and fluorescent microscopy. Some future directions of the IFC technology are described and discussed.

Key words Imaging flow cytometry, Flow cytometry, Quantitative morphology, Spectral compensation, Autofluorescence, Automated image analysis

1 Introduction

The history of imaging in the biomedical sciences began with the invention of microscope in seventeenth century. Over the last 50 years, microscope technology has advanced by the development of sophisticated optics, digital cameras and different modes of observation that ignited a new era in cell biology, diagnostics, and related areas. Currently, the advanced features available in automated fluorescent and confocal microscopes have made them indispensable tools for cell and tissue analysis.

The next step in cellular analysis technology was invention of flow cytometer in the 70th years of last century. Flow cytometry allowed for the first time analysis of the large cell populations, however, the flow cytometer does not give real images of the analyzed cells. The technique combining features of advanced fluorescent microscopy and analytical tools of flow cytometry was laser scanning cytometer (LSC) introduced in late 90th. LSC software allows multiparametric analysis of individual cells attached to the surface of microscopic slides or multiwell plate. However this

Natasha S. Barteneva and Ivan A. Vorobjev (eds.), *Imaging Flow Cytometry: Methods and Protocols*, Methods in Molecular Biology, vol. 1389, DOI 10.1007/978-1-4939-3302-0_1,

technique was still limited by the number of cells analyzed and did not allow direct analysis of live cells in suspension.

The Imagestream system has been developed as the next stage of hybrid technique, based on direct imaging of thousands of cells in suspension and sophisticated analysis of the obtained datasets. Nowadays imaging flow cytometry is a big advance in the field cell analysis with regards to novel applications and method complementarity to conventional flow cytometry and fluorescent microscopy. Advances in instrumentation for imaging cytometry are fueling a renaissance in the fluorescence-based analysis of cellular morphology and cellular heterogeneity.

Quantitation is becoming highly important in the area of morphological parameters measurement and the measurement of fluorescent and chromogenic intensities as the questions being asked become increasingly complex and require multimodal and multicolor cellular analysis.

The major advantage of microscopy is high spatial resolution (down to 0.2 μm using high-end optics and detectors). Sensitivity of modern microscopes allows identification of a single molecule of fluorochrome. However, primary objects for microscopic analysis are cells firmly attached to the substrate or fixed prior to observation. Current techniques in microscopy available to quantify fluorescent images rely on MetaMorph (Molecular Devices, Sunnyvale, CA), ImagePro (Media Cybernetics Inc.), and open-source software such as Image J (NIH) and CellProfiler (Broad Institute, MA) and similar software packages providing script-based manual analysis. One of the key problems in automated microscopic analysis is image segmentation [1]. Automated image analysis in microscopy is under development for last 40 years, however is still limited by the necessity of image segmentation. Lack of capabilities of automated image analysis makes it almost impossible to manipulate with large datasets of thousands of cells and usually microscopic analysis includes not more than tens or (rarely) several hundreds of events. Another restriction of fluorescent microscopy is a problem of spectral overlap of fluorescent probes that limits currently routine observations with 3–4 fluorescent dyes. Simultaneous use of larger number of fluorescent probes is possible technically; however it requires development of specialized software tools and high-skilled personnel for image analysis.

The major advantages of flow cytometry are (1) high-throughput analysis for up to millions of events and (2) capability of simultaneous use of numerous fluorescent markers [2]. The major drawbacks of flow cytometry are (1) lower sensitivity compared to the microscope—only simultaneous signal coming from 20 to 100 molecules could be recorded and (2) lack of spatial information on the distribution of a probe on the cells surface and/or in the cell volume.

The current dogma of flow cytometry states that it is based on the analysis of single events. However, in some cases this is not true, for example, doublet discrimination based on cellular width (FCS-W) does not allow separation of large activated cells from a pair of small cells. Besides, in the cases when live cells could be associated with dead cells and apoptotic bodies [3], adherent platelets/platelet fragments [4], or extracellular vesicles (EVs) capabilities of flow cytometry are limited. Also this method does not allow the analysis of multiple events and cellular interactions—issues becoming increasingly popular in modern research.

1.1 Flow Cytometry: Problems in Data Interpretation

Under optimal conditions the quantification of molecular target, fluorescent intensity and its spatial distribution are expected to reflect the true distribution of molecular target. However, the labeling can be misrepresentative of distribution and quantity of target molecules because the fluorescent counts and distribution will include specific and nonspecific binding of antibodies and dyes, and multicolor combinations will take in consideration spectral compensation and nonspecific background. High level of multiplexing and multicolor cytometry have been a goal of instruments development in cytometry leading to instruments with as much as 17 fluorescent channels [2] and recently to the development of mass cytometry (CyTOF) (up to 52 parameters [5]). However, multicolor fluorescent staining may be an important potential source of bias. Adding fluorescent parameters into the panel leads to increase of background and diminishes signal-to-noise ratio because of the compensation.

In multicolor panels because of significant spillover between several channels, true MFI ($MFI_{compensated}$) might be significantly less than originally measured by the instrument, while background staining remains the same or increases. As a result staining index could become rather low. The following recommendations are useful to minimize the problem: (1) minimize compensation by using spectrally separated fluorophores; (2) use multiple lasers to spread your fluorochromes between them. Take into account that compensation is always higher for the same pair of dyes when excited by one laser than interlaser compensation; (3) try not to use tandem dyes—they always require significant compensation with the donor dye channel.

1.2 Autofluorescence

Autofluorescence is a significant problem in conventional flow cytometry [6], particularly when cells expressing green fluorescent protein or labeled with a green fluorophore-conjugated AB (such as FITC and Alexa 488), and may lead to false-positive results [7]. Though using a narrow-band bandpass filters [8] helps to minimize autofluorescence problem in conventional flow cytometry settings, often morphological analysis and more rigorous methods such as IFC and/or fluorescent microscopy need to be employed to characterize cell population and increase accuracy of distinguishing false-positive and false-negative events.

The *quantitative imaging flow cytometry* (IFC) is a hybrid technology, which is using advantages of microscopy and cytometry in order to overcome limitations related to these methods. IFC allows overcoming the subjective bias and limited statistics of microscopy and absence of spatial distribution and morphological features associations for fluorescent parameters of flow cytometry.

IFC generates of galleries of cell images, has spatial resolution down to 0.3 μm (bright field) and allows rather rapid analysis of thousands to tens of thousands events. It has sensitivity in a range of 20–100 fluorescent molecules that is the same as in conventional flow cytometry. In IFC the problem of image segmentation is mainly solved by putting cell suspension into the stream, where cell-cell average distance is much larger than cell diameter and thus single events are separated by the instrument during data acquisition.

2 Fluorescent Probes for IFC

Fluorescent probes and ways of labeling in IFC are the same as for conventional flow cytometry and selection of dyes might be different from fluorescent microscopy. There are a number of methods by which a fluorescent probe can be generated and introduced in the cell, including covalent linking antibodies with fluorophore of the choice, genetic manipulation to label a protein with fluorophore [9], use of fluorescent ligands [10], and fluorescent dyes.

Labeling reagents for IFC include a wide range of macromolecules (proteins and antibodies, miRNA, etc.) conjugated with fluorochromes differed by their molecular weight and volume. Steric hindrance may be a problem when labelling antigens that are in close proximity on the surface of cell or particle [11]. Another factor influencing the labeling of target molecules in cell is a labeling efficiency (LE), i.e., a number of labeling molecules per target molecule. IFC allows working with alive and fixed cells. Labeling efficiency can be influenced by several factors such as fixation, permeabilization (for intracellular staining), molecular size of fluorochromes (difficulties to reach intracellular compartments because of significant molecular weight/size), different penetration in subcellular compartments, and specifics of labelling protocols [12]. Because of the importance of immunophenotyping in conventional flow cytometry this multimarker strategy that utilizes traditional fluorochromes (such as FITC, PE, PERCP, APC) has been replicated in IFC. However, IFC differs from traditional fluorescent microscopy in that dyes and fluorochromes might be less prone to photobleaching due to a short time of cell exposure at the laser interrogation point. Therefore, traditional flow cytometry fluorochromes and panels can be successfully used for labeling samples for IFC. These methods form basis for minimally biased multicolor analysis in IFC and combined with feature extraction for bright field and fluorescent images allow discrimination of cell types not feasible for standard cytometry [13].

3 Spectral Compensation in IFC

An important aspect of IFC is a user-friendly generation for fluorescent images correct spectral compensation (manual or automatic) applying a set of singly stained spectral control samples and control for autofluorescence. It is performed on the pixel-by-pixel basis allowing both online and off-line compensation to be performed as in standard flow cytometry. However, band-pass filters used in IFC have larger transmittance window compared to standard flow cytometry and so far the maximal number of simultaneously used channels is lower.

4 Statistics in IFC

To analyze cellular subsets reproducibly, measuring sufficient cell numbers (hundreds and thousands) is required. To be reliably identified by statistical methods a subpopulation has to include 10–25 clustered events with identical features [14, 15]. Modern IFC instrumentation allows collection of 10–20 thousand events in a single file and up to 100 thousand events could be collected within the reasonable time. Thus selection of rare events (0.1–1 % of the population) becomes statistically significant.

5 Software and Feature Extraction

IFC, especially because it combined with advanced software is capable to provide multi-marker, per-cell quantitative data, combining morphological features and particular morphological subregions on images with fluorescent intensity quantitation. IFC aided the development of traditional flow cytometry methods by addressing the following major issues: (1) adding bright-field parameter to cells in laminar flow allows the separation and independent quantification of multiple fluorescently labeled molecules in the same or in different cellular compartments; (2) high-speed analysis of cells in stream enables accurate spatial distribution of fluorescent intensities in hundreds and thousands of images of particles, cells, and cell aggregates providing statistical means to work with rare cell subpopulations; (3) creating combinations of new parameters, combining of cellular masks, as well as automated selection of statistically significant morphological cellular features.

In IFC size and shape of cells (aspect ratio or ratio of cell diameters) is a widely used parameter for discrimination of single, double and multiple events (Fig. 1) which is possible to further control via looking through image galleries. Besides, IDEAS software allows automated analysis of more than 85 morphological characteristics

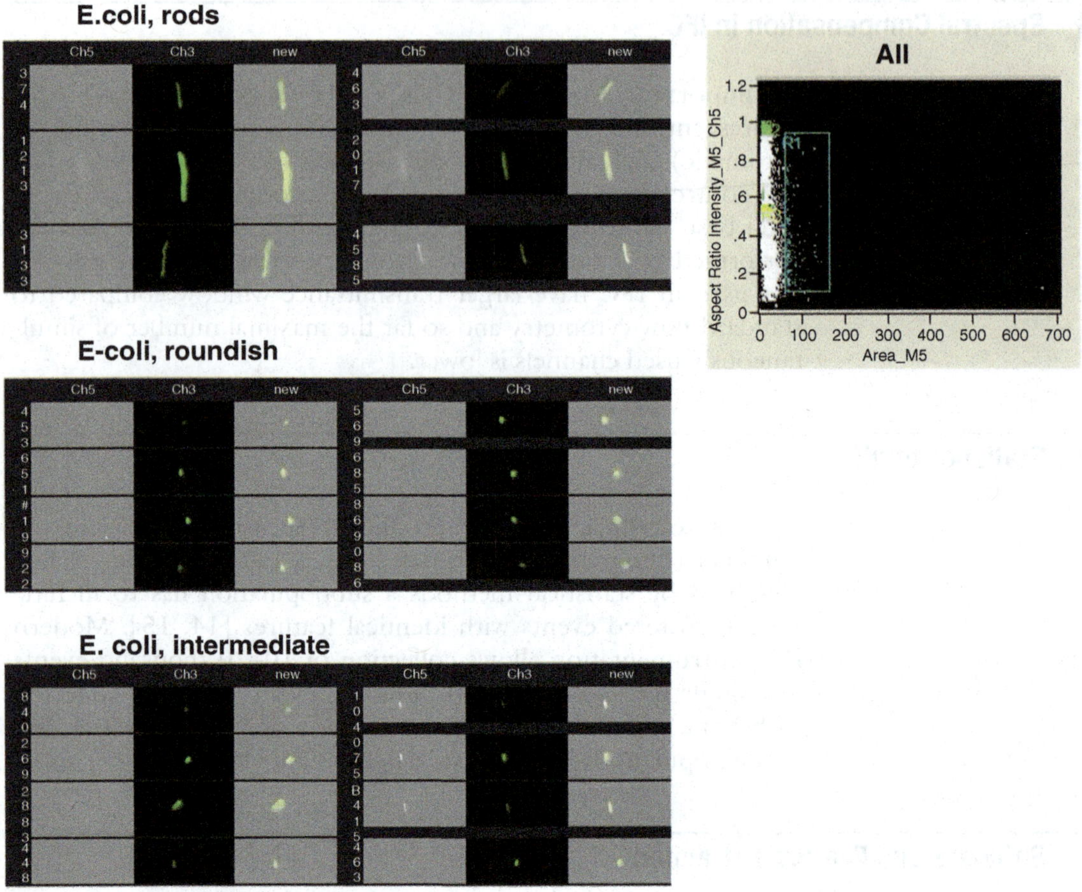

Fig. 1 Size parameter in *E. coli* analysis by imaging flow cytometry. Examples of rods, roundish and intermediate-sized bacteria on images (panels) and on dotblot (area vs. aspect ratio intensity)

that could be selected from the list and applied to bright-field and fluorescent images. The recent breakthrough in IFC represented by Feature Finder option (in IDEAS software) which allows computer-aided pattern recognition of statistically significant differences in morphological and/or fluorescence cellular features based on the comparison of two panels of images chosen by researcher. This is one of the most important and user-friendly recent developments in imaging analysis software.

However, IFC has a number of limitations: (1) it does not allow interrogation of individual cells repeatedly creating time-lapse series; (2) the amount of collected light is limited, because of the short time in which cell pass the flow cell (typically milliseconds). Though comparing with a conventional cytometry the amount of collected light is larger due to lower flow rate and time delay approach (collecting light over longer interval of time) using custom-modified charge-coupled device (CCD) camera for capturing of images; (3) because of the lower flow rate the number of events collected per unit of time is lower than in conventional flow cytometry. Besides necessity of image analysis of each cell makes generation of large data sets inconvenient.

6 Examples of IFC Applications

Recent studies show that IFC allows to localize and quantitate molecules (peptides, proteins, protein complexes, nucleic acids (DNA, RNA, miRNA), glycolipids, molecular assemblages, etc.) not only in the single cells, but also in multicellular clusters and small aggregates and to map their spatial distribution to the cellular compartments. Single cell measurements allow revealing information hidden in population averages.

Using fluorescent molecules as markers allows IFC to perform analysis of host-pathogen interaction, quantification of intracellular parasites and has a potential of providing high-throughput analysis of host-protein interactions and assessment of drugs for parasitic and microbial infections [16, 17].

IFC is extremely helpful in the cases when live cells could be associated with dead cells and apoptotic bodies [3], adherent platelets/platelet fragments [4], or extracellular vesicles (EVs). In these situations use of conventional flow cytometric techniques and excluding events only by FSC/SSC parameters may result in the interpretation of negatively stained cells as positive ones. This was shown for apoptotic markers [18] and/or the CD41a^{+} expression [19]. CD41a is a glycoprotein (Ca^{++}-dependent heterodimer) that serves as a receptor for adhesion proteins on the surface of platelets and mediate cell-to-substrate and cell-to-cell adhesion. Many cancer cells floating in the blood vessels are temporarily coated with platelets, possible to escape immune surveillance [19, 20] and their analysis by IFC is the most promising approach.

IFC is also very helpful in distinguishing false-positive events in cell death evaluation [21] and during analysis of activated cells. The cellular suspensions always have level of spontaneous apoptosis >0 % and constantly produce extracellular vesicles (EVs) in response to apoptotic and activation stimuli. The EVs are lipid vesicles produced by all cells, and can carry nucleic acids and a number of proteins originated from cell-progenitors. Therefore EVs can be positively stained with DNA and lipid dyes, and/or fluorochrome-conjugated antibodies and attach to negatively stained cells. Annexin V^{+} staining is a marker of significant subset of EVs (microparticles, size range 100–1000 nm). Only IFC allows verification and exclusion of a subset of cells with EVs attached to their surface and obtaining proper statistics on the cell populations.

7 Conclusions

While IFC boasts over a 30-year history, its development within last decade (Imagestream-based IFC with IDEAS software including the advantages highlighted in this review) is indeed exploding. The applications of IFC are extensive and include but not limited to immunophenotypic multiparametric analysis combined with morphometric assessment of cells. An interesting road lies ahead when the automated search of morphological differences between cellular structures will become easy available and useful tool for future research and drug discovery.

Acknowledgements

The work was supported in part by NIH S10 RR023459 grant, Harvard Pilot grant, and the Russian Foundation for Basic Research grants 13-04-40189-H and 11-04-01749a. We are also grateful to Aleksandra Gorelova (Harvard University) for help with the editing and preparation of the manuscript.

References

1. Oheim M (2011) Advances and challenges in high-throughput microscopy for live-cell subcellular imaging. Expert Opin Drug Discov 6:1299–1315. doi:10.1517/17460441.2011.637105
2. Perfetto SP, Chattopadhyay PK, Roederer M (2004) Seventeen-colour flow cytometry: unravelling the immune system. Nat Rev Immunol 4:648–655
3. Henery S, George T, Hall B, Basiji D, Ortyn W, Morrisey P (2008) Quantitative image based apoptotic index measurement using multispectral imaging flow cytometry: a comparison with standard photometric methods. Apoptosis 13:1054–1063. doi:10.1007/s10495-008-0227-4
4. Ouk C, Jayat-Vignoles C, Donnard M, Feuillard J (2011) Both CD62 and CD162 antibodies prevent formation of CD36-dependent platelets, rosettes, and artefactual pseudoexpression of platelet markers on white blood cells: a study with ImageStream®. Cytometry A 79:477–484. doi:10.1002/cyto.a.21050
5. Di Palma S, Bodenmiller B (2015) Unraveling cell populations in tumors by single-cell mass cytometry. Curr Opin Biotechnol 31:122–129. doi:10.1016/j.copbio.2014.07.004
6. Hulspas R, O'Gorman MRG, Wood BL, Gratama JW, Sutherland DR (2009) Considerations for the control of background fluorescence in clinical flow cytometry. Cytometry B Clin Cytom 76:355–364. doi:10.1002/cyto.b.20485
7. Li F, Yang M, Wang L, Williamson I, Tian F, Qin M, Shah PK, Sharifi BG (2012) Autofluorescence contributes to false-positive intracellular Foxp3 staining in macrophages: a lesson learned from flow cytometry. J Immunol Methods 386:101–107. doi:10.1016/j.jim.2012.08.014
8. Vorobjev IA, Buchholz K, Prabhat P, Ketman K, Egan ES, Marti M, Duraisingh MT, Barteneva NS (2012) Optimization of flow cytometric detection and cell sorting of transgenic *Plasmodium* parasites using interchangeable optical filters. Malar J 11:312. doi:10.1186/1475-2875-11-312
9. Stepanenko OV, Shcherbakova DM, Kuznetsova IM, Turoverov KK, Verkhusha VV (2011) Modern fluorescent proteins: from chromophore formation to novel intracellular applications. Biotechniques 51:313–314. doi:10.2144/000113765
10. Kovalenko EI, Ranjbar S, Jasenosky LD, Goldfeld AE, Vorobjev IA, Barteneva NS

(2011) The use of HaloTag-based technology in flow and laser scanning cytometry analysis of live and fixed samples. BMC Res Notes 4:340. doi:10.1186/1756-0500-4-340

11. Johanasson U, Macey M (2011) Pitfalls in the use of multicolour flow cytometry in haematology. J Clin Pathol 64:561–563
12. Mayhew TM (2011) Mapping the distributions and quantifying the labelling intensities of cell compartments by immunoelectron microscopy: progress towards a coherent set of methods. J Anat 219:647–660. doi:10.1136/jcp.2010.085183
13. George TC, Basiji DA, Hall BE, Lynch DH, Ortyn WE, Perry DJ, Seo MJ, Zimmerman CA, Morrissey PJ (2004) Distinguishing modes of cell death using the Imagestream multispectral imaging flow cytometer. Cytometry A 59:237–245
14. Kraan J, Gratama JW, Haioun C, Orfao A, Plonquet A, Porwit A, Quijano S, Stetler-Stevenson M, Subira D, Wilson W (2008) Flow cytometric immunophenotyping of cerebrospinal fluid. Curr Protoc Cytom Chapter 6: Unit 6.25, Wiley, New York
15. de Graaf MT, de Jongste AHC, Kraan J, Boonstra JG, Sillevis Smitt PA, Graatama JW (2011) Flow cytometric characterization of cerebrospinal fluid cells. Cytometry B 80:271–281. doi:10.1002/cyto.b.20603
16. Muskavitch MA, Barteneva N, Gubbels MJ (2008) Chemogenomics and parasitology: small molecules and cell-based assays to study infectious processes. Comb Chem High Throughput Screen 11:624–646
17. Tree JA, Flick-Smith H, Elmore MJ, Rowland CA (2014) The impact of "omic" and imaging technologies on assessing the host immune response to biodefense agents. J Immunol Res 2014, Article ID 237043. doi: 10.1155/2014/237043
18. Wlodkowic D, Telford W, Skommer J, Darzinkiewicz Z (2011) Apoptosis and beyond: cytometry in studies of programmed cell death. Methods Cell Biol 103:55–98
19. Betz SA, Foucar K, Head DR, Chen I-M, Willman CL (1992) False-positive platelet glycoprotein IIb/IIIa expression in myeloid leukemias secondary to platelet adherence to blasts. Blood 79:2399–2403
20. Placke T, Orgel M, Schaller M, Jung G, Rammensee H-G, Kopp H-G, Salih HR (2012) Platelet-derived Class I confers a pseudonormal phenotype to cancer cells that subverts the antitumor reactivity of natural killer immune cells. Cancer Res 72:440–448. doi:10.1158/0008-5472.CAN-11-1872
21. Rieger AM, Hall BE, Luong LT, Schang LM, Barreda DR (2010) Conventional apoptosis assays using propidium iodide generate a significant number of false positives that prevent accurate assessment of cell death. J Immunol Methods 358:81–92. doi:10.1016/j.jim.2010.03.019

Chapter 2

Principles of Amnis Imaging Flow Cytometry

David A. Basiji

Abstract

Over the last decade imaging flow cytometry (IFC) has become an established technique, as evidenced by its use in over 500 peer-reviewed scientific articles. Nevertheless, it is still an emerging technique with an installed base of less than 5 % that of conventional flow cytometers. In parallel with its adoption, the technology has evolved rapidly, increasing in speed, sensitivity, and real-time data analysis capacity by over an order of magnitude since its introduction. This chapter summarizes IFC's basic principles of operation and describes the current state of the art.

Key words Imaging, Flow cytometry, Time delay integration

1 Introduction

Imaging flow cytometers are distinguished from conventional flow cytometers by their ability to produce images of cells directly in fluid flow. The first functional imaging flow cytometers were developed by Leon Wheeless and colleagues at the University of Rochester in the late 1970s, but were limited by the detector technologies available at the time [1]. Today, imaging flow cytometers with modest image resolution and limited fluorescence sensitivity are available from a number of commercial suppliers and are typically optimized for the analysis of urine sediment, industrial particulates, and marine microorganisms.

This chapter describes the Amnis imaging flow cytometers, which are designed to augment and in certain cases replace conventional flow cytometers and fluorescence microscopes in research labs by virtue of their diffraction-limited image resolution, high-speed operation, and excellent fluorescence sensitivity. At the time of this writing, the Amnis systems can produce up to 12 images per cell at 60× magnification using a combination of transmitted light, scattered light, polarized light, and fluorescence. With operating speeds of several thousand cells per second and sensitivity equal to or better than conventional flow cytometers, these instruments

Natasha S. Barteneva and Ivan A. Vorobjev (eds.), *Imaging Flow Cytometry: Methods and Protocols*, Methods in Molecular Biology, vol. 1389, DOI 10.1007/978-1-4939-3302-0_2, © Springer Science+Business Media New York 2016

combine the analytical power of microscopy with the rare cell detection and statistical power of conventional flow cytometry [2]. Recent applications of Amnis IFC described in the literature include immunology [3–6], oncology [7–11], hematology [12–15], cell biology [16–18], infectious disease [19–23], transplantation medicine [24], biofuel research [25], fertility [26], and others.

2 Time Delay Integration CCD Operation

High fluorescence sensitivity is a requirement for effective IFC, allowing faint probes to be detected, facilitating image segmentation, and increasing the accuracy of fluorescence crosstalk compensation. The Amnis platforms utilize charge-coupled device (CCD) cameras that are read out using time delay integration (TDI), a technique originally developed for imaging sub-micron defects on semiconductor wafers, aerial reconnaissance, and other applications involving objects moving in one axis. In TDI, the photocharges stored in each image pixel are shifted down the chip as each row of pixels is read sequentially off the bottom of the chip. As a result, the latent images stored in the CCD continuously move from the top of the sensor to the bottom. If the photocharge movement is synchronized with the motion of the object via control of the TDI line readout rate, the effect is akin to panning the camera to track the object so no image streaking takes place as it traverses the field of view. The benefit of this approach compared to conventional flow cytometry or strobed illumination imaging is that the signal integration times with TDI are increased by several orders of magnitude, typically from the microsecond scale to the millisecond scale. However, because there is no readout noise penalty with TDI, higher net sensitivity is achieved. In comparison to conventional flow cytometry, TDI data collection is accomplished with a 100 % duty cycle, so every cell that flows through the system is imaged and there is no need for a separate data acquisition trigger.

The current generation ImageStream^X Mark II system employs a custom back-thinned CCD of 256 rows by 768 columns. At a nominal system magnification of 40×, the pixel size in object space is 0.5 × 0.5 μm and the field of view is 128 μm high. Each pixel is digitized with 12-bit resolution but since typical mammalian cell images consist of hundreds of pixels, the effective dynamic range exceeds 5 decades (Fig. 1).

3 Velocity Detection and Autofocus

In order to effectively image using TDI, the fluidic system must operate without pulsatility and prevent uncontrolled cell rotation or translation across the field of view. The ImageStream system

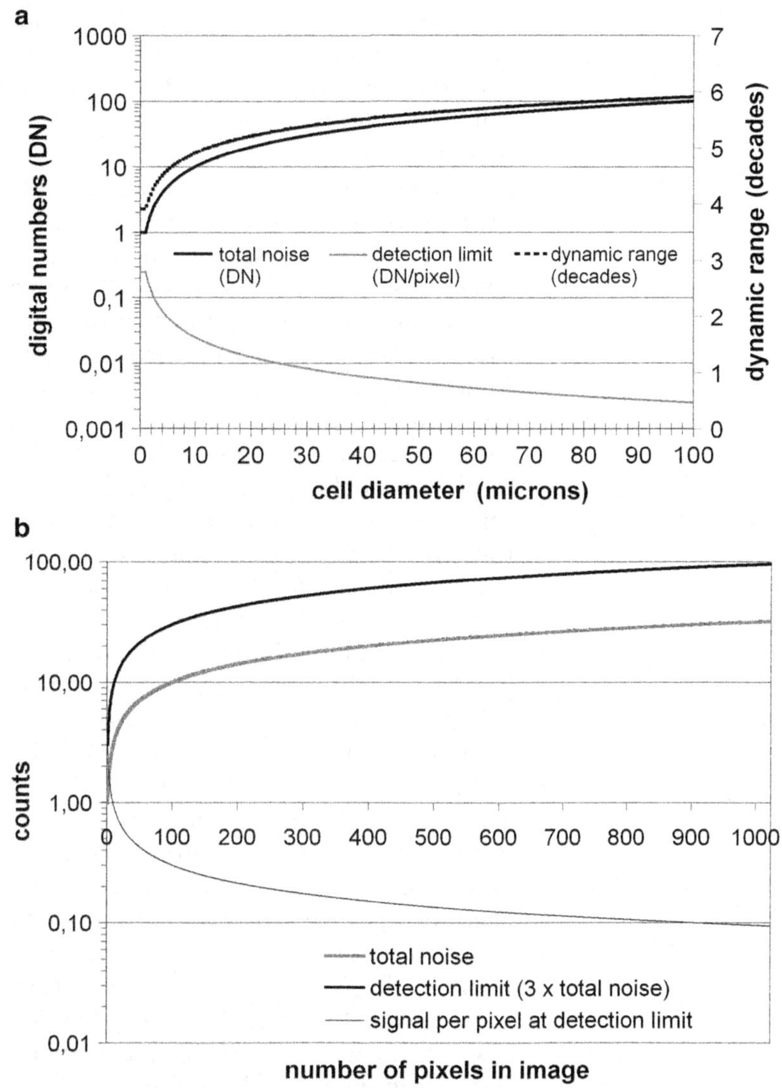

Fig. 1 Fluorescence detection limit and dynamic range as a function of cell size. Because each cell image covers multiple pixels, the sensitivity and dynamic range of cell measurements in imaging flow cytometry are both dependent on cell size. The Amnis systems use the probe-free bright-field and dark-field images to define the location and boundaries of the corresponding fluorescence images, thereby eliminating background noise from pixels within the field of view that do not contain cell imagery. The graph (**a**) shows the total noise, the detection limit (total noise divided by image size), and dynamic range as a function of cell size, assuming ~×40 magnification. Note that with objects larger than 10 microns in diameter, imaging flow cytometry can exceed five decades of dynamic range and with objects larger than 25 microns, IFC can detect cellular fluorescence at levels below 0.1 counts per pixel, well below the traditional 1 count per pixel detection floor of conventional imaging systems (**b**). Conversely, with bacteria and other objects just a few microns in diameter, the use of bright-field and dark-field imagery for background rejection can increase sensitivity by an order of magnitude relative to conventional flow cytometry

employs precision pumps and an in-line air chamber for fluidic system compliance, resulting in very smooth sheath flow. In addition, the flow cell is designed and manufactured to ensure that the cells are hydrodynamically focused at the center of the cuvette, thereby balancing the shear forces on either side of the cells to prevent rotation. Precise centration of the cells also assures a consistent plane of focus from cell to cell.

Even with non-pulsatile and centered flow, TDI detection still requires the precise tracking of cells as they transit the field of view. If the cumulative tracking error over the height of the detector exceeds one quarter of a pixel, image contrast will degrade. In order to achieve the necessary tracking accuracy, the ImageStream system employs a dedicated velocity detection subsystem that provides closed loop control of the TDI readout rate. The principle is similar to watching a car's headlights through a picket fence, as the car moves faster, the frequency at which the lights appear to flash through the fence is proportional to the car's speed. The Amnis velocity detection subsystem includes an infrared laser source to generate side-scattered light from objects in the flow core, a "hot mirror" to reflect the infrared scatter out of the primary imaging path, a Ronchi ruling of alternating opaque and transparent bars oriented orthogonal to the axis of cell motion that modulates the amplitude of the scattered light, and a detector and associated signal processing circuitry that Fourier transforms the signal and accurately identifies the fundamental modulation frequency. The modulation frequency is proportional to the speed of the flowing objects and is used to dynamically adjust the TDI readout rate to keep the detector synchronized with the motion of the cells.

In practice, two Ronchi rulings and detectors are employed behind a 50/50 beam splitter. The rulings are spaced such that their conjugate images lie on either side of the center of the flow cuvette. In this way, an auto focus score can also be generated by comparing the relative amplitudes of the signals from each velocity detection leg, which is then used to correct focus by moving the objective in the *Z*-axis.

4 Multispectral Imaging Optics

Light collection in the ImageStream system is via a microscope objective and the images cover a spectral range of 420–800 nm. Prior to projection on the TDI detector, the collected light passes through a spectral decomposition element that generates six spectral images of each object simultaneously. The spectral decomposition element consists of a stack of five dichroic long-pass filters with progressively longer cut-on wavelengths mounted in front of a backing mirror. The entire assembly is mounted in aperture space with each filter reflecting a different spectral band at a different angle in the horizontal plane. A secondary imaging lens assembly then focuses this light at the TDI camera, forming six spectral images arrayed side by side on the detector. The original ImageStream 100 employed a single camera but the ImageStream^{X} and later versions can accept up to two cameras, each with its own spectral decomposition element, to form a total of 12 images per cell. The fields of view of the two cameras are staggered vertically so the cells are first imaged on one camera and then, after a few milliseconds, on the second camera. Different lasers are steered towards the upper and lower fields of view, thereby suppressing cross talk between cameras in a fashion analogous to the use of different pinholes in a conventional flow cytometer.

5 Real-Time Data Analysis

Imaging flow cytometry makes it possible to image populations of rare cells at frequencies at or below 1:100,000. However, with image data rates exceeding 500 megabytes per second, it is critically important to analyze cells in real time so that only cell data of interest are stored. Expanded real-time data analysis also allows IFC to more closely match the workflow of conventional flow cytometry.

The original ImageStream 100 system, which was introduced in 2005, produced 60 megabytes of data per second and was capable of detecting in the presence of objects in flow, performing basic image segmentation from background, and calculating a limited set of morphometric and photometric features. However, all image registration and fluorescence compensation steps were performed post-acquisition, as well as the calculation of ~40 quantitative features per image.

At the time of this writing, the third-generation ImageStream^{X} Mark II is capable not only of object detection and image segmentation, but also real-time image registration, fluorescence compensation, and the calculation of a much broader and more sophisticated set of image features, despite the fact that it operates at nearly ten times the data rate of the original system. Thanks to Moore's law, subsequent generations of instrumentation should be capable of the real-time determination of co-localization, spot counting, and other sophisticated image processing tasks at rates fast enough to enable cell sorting.

6 Post-acquisition Data Processing

Amnis data sets routinely contain over a million digital images, presenting challenges both in ensuring data integrity and in managing data analysis. Off-line analysis of Amnis data is preceded by data decompression, correction for gain and offset variations in the CCD detector, normalization for velocity variation from object to object, and (if not performed at the time of data acquisition) spatial registration of the images of each object and spectral crosstalk correction. Registration is performed to sub-pixel accuracy via interpolation using spatial offset coefficients measured during daily instrument calibration. A spectral crosstalk correction matrix, calculated from singly labeled fluorescence cell images as in conventional flow cytometry, is applied independently to every pixel of image data to improve fluorescence quantitation and image analysis.

After crosstalk compensation, approximately 100 photometric and morphometric feature calculations are performed for each image. In an application employing all 12 images per cell, approximately 1200 quantitative analysis features are available. The breadth of this "feature space" allows the creation of biologically meaningful derivative features by the algebraic combination of existing features and/or masks. For example, the nuclear-to-cellular area ratio can be calculated by taking the quotient of the nuclear fluorescence image area and the bright-field cellular area.

Amnis data allows the linkage of cell imagery with the univariate histograms and bivariate scatter plots used in standard flow cytometry data analysis. For example, a dot representing a cell on a scatter plot can be selected to inspect its associated imagery. Alternatively, a cell image can be selected and the corresponding dot highlighted in every scatter plot containing that cell. This functionality extends to populations, which are defined on a scatter plot or histogram using graphical drawing tools. Once a population is defined, its members can be visually inspected as a gallery in a "virtual cell sort." Conversely, a population can be defined by the inspection and selection of individual cell images and then highlighted in various scatter plots to determine the utility of different features for the definition of "like" cells.

Finally, correlated multispectral imagery allows the use of image cross-correlation algorithms to determine the degree of co-localization of probes in different images of the same cell. This capability is useful for numerous applications, particularly in cell signaling, and can substitute for Forster resonance energy transfer (FRET) in many applications.

7 Multimode Image Correlation for Improved Detection Limits

The simultaneous acquisition of dark-field and bright-field along with fluorescence imagery provides a means for substantially improving the fluorescence detection limit. In any digital imaging system where the camera has one count of dark noise per pixel, the detection limit will be approached when overall pixel values are within a few counts of zero. At these signal levels, image segmentation algorithms become unreliable, despite the fact that there may be hundreds of pixels in the image and therefore hundreds of counts of signal overall.

In the Amnis systems, the bright-field and dark-field images offer a means to enhance the detection of extremely faint fluorescence from the same object. During image segmentation the logical OR of the bright-field and dark-field masks is applied to the fluorescence image(s) of the object, thereby defining the boundaries of any faint fluorescence that may be present in or on the cell. In order to detect the fluorescence signal, it must be isolated from the background, which can be determined by using the pixels outside the mask. The mean value of the background per pixel outside the mask is multiplied by the area of the mask and subtracted from the total signal-plus-background within the mask. The result of this calculation is a net total signal, which may average significantly less than one count per pixel in larger cell images, and can greatly enhance sensitivity relative to conventional flow cytometry of smaller objects like bacteria.

Both sensitivity and dynamic range are a function of image size in IFC. Because the total noise scales with the square root of the number of image pixels, for a given fluorescence signal, sensitivity will go down as the image size increases due to higher magnification or larger cell sizes. However, the dynamic range goes up as the fluorescence is distributed over more pixels, since each pixel is independently digitized. Figure 1 shows the total noise, the detection limit (total noise divided by image size), and dynamic range as a function of object size, assuming ~40× magnification. Note that with objects larger than 25 μm in diameter, imaging flow cytometry can detect cellular fluorescence at levels below 0.01 counts per pixel.

8 Conclusions

The speed, sensitivity, and rich feature set offered by IFC allow a wide range of image-based cell analysis applications to be performed with statistical rigor, even when the cells of interest are present at extremely low concentrations. Thanks to these characteristics, the current Amnis analytical platforms are proliferating in research labs and will soon be in use for clinical applications. Advances in TDI camera technology and the ongoing increase in computer processing power will also enable image-based cell sorting at thousands of cells per second, allowing IFC to be used as a front end for single-cell sequencing and other preparative applications of significance.

References

1. Kay DB, Cambier JL, Wheeless LL Jr (1979) Imaging in flow. J Histochem Cytochem 27: 329–334
2. Ortyn WE, Hall BE, George TC, Frost K, Basiji DA, Perry DJ, Zimmerman CA, Coder D, Morrisey PJ (2006) Sensitivity measurement and compensation in spectral imaging. Cytometry A 69:852–862
3. Bezbradica JS, Rosenstein RK, DeMarco RA, Brodsky J, Medzhitov R (2014) A role for the ITAM signaling module in specifying cytokine-receptor functions. Nat Immunol 15:333–342. doi:10.1038/ni.2845
4. Bonaccorsi I, Morandi B, Antsiferova O, Costa G, Oliveri D, Conte R, Pezzino G, Vermiglio G, Anastasi GP, Navarra G, Münz C, DiCarlo E, Mingari MC, Ferlazzo G (2014) Membrane transfer from tumor cells overcomes deficient phagocytic ability of plasmacytoid dendritic cells for the acquisition and presentation of tumor antigens. J Immunol 192:824–832. doi:10.4049/jimmunol.130103
5. Zhao W, Minderman H, Russell MW (2014) Identification and characterization of intestinal antigen-presenting cells involved in uptake and processing of a nontoxic recombinant chimeric mucosal immunogen based on cholera toxin using imaging flow cytometry. Clin Vaccine Immunol 21:74–84. doi:10.1128/CVI.00452-13
6. Waysbort N, Russ D, Chain BM, Friedman N (2013) Coupled IL-2-dependent extracellular feedbacks govern two distinct consecutive phases of CD4 T cell activation. J Immunol 191:5822–5830. doi:10.4049/jimmunol.1301575
7. Ryszawy D, Sarna M, Rak M, Szpak K, Kedracka-Krok S, Michalik M, Siedlar M, Zuba-Surma E, Burda K, Korohoda W, Madeia Z, Czyz J (2014) Functional links between Snail-1 and Cx43 account for the recruitment of Cx43-positive cells into the invasive front of prostate cancer. Carcinogenesis 35:1920–1930. doi:10.1093/carcin/bgu033
8. Kefas B, Floyd DH, Comeau L, Frisbee A, Dominquez C, Dipierro CG, Guessous F, Abounader R, Purow B (2013) A miR-297/hypoxia/DGK-α axis regulating glioblastoma survival. Neuro Oncol 15:1652–1663
9. Franzen CA, Simms PE, Van Huis AF, Foreman KE, Kuo PC, Gupta GN (2014) Characterization of uptake and internalization of exosomes by bladder cancer cells. Biomed Res Int 2014:619829. doi:10.1155/2014/619829
10. Leigh ND, Bian G, Ding X, Liu H, Aigun-Sunar S, Burdelya LG, Gudkov AV, Cao X (2014) A flagellin-derived toll-like receptor 5 agonist stimulates cytotoxic lymphocyte-mediated tumor immunity. PLoS One 9, e85587. doi:10.1371/journal.pone.0085587
11. Clarke AJ, Ellinghaus U, Cortini A, Stranks A, Simon AK, Botto M, Vyse TJ (2014) Autophagy is activated in systemic lupus erythematosus and required for plasmablast development. Ann Rheum Dis 74:912–920. doi:10.1136/annrheumdis-2013-204343
12. van Beers EJ, Samsel L, Mendelsohn L, Salyed R, Fertrin KY, Brantner CA, Daniels MP, Nichols J, McCoy JP, Kato GJ (2014) Imaging flow cytometry for automated detection of hypoxia-induced erythrocyte shape change in sickle cell disease. Am J Hematol 89:598–603. doi:10.1002/ajh.23699
13. Niswander LM, McGrath KE, Kennedy JC, Palis J (2014) Improved quantitative analysis of primary bone marrow megakaryocytes utilizing imaging flow cytometry. Cytometry A 85:302–312. doi:10.1002/cyto.a.22438
14. Foudi A, Kramer DJ, Qin J, Ye D, Behlich AS, Mordecai S, Preffer FI, Amzallag A, Ramaswamy S, Hochedlinger K, Orkin SH, Hock H (2014) Distinct, strict requirements for Gfi-1b in adult bone marrow red cell and platelet generation. J Exp Med 211:909–927. doi:10.1084/jem.20131065
15. Malik J, Kim AR, Tyre KA, Cherukuri AR, Palis J (2013) Erythropoietin critically regulates the terminal maturation of murine and human primitive erythroblasts. Haematologica 98:1778–1787. doi:10.3324/haematol.2013.087361
16. Ma H, Reyes-Gutierrez P, Pederson T (2013) Visualization of repetitive DNA sequences in human chromosomes with transcription activator-like effectors. Proc Natl Acad Sci U S A 110:21048–21053. doi:10.1073/pnas.1319097110
17. Barteneva NS, Ponmarev ED, Tsytsykova A, Armant M, Vorobjev IA (2014) Mitochondrial staining allows robust elimination of apoptotic and damaged cells during cell sorting. J Histochem Cytochem 62:265–275. doi:10.1369/0022155413520404
18. Dias da Silva D, Carmo H, Lynch A, Sylva E (2013) An insight into the hepatocellular death induced by amphetamines, individually and in combination: the involvement of necrosis and apoptosis. Arch Toxicol 87:2165–2185. doi:10.1007/s00204-013-1082-9
19. Bergquist M, Nurkkala M, Rylander C, Kristiansson E, Hedenstierna G, Lindholm S

(2013) Expression of the glucocorticoid receptor is decreased in experimental *Staphylococcus aureus* sepsis. J Infect 67:574–583. doi:10.1016/j.jinf.2013.07.028

20. Singh MV, Davidson DC, Jackson JW, Singh VB, Silva J, Ramirez SH, Maggirwar SB (2014) Characterization of platelet-monocyte complexes in HIV-1-infected individuals: possible role in HIV-associated neuroinflammation. J Immunol 192:4674–4684. doi:10.4049/jimmunol.1302318
21. Caccuri F, Giagulli C, Reichelt J, Martorelli D, Marsico S, Bugatti A, Barone J, Rusnati M, Guzman CA, Dolcetti R, Caruso A (2014) Simian immunodeficiency virus and human immunodeficiency virus type 1 matrix proteins specify different capabilities to modulate B cell growth. J Virol 88:5706–5717. doi:10.1128/JVI.03142-13
22. Németh T, Tóth A, Hamari Z, Falus A, Éder K, Vágvölgyi C, Guimaraes AJ, Nosanchuk JD, Gácser A (2014) Transcriptome profile of the murine macrophage cell response to *Candida parapsilosis*. Fungal Genet Biol 65:48–56. doi:10.1016/j.fgb.2014.01.006
23. Lee YQ, Goh AS, Ch'ng JH, Nosten FH, Preiser PR, Pervaiz S, Yada SK, Tan KS (2014) A high-content phenotypic screen reveals the disruptive potency of quinacrine and 3′,4′-dichlorobenzamil on the digestive vacuole of *Plasmodium falciparum*. Antimicrob Agents Chemother 58:550–558. doi:10.1128/AAC.01441-13
24. Maguire O, Tornatore KM, O'Loughlin KL, Venuto RC, Minderman H (2013) Nuclear translocation of nuclear factor of activated T cells (NFAT) as a quantitative phamacodynamic parameter for tacrolimus. Cytometry A 83:1096–1104. doi:10.1002/cyto.a.22401
25. Trentacoste EM, Shrestha RP, Smith SR, Glé C, Hartmann AC, Hildebrand M, Gerwick WH (2013) Metabolic engineering of lipid catabolism increases microalgal lipid accumulation without compromising growth. Proc Natl Acad Sci U S A 110:19748–19753. doi:10.1073/pnas.1309299110
26. Kennedy CE, Krieger KB, Sutovsky M, Xu W, Vargovič P, Didion BA, Ellersieck MR, Hennessy ME, Verstegen J, Oko R, Sutovsky P (2014) Protein expression pattern of PAWP in bull spermatozoa is associated with sperm quality and fertility following artificial insemination. Mol Reprod Dev 81:436–449. doi:10.1002/mrd.22309

Chapter 3

Ultrafast Microfluidic Cellular Imaging by Optical Time-Stretch

Andy K.S. Lau, Terence T.W. Wong, Ho Cheung Shum, Kenneth K.Y. Wong, and Kevin K. Tsia

Abstract

There is an unmet need in biomedicine for measuring a multitude of parameters of individual cells (i.e., high content) in a large population efficiently (i.e., high throughput). This is particularly driven by the emerging interest in bringing Big-Data analysis into this arena, encompassing pathology, drug discovery, rare cancer cell detection, emulsion microdroplet assays, to name a few. This momentum is particularly evident in recent advancements in flow cytometry. They include scaling of the number of measurable colors from the labeled cells and incorporation of imaging capability to access the morphological information of the cells. However, an unspoken predicament appears in the current technologies: higher content comes at the expense of lower throughput, and vice versa. For example, accessing additional spatial information of individual cells, imaging flow cytometers only achieve an imaging throughput ~1000 cells/s, orders of magnitude slower than the non-imaging flow cytometers. In this chapter, we introduce an entirely new imaging platform, namely optical time-stretch microscopy, for ultrahigh speed and high contrast label-free single-cell (in a ultrafast microfluidic flow up to 10 m/s) imaging and analysis with an ultra-fast imaging line-scan rate as high as tens of MHz. Based on this technique, not only morphological information of the individual cells can be obtained in an ultrafast manner, quantitative evaluation of cellular information (e.g., cell volume, mass, refractive index, stiffness, membrane tension) at nanometer scale based on the optical phase is also possible. The technology can also be integrated with conventional fluorescence measurements widely adopted in the non-imaging flow cytometers. Therefore, these two combinatorial and complementary measurement capabilities in long run is an attractive platform for addressing the pressing need for expanding the "parameter space" in high-throughput single-cell analysis. This chapter provides the general guidelines of constructing the optical system for time stretch imaging, fabrication and design of the microfluidic chip for ultrafast fluidic flow, as well as the image acquisition and processing.

Key words Imaging flow cytometry, Time-stretch microscopy, Ultrafast label-free imaging, Quantitative phase imaging, Microfluidic fabrication, Optofluidics

1 Introduction

The pressing need for progressively larger size of biomedical dataset has transformed biomedicine into the era of "Big Data". While it is vaguely defined across different disciplines, this Big-Data

Natasha S. Barteneva and Ivan A. Vorobjev (eds.), *Imaging Flow Cytometry: Methods and Protocols*, Methods in Molecular Biology, vol. 1389, DOI 10.1007/978-1-4939-3302-0_3, © Springer Science+Business Media New York 2016

concept has indeed long been the unspoken pursuit in the biomedical field [1, 2]. One of the rationales is the bigger the data size (i.e., measurable parameters), the higher degree the statistical accuracy and thus the higher confidence decision making in diagnostics. It could also lead to discovery of new insights because of the access to the unprecedented data space. A notable example is the trend in the development of flow cytometry. The number of measurable parameters in a classical flow cytometer is primarily limited by the number of laser sources equipped in the system, as well as by the number of fluorescent labels used in the test (~3–4 colors in typical flow cytometry). There has been a continuing move toward higher number of measurable parameters by integrating more lasers together with multiple labels, such as the development of 17-color flow cytometers [3]. Apart from scaling the measurement colors, improved measurement accuracy can also be achieved by capturing images of the cells. Imaging reveals morphological and structural details of single cells. Such information can be harnessed for better cellular identification and thus high-confidence statistical data can be obtained. It is of great importance for single-cell analysis, which is to assess cell-to-cell differences to elucidate the variation in disease processes within a large population of cells, for the purpose of differentiating and quantifying either normal or abnormal early progenitors from differentiated cells. Such phenomenon is commonly encountered in both stem cells and cancer research. As the traditional flow cytometers are unable to image the individual cells with high throughput, there has been a rising demand for incorporating imaging capability to flow cytometry—imaging flow cytometers [4].

However, gaining access to image information compromises the measurement throughput. This is because of the intrinsic trade-off between imaging speed and sensitivity appeared in image sensors (i.e., charge-coupled devices (CCDs) or complementary metal oxide semiconductor (CMOS)) of optical microscope systems—higher imaging speed, thus throughput, is impeded by the loss of detectable light signal [5]. This explains why the current state-of-the-art imaging flow cytometers can only reach an imaging throughput of few thousands cells/s, compared with the throughput of ~100,000 cells/s of the classical non-imaging flow cytometers [4]. Realizing image-based cellular bioassays, which can access detailed morphological information of the individual cells and deliver multiparametric cytometry without sacrificing the throughput, is hence of significant value. To achieve it, an ultrafast imager is the key.

While the vast majority of the existing optical imaging modalities still rely on the use of CCD or CMOS cameras, an entirely new imaging concept, based on a technique called optical time-stretch, has recently emerged to achieve ultrafast real-time continuous imaging at a frame rate or line-scan rate as high as tens of millions Hz—orders of magnitude faster than any CCD/CMOS cameras [6–8].

Originally adopted in high-speed fiber-optics communication applications, optical time-stretch (also known as dispersive Fourier transform) is now employed for imaging [6–10]. The central idea lies in ultrafast retrieval of image information encoded in the spectrum of a broadband optical pulse by converting it into a serial temporal data format in real time. Such mapping is done by stretching the pulse via group velocity dispersion (GVD) in a dispersive medium—e.g., a dispersive optical fiber.

To be precise, an additional mapping step is necessary before the time-stretch process—i.e., mapping the spatial information of the specimen to the spectrum of the pulse by the use of diffractive optics. This process is called spectral-encoding [6, 11, 12]. A single diffraction grating is used in the 1-D line-scan mode (particularly useful for unidirectional flow imaging), whereas an orthogonally oriented pair of a diffraction grating and a virtually imaged phased array is employed in the 2-D mode [6]. The time-stretched pulse, now stamped with image or spectral information, is captured by a high-speed single-pixel photodetector and analog-to-digital converter, instead of a CCD/CMOS imager.

One key strength of optical time-stretch is its time-domain operation using a single-pixel photodetector and a high-speed electronic digitizer to acquire high-resolution images in real time. The imaging frame rate is governed by the repetition rate of the laser pulse train (typically on the order of 10 MHz for most of the femtosecond-picosecond lasers). Another key strength is its enhanced sensitivity. During the time-stretch process, optical amplification can also be incorporated to overcome the fundamental trade-off between optical loss and GVD, circumventing the trade-off between sensitivity and speed. This is particularly useful if the captured optical image signal is buried below the noise floor of the photodetector. Such feature of *optical* image amplification is absent in any CCD/CMOS-based imaging systems.

Time-stretch imaging has been found particularly pertinent to microparticle imaging in microfluidic flow, with an imaging throughput comparable to conventional non-imaging flow cytometry (Fig. 1). It has been proven to be able to perform high-throughput image-based rare cancer cell screening, which is an ideal complementary tool to typical multiparametric flow cytometry [7]. In this chapter, two advanced modalities of time-stretch imaging, namely asymmetric-detection time-stretch optical microscopy (ATOM) and interferometric time-stretch (iTS) microscopy will be introduced [13]—in the context of imaging microparticles or cells in ultrafast flow in the microfluidic channel. These platforms could have great potential for enabling high-throughput multiparametric imaging flow cytometry. As the entirely new optical imaging concepts, the chapter will specifically focus on the details of system construction (both the optics and microfluidic platforms), imaging procedures, as well as image reconstruction methods.

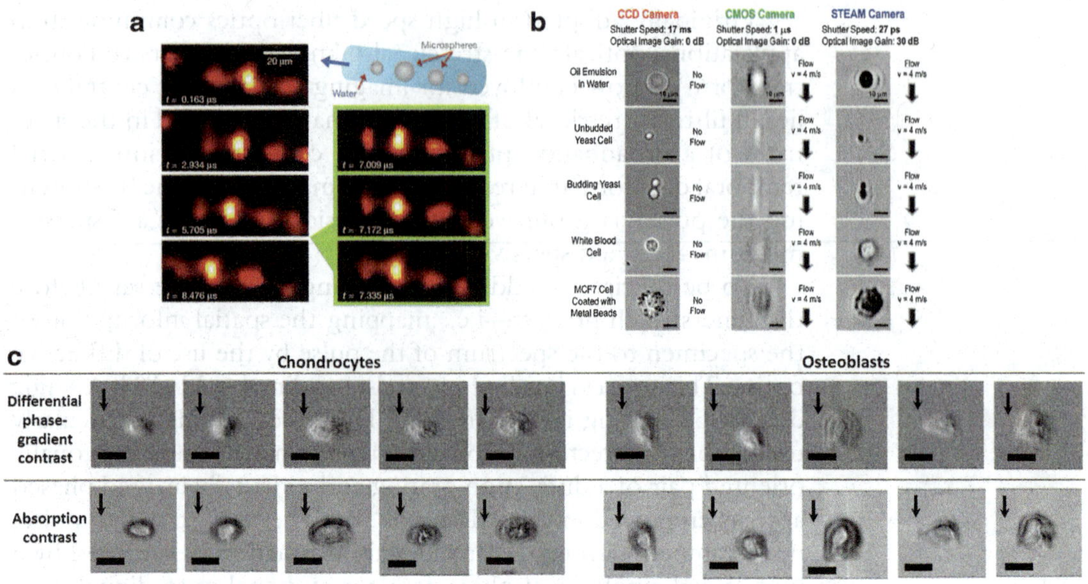

Fig. 1 Development of time-stretch imaging in ultrafast microparticle and single-cell imaging in microfluidic flow. (**a**) An ultrafast microfluidic flow of water-suspended metal microspheres (sphere diameters ranging from 10 to 30 μm) was established in a hollow fibre with an inner diameter of 50 μm. On the left, one out of every 17 snapshots is shown, for clarity. On the right, consecutive snapshots with the full temporal resolution of 163 ns and shutter speed of 440 ps are shown [6]. (**b**) Performance of the time-stretch microscopy at 1.5 μm and the comparison with a conventional CCD camera and a state-of-the-art CMOS camera [7]. (**c**) A new generation of time-stretch imaging, called ATOM, capturing images of label-free chondrocytes and osteoblasts at 1.0 μm, a wavelength window more favorable for biological imaging. Internal details of individual cells can be clearly revealed, e.g., nucleus and vesicle-like structures. ATOM can also provide two different image contrasts simultaneously, namely absorption and differential phase-gradient contrasts (see the text for details). The arrows indicate the flow direction of the cells (at a speed on the order of m/s). Scale bars represents 20 microns

1.1 Asymmetric-Detection Time-Stretch Optical Microscopy

Time-stretch microscopy had mostly been operated in bright-field (BF) imaging mode in the longer wavelength range (~1500 nm), and is thus incapable of revealing high-contrast and high-resolution morphology of the transparent cells—hindering accurate cell recognition and screening. As a result, effective use of time-stretch imaging had been limited to microparticle or cell screening in high-speed flow with trivial size and shape differences, especially when the targeted cells are labeled with contrast agents [6, 7]. Similar to the classical optical imaging modalities, the time-stretch image quality, which is typically characterized by image resolution and image contrast, is still compromised by the higher imaging speed. To enlarge the scope of time-stretch imaging applications, particularly in biomedical diagnostics, we have recently introduced a new generation of time-stretch imaging, dubbed ATOM, for obtaining label-free, high-contrast images of the transparent cells at ultrahigh speed, and with sub-cellular resolution.

ATOM relies on the same spectral-encoding strategy as its BF time-stretch imaging counterpart. The key difference is that ATOM generates the enhanced phase-gradient contrast in the time-stretch image based on a simple asymmetric detection scheme after the spectral-encoding step (*see* Fig. 4). The phase-gradient contrast results in three-dimensional (3D) appearance in the image, resembling the contrast-enhancement effect in Schlieren imaging [8]. Moreover, by time-multiplexing two ATOM images with opposite phase-gradient contrasts, we further obtain two different contrasts from the same specimen: one with differential (enhanced) phase-gradient contrast and another with absorption contrast, simultaneously (Fig. 2). This method decouples the phase-gradient information from absorption, resulting in further enhancement of the image contrast. Together with operating in the 1 μm wavelength regime, a more favorable spectral window for biophotonic applications as opposed to the telecommunications wavelength band (~1500 nm) used by most of the previously reported time-stretch imaging systems, the ATOM system is able to achieve higher diffraction-limited resolution and high-contrast cellular time-stretch imaging. Similar to the original time-stretch imaging, the detection sensitivity in ATOM is not compromised by the high-speed operation because of the in-line optical image amplification—a feature rarely implemented in typical optical microscopy. We have demonstrated the unique capability of ATOM in visualizing the detailed cellular morphology (e.g. normal human blood cells from fresh blood and human leukemic monocytes) without contrast agent in ultrafast microfluidic flow (up to ~10 m/s), which is yet to be demonstrated in the existing time-stretch imaging modality (Fig. 2) [8]. The achieved flow speed here is equivalent to an imaging throughput of ~ 100,000 cells/s. ATOM thus represents a significant advancement in bringing the essential imaging metrics—high resolution and high contrast—to high-speed time-stretch imaging, making it a genuinely appealing platform for realizing high-throughput image-based cellular assays.

1.2 Interferometric Time-Stretch Microscopy

The functional and structural information of the biological cells and tissues can commonly be inferred by the correlated image contrasts. While exogenous fluorescent labels are the well-acclaimed and thus the prevalent contrast agent used in many applications, they are however not always ideal in view of the complication introduced by cytotoxicity and photobleaching. In contrast, endogenous image contrast, e.g., absorption and scattering, could sometimes serve as the effective intrinsic biomarker without the need for label or stain and the associated laborious sample preparation procedures. Among different label-free optical imaging modalities, quantitative phase-contrast imaging (QPI) is proven to be an effective approach to scavenge different useful intrinsic information about the cells and tissues. By mapping the optical phase

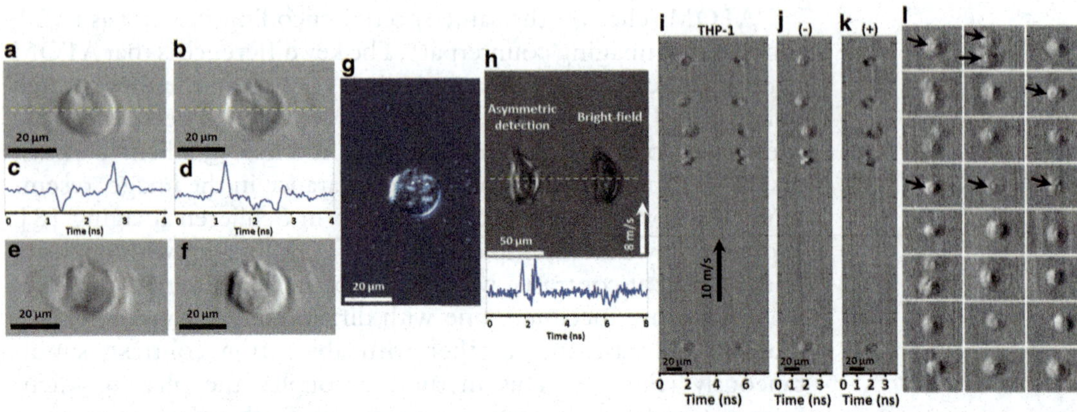

Fig. 2 High-contrast and dual-contrast cellular imaging in flow by ATOM. (**a**, **b**) Two single-angle ATOM images of a normal hepatocyte (MIHA) fixed on a glass slide, which show the opposite phase-gradient contrasts, respectively. (**c**, **d**) The corresponding line profiles (*yellow dotted lines*) of the single-angle ATOM images in a, b, respectively. Each line scan of the image is captured within ~4 ns. (**e**) By calculating the sum of the two single-angle ATOM images (i.e., a + b), an image with absorption contrast can be revealed. (**f**) Dual-angle ATOM image with differential (enhanced) phase-gradient contrast can be obtained by calculating the difference of the two single-angle ATOM images (i.e., a − b). (**g**) White-light DIC image of the same MIHA cell for comparison. (**h**) Demonstration of the contrast enhancement in ATOM by comparing the single-angle asymmetric detection scheme (i.e. off-axis fiber coupling) and the bright-field (BF) scheme (i.e. on-axis fiber coupling), in the context of high-speed flow imaging (8 m/s) of the MIHA cells in a PDMS microfluidic channel. In the time-multiplexed ATOM image, it shows the image replica of the same cell, as if they are flowing in parallel: one is captured by asymmetric detection (*left*) whereas another is captured by on-axis detection, i.e., BF time-stretch image (*right*). The bottom inset is the corresponding line profiles (*yellow dotted lines*) of ATOM image and BF time-stretch image shown. Note that in one single-shot line-scan, both phase-gradient and BF contrasts of the same cell are captured within ~8 ns. (**i**) Time-multiplexed single-angle ATOM images of the stain-free THP-1 cells flowing at a speed of 10 m/s. (**j**) Dual-angle (differential) ATOM image obtained by subtraction of the two opposite-contrast images (the left and right group of cells) in (**i**). (**k**) Dual-angle (absorption) ATOM image obtained by addition of the two opposite-contrast images (the left and right group of cells) in (**i**). More dual-angle (differential) ATOM images of THP-1 cells are shown in (**l**). The enhanced contrast enables visualization of the nuclei of the cells (indicated by the *arrows*)

shift across the transparent cells or tissues, QPI can provide not only high image contrast non-invasively, but also quantitative evaluation of cellular information (e.g., cell volume, mass, refractive index, stiffness, membrane tension) at nanometer scale based on the optical phase [14, 15]. The scattering properties of the cells/tissues can also be extracted from the optical phase [16–19]. All information derived from QPI can then be used as the intrinsic biomarkers for cellular identification and understanding the corresponding physiological information, such as disease state of cells/tissues. Notably, refractive index of cell nuclei serves a distinct indicator for label-free detection of cancer cells with high sensitivity [20, 21].

Motivated by that time-stretch imaging is intrinsically compatible with label-free phase-contrast imaging, it has been demonstrated iTS microscopy for quantitative phase contrast tissue and

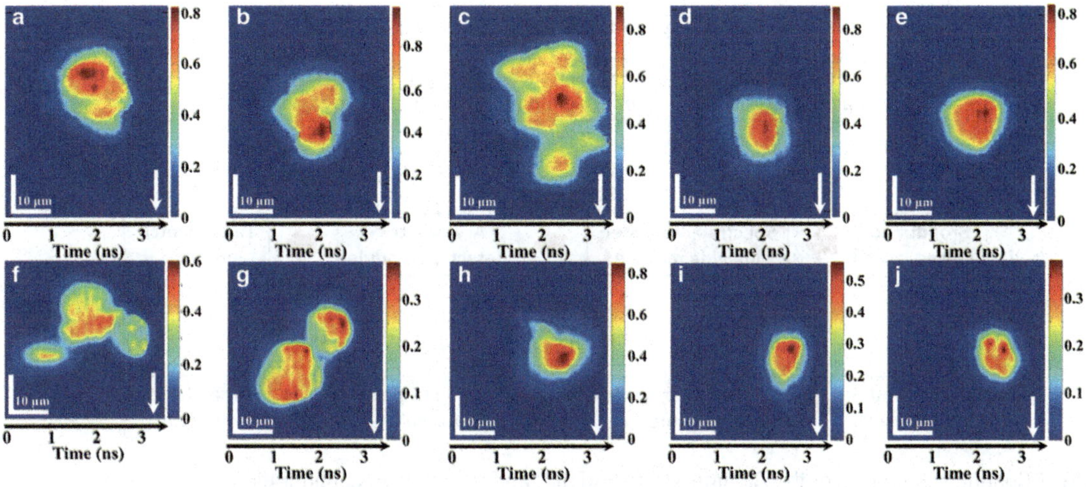

Fig. 3 Imaging of mammalian cells by iTS microscopy in ultrafast flow: (**a–e**) iTS microscopic images of stain-free HeLa cells flowing at a speed of 8 m/s. (**a–c**) Clusters of HeLa cells can be easily distinguished under iTS microscopy with subcellular resolution (~1.2 μm). (**d–e**) Single HeLa cells flowing at ultrahigh speed. (**f–j**) iTS microscopic images of stain-free MIHA cells flowing at a speed of 0.4 m/s. (**f–g**) Clusters of MIHA cells can be easily distinguished under iTS microscopy with subcellular resolution. (**h–j**) Single MIHA cells flowing at 0.4 m/s. All color bars represent the phase in radian. *Arrows* indicate the flow direction

cellular imaging at an ultrahigh imaging scan rate as high as tens of MHz, with nanometer precision [13, 15]. Again, iTS microscopy employs the same spectral-encoding imaging scheme as ordinary BF time-stretch imaging. The key difference is that an additional interferometer (e.g., Michelson interferometer) is required in iTS microscopy such that the spectrally-encoded pulsed beam from the sample arm is interfered with its uncoded replica from the reference arm. The interfered pulses then undergo optical time-stretch which generates the temporal interferograms. The quantitative phase information can then be extracted from the interferograms by the phase-retrieval and unwrapping algorithms routinely used in classical QPI (e.g., Goldstein's algorithm) (Fig. 3). It should be emphasized again that the quantitative phase information has long been known as valuable biomarkers of label-free cellular bioassay [13]. However, the use of these intrinsic biomarkers for large-scale screening has almost been impossible because of the limited imaging throughput of the existing technologies.

2 Materials and System Preparation

In general, a time-stretch imaging system consists of three main modules: a broadband pulsed laser source, a spectral-encoded imaging module and a time-stretch module (Fig. 4). This configuration essentially refers to the BF imaging mode of time-stretch imaging.

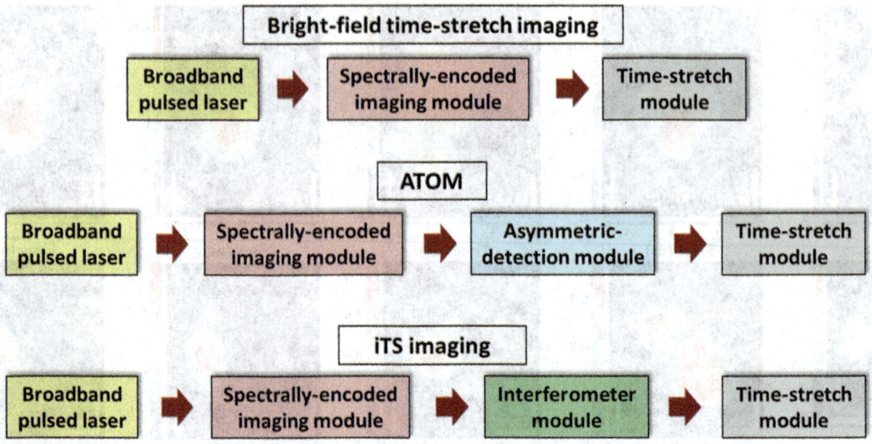

Fig. 4 Generic schematics of the BF time-stretch, ATOM, and iTS imaging configurations

In this chapter, we focus on two other imaging modalities: ATOM and iTS microscopy, each of which has an additional module in the system. In ATOM, an asymmetric-detection module is added to enable enhanced dual contrast of the images. In iTS microscopy, an interferometer module is inserted to obtain the quantitative phase information of the image (*see* Fig. 4). In this section, we describe the enabling components required for each of these modules.

2.1 Equipment and Materials for Microfluidic Chip Fabrication

In both ATOM and iTS microscopy, cellular imaging is performed in the microfluidic platform. Here below are the key materials and equipment for microfluidic channel fabrication:

1. Polydimethylsiloxane (PDMS).
2. Silicon wafer.
3. Incubator (Thermo Scientific, Heratherm, IGS60).
4. Expanded plasma cleaner (Harrick Plasma, PDC-002).
5. Cover slips.
6. Polymethyl methacrylate (PMMA).
7. Maskless soft lithography machine (Intelligent Micro Patterning, LLC SF-100 XCEL).
8. Photoresist (MicroChem Corp. SU-8 2025).
9. Spin coater (Apex Instruments Co. spin NXG-P1).
10. Hot plate (IKA, C-MAG HP 7).
11. SU-8 developer.
12. Isopropyl alcohol.
13. PDMS precursor (SYLGARD 184 Silicone Elastomer): 18 g Sylgard 184 Silicone elastomer oligomer and 2 g cross-linker were mixed together by an agitator.

14. Biopsy punch (Miltex 33-31AA).
15. Plastic tubing (Scientific Commodities, Inc. BB31695-PE/2).

2.2 Essential Components for Spectral Encoding in Both ATOM and iTS Systems

As described in Subheading 1, spectral-encoding is the step in which the spatial information of the cells is optically encoded into the wavelength spectrum of the laser pulse. Here below is the list of the key components for configuring this module:

1. A broadband pulsed laser, which is preferably to be optical fiber-based laser, e.g., mode-locked fiber laser, or supercontinuum (SC) fiber laser, for the sake of system compactness, in contrast to the use of the bulky solid-state pulse lasers (e.g., Ti:Sapphire laser, or Nd:YAG laser).
2. Fiber collimator ×2.
3. Beam splitter ×1.
4. Diffraction grating ×1.
5. Objective lens ×2.
6. Mirror ×1.
7. Compatible kinematic mounts, post holders, and posts for assembling the optics stated in **items 1–6**.

2.3 Essential Components for Enabling ATOM Operation

The following components are required for setting up the asymmetric detection module of the spectrally-encoded optical pulsed beam:

1. Beam splitter ×1.
2. Several mirrors for beam propagation alignment.
3. Knife edge.
4. Fiber collimator ×2.
5. Linear translation stage ×2.
6. Fiber coupler ×1.
7. Compatible kinematic mounts, post holders, and posts for assembling the optics stated in **items 1–6**.

2.4 Essential Components for Enabling iTS Operation

These are the components required for constructing a Michelson interferometer of the iTS microscope. The input beam to the interferometer is spectrally-encoded with the spatial information of the specimen:

1. Linear translation stage ×1.
2. Several mirrors for optical beam propagation alignment within the interferometer.
3. Compatible kinematic mounts, post holders, and posts for above optics.

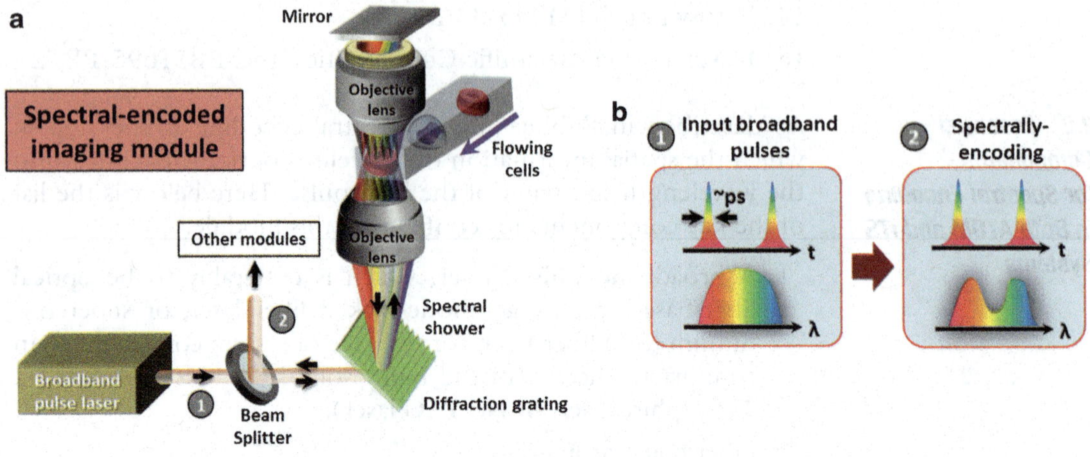

Fig. 5 (**a**) Schematic of the spectrally-encoded imaging module in time-stretch imaging. (**b**) Evolutions of the wavelength spectrum and the temporal waveforms in spectrally encoded imaging

2.5 Key Components for Time-Stretch Operation

These are the components to perform wavelength-to-time mapping of the spectrally-encoded image data (either the ATOM or iTS signal) and thus enable ultrafast image capture:

1. Optical fiber module for time-stretch (with a total GVD value such that the spatial resolution is limited by the diffraction limit, but not by the dispersion process nor the electronic bandwidth of the detection.)
2. A multistage optical fiber amplifier with a total gain high enough to compensate the loss due to dispersion (*see* **Note 4.1.5**).
3. High-speed single-pixel photodetector (8 GHz, Picometrix).
4. A real-time oscilloscope with a recommended sampling rate greater than 10 GS/s and electrical bandwidth more than 10 GHz (e.g., Agilent, 16 GHz, 80 GS/s) or any high-speed data-acquisition (DAQ) system with high sampling rate of >10 GS/s).

3 Methods

3.1 Construction of the Spectral-Encoded Imaging Module

The general schematic of a spectral-encoded imaging module is shown in Fig. 5. A fiber-based pulsed laser (*see* **Note 4.1.10**) is first launched to free space from a fiber collimator and passes through a beam splitter and then a diffraction grating (e.g., a transmission holographic grating as shown in Fig. 5) to perform spatial dispersion. This generates a one-dimensional (1-D) line-scan beam called, spectral shower. This spectral shower maps different wavelengths of the light source into 1-D spatial coordinates. It is then focused by an objective lens for illumination. Another identical

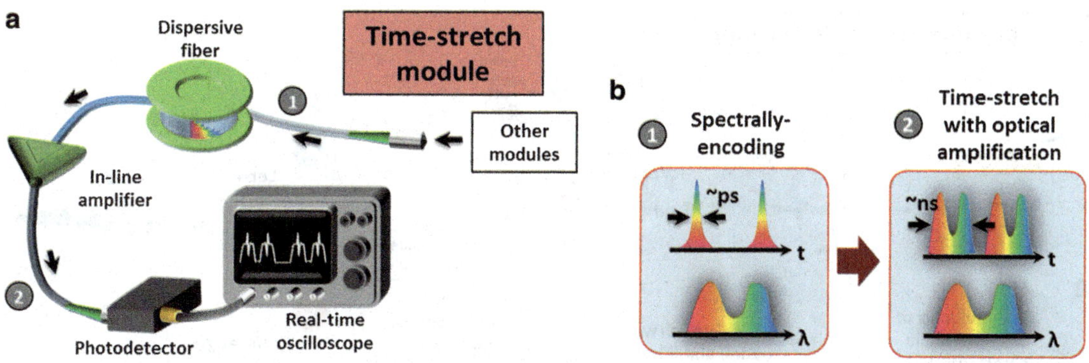

Fig. 6 (**a**) Schematics of the time-stretch module in time-stretch imaging. (**b**) Evolutions of the wavelength spectrum and the temporal waveforms during the time-stretch process

objective lens and a mirror are added behind the sample in order to operate spectral-encoding imaging in a double-pass transmission mode, which is favorable especially for transparent biological cells in flow. The double-passed spectral shower that encodes the spatial information of the sample, is then restored back to original pulsed beam by the same grating. This encoded pulsed beam is then routed by the same beam splitter toward the optical fiber through a fiber collimator. With the above settings, the sample information is now mapped into the spectrum of the pulsed laser, which can be detected using any ordinary optical spectrometer to perform spectral-encoding imaging. However, a time-stretch module is employed for high-speed "readout" of encoded wavelength spectrum in time domain, which is introduced in Subheading 3.2.

3.2 Construction of the Time-Stretch Module

After all the aforementioned imaging techniques (BF, ATOM, or iTS), a spectrally-encoded pulse can be detected in an ultrafast manner with the use of a time-stretch module. The spectrally-encoded light pulse is coupled into the dispersive fiber in which the encoded wavelength spectrum information is transformed into a serial temporal waveform by the process of GVD. The time-stretch signal is also amplified by an in-line fiber-based optical amplifier (*see* **Note 4.1.5**) in order to compensate the optical loss due to free-space coupling as well as the fiber attenuation. Without such optical (image) gain, the time-stretch signal could be too weak to be detectable. Finally, the signal is detected by a high-speed photo-detector and a real-time oscilloscope or any real-time DAQ system (Fig. 6).

3.3 Construction of the Asymmetric Detection Module for the ATOM System

To construct an asymmetric detection module in an ATOM system, the spectrally-encoded beam routed from the beam splitter (described in Subheading 3.1 and Fig. 5) is further split into two identical replicas by an additional beam splitter (Fig. 7a). Each of the two replicas is then partially blocked by a knife edge in its light path.

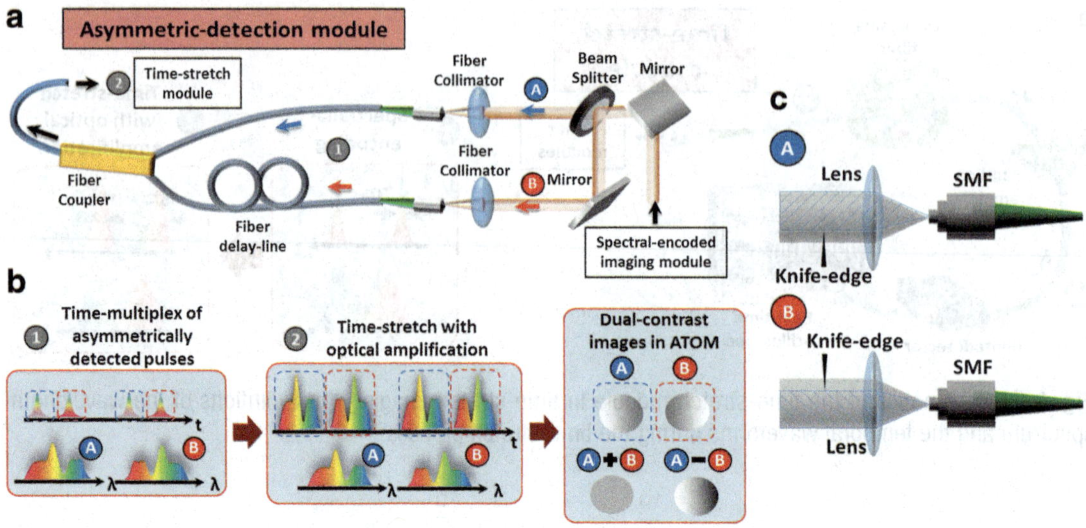

Fig. 7 (**a**) Schematics of the asymmetric detection module in the ATOM system. (**b**) Evolutions of the wavelength spectrum (*bottom*) and the temporal waveforms (*top*) during the time-stretch process. (**c**) The knife-edge configuration for dual-angle ATOM detection

The consequence is that an asymmetric light-cone is then formed and coupled to the fiber through a fiber collimator (Fig. 7c). To enhance the phase-gradient contrast, the knife edge in each light path should oppositely block the light cone, such that the detectable intensity level of the two separate paths is the same. These two separately coupled light beams are then time-multiplexed by a fiber-delay line and a fiber coupler (*see* **Note 4.1.11**), such that one is time-delayed with respect to the other (Fig. 7b).

Furthermore, one can access different phase-gradient contrasts simultaneously within each line scan (i.e., each laser repetition period) by this time-multiplexing scheme. It should be emphasized that we can simultaneously obtain two separate time-stretch images of the same specimen, with two different contrasts in a single run of ATOM measurement: one with differential phase-gradient contrast (subtraction of the two image signals) and another with absorption contrast (summation of the two image signals) (Fig. 7b). This method is based on using the two opposite asymmetric fiber-coupling angles, dual-angle ATOM, and thus allows the access to the absorption information of the specimen to be decoupled from the phase-gradient information. It should be emphasized again that such time-multiplexing scheme results in no compromise on the imaging speed as long as the total duration of two replicas does not exceed one line-scan period (i.e., total duty cycle should be kept <100 %). Thus, ultrafast operation can be maintained, a (i.e., at the line-scan rate of >MHz).

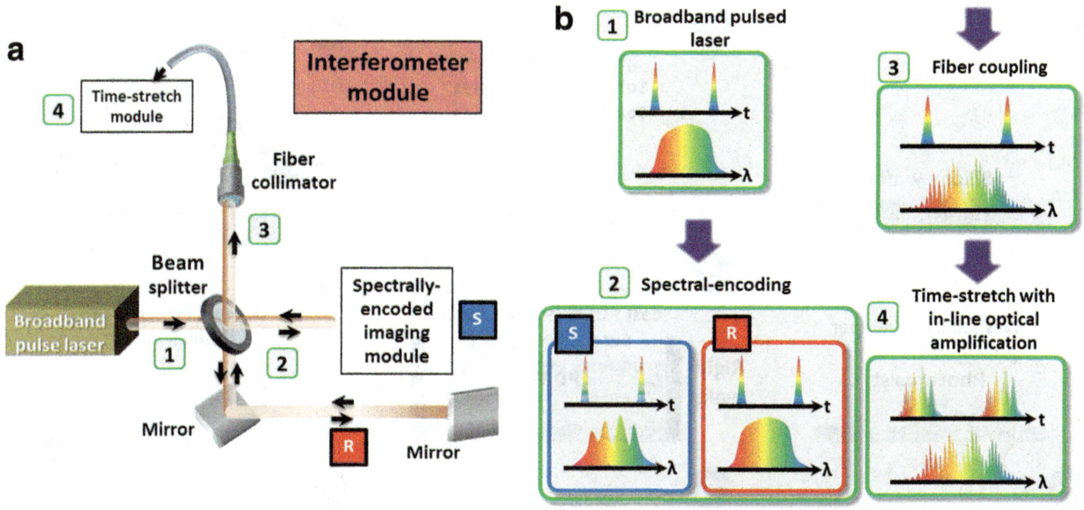

Fig. 8 (**a**) Schematics of the interferometer in the iTS microscope. (**b**) The evolutions of the wavelength spectrum (interferogram) (*bottom*) and temporal waveforms (*top*) from the source to the output of the time-stretch module

3.4 Construction of the Interferometer Module for the iTS Microscope

In order to capture the quantitative phase information of the specimens, iTS microscopy needs an interferometer (specifically, it is in the Michelson interferometer configuration in this case) combined with the spectrally-encoded imaging module (Fig. 8a). This can be done easily in free-space by splitting the broadband pulsed beam into two arms: one to the spectrally-encoded imaging module (called sample arm, S, in Fig. 8) whereas another to a plane mirror (called reference arm, R, in Fig. 8). The quantitative phase image information is obtained from the combined beam of the back-reflected beam from the reference arm and the spectrally-encoded beam from the sample arm. Such beam combination (by using the same beam splitter), which is essentially an effect of interference, gives rise to an interferogram in the wavelength spectrum of the combined beam. In order to generate proper interferogram for phase value extraction, we should carefully tune the path length in the reference arm using a translation stage, which is mounted with the mirror, to be roughly identical to the path length in the sample arm (*see* **Note 4.1.6** for more details about the proper alignment). Note that the spectrally-encoded interferogram can be detected by a classical optical spectrometer, which is a good reference for verifying the wavelength-to-time mapping, but only for calibration purpose. The interferogram is then detected in ultrafast manner by using the time-stretch module—the spectral interferogram is then mapped into a temporal interferogram (Fig 8b).

3.5 Microfluidic Channel Fabrication

The PDMS microfluidic channel platform (chip) has to be designed and fabricated in a way that the balance between the inertial lift force and the viscous drag force is achieved for manipulating the

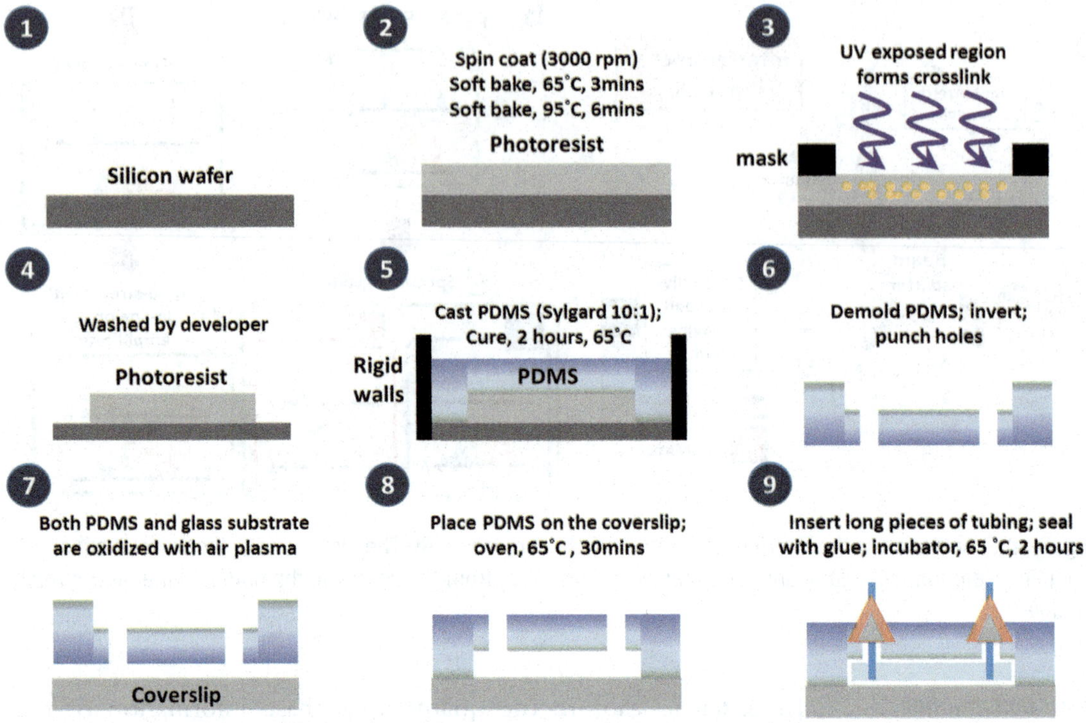

Fig. 9 Fabrication procedures of microfluidic channel for time-stretch imaging [8]

positions of the individual cells and focusing them in ultrafast flow inside the channel [22]. This microfluidic technique is essential for ensuring robust imaging by ATOM at the record high microfluidic flow speed (as high as ~10 m/s). Detailed fabrication steps are described as following (Fig. 9):

1. The microchannel is fabricated by curing PDMS on a silicon mold. The silicon mold is prepared by soft lithography.
2. First, a layer of photoresist is coated on a silicon wafer using a spin coater.
3. The photoresist is soft baked at 65 and 95 °C for 3 and 6 min on a hot plate, respectively.
4. The silicon wafer and photoresist are then cooled under ambient temperature before being exposed to ultraviolet (UV) light.
5. A maskless soft lithography machine is used to transfer a pattern defined by a computer-aided design (CAD) software onto the photoresist. Four seconds of exposure time is chosen. The exposed photoresist is post-baked for 1 and 6 min at 65 and 95 °C on the hot plate, respectively.

6. Afterwards, the photoresist undergoes a developing step by immersing the photoresist-coated silicon wafer in SU-8 developer for 5 min. The silicon wafer is rinsed with isopropyl alcohol followed by a drying step of either nitrogen blow or air blow over the wafer.
7. The PDMS precursor is mixed with the corresponding curing agent with a ratio of 10:1 before pouring onto the silicon mold.
8. A PMMA block is designed and placed on the silicon wafer to form a thin PDMS layer in the imaging section of the channel.
9. The PDMS is cured in an oven at 65 °C for 2 h. Then, the PDMS is demolded and holes are punched using a biopsy punch.
10. The PDMS and a cover slip are cleaned before they are exposed to oxygen plasma. This process makes the surface hydrophilic and enhances the bonding strength between the cover slip and the PDMS block.
11. Afterwards, the whole device is placed in an oven at 65 °C for 30 min for better bonding performance.
12. Lastly, sections of plastic tubing are inserted into the punched holes with PDMS glue applied at the gap between the tubing and the PDMS block to prevent water leakage.

3.6 Key Microfluidic Imaging Operation Steps in BF Time-Stretch Microscopy, ATOM, or iTS Microscopy

1. The microfluidic chip is inserted into the specimen platform, which consists of a microscope clamping platform and a translation stage—essentially similar to the specimen stage of conventional light microscope.
2. Move the upper objective lens shown in Fig. 5 slowly and carefully to align it respect to the lower objective lens as well as other optical components in the spectrally-encoded imaging module in order to obtain the highest power collected by the fiber (*see* **Note 4.1.7**).
3. Find the predefined imaging region on the chip (which should be defined during the chip design step) by moving the translation stage (*see* **Note 4.1.8**).
4. Having located the optimal position, further fine-adjust the in-plane position of the microfluidic chip such that the spectral shower illuminates the entire microfluidic channel width.
5. Adjust the vertical position of the microfluidic chip to ensure an in-focus condition. This can be confirmed by maximizing the extinction ratio of the time-stretch waveform intensity across the channel wall/boundary. After this step, the system is ready for microfluidic imaging.

6. Start to slowly fill in solution (preferably at a flow rate <1 m/s) used for cell suspension. A small rise in the detected optical signal power can be observed. This is due to the reduction of refractive index contrast between PDMS and the channel core (changing from air channel to liquid channel) (*see* **Note 4.1.9**).
7. Once the channel has been completely filled up with the fluid, re-adjust the position of the upper objective lens to optimize the in-focus condition.
8. Fill up the syringe with the cell samples, pipette up and down the sample before fill-up to ensure that cells are all suspended in the syringe.
9. Start to flow the cells at a slower flow rate. Once the fluid flow is observed from the outlet, the flow rate can be increased.
10. Continuously detect and save the temporal waveform data (which is encoded with ATOM, iTS or BF time-stretch image information) using the oscilloscope or real-time DAQ system for subsequent off-line image processing and image reconstruction.

3.7 Image Processing: BF Time-Stretch Imaging and ATOM

All images are processed and generated by a custom-designed MATLAB program off-line (Fig. 10).

1. Read all the detected pulses in a single temporal data trace (Fig. 10a).
2. Find out the number of pixels for each laser repetition period.
3. Digitally segment individual periods and stack each period along the flow direction to form a 2D raw image (Fig. 10b).
4. The final 2D time-stretch image can be obtained by numerically normalizing it by the background pulse in the raw image (i.e., the input spectral shape of the laser source).
5. In the case of ATOM, each raw line-scan contains two asymmetrically-detected time-stretch waveform replicas (segments A and B in Fig. 10). By segmenting the line scan data into two parts (see the dashed line in Fig. 10c), the differential phase-gradient and absorption contrasts can be further obtained by calculating the difference and sum of the two asymmetrically detected time-stretch waveform replicas (having opposite phase-gradient contrasts), respectively (Fig. 10c–f).

3.8 Image Processing: iTS Microscopy

1. Following the same digital line-scan stack procedure mentioned in Subheading 3.7, **steps 1–3**, a 2D interferogram (by stacking the raw 1D interferograms), each single interferogram is high-pass filtered to remove the DC offset (Fig. 11a).
2. The high-pass-filtered interferogram is then Hilbert-transformed to obtain the wrapped phase from a complex analytical signal (Fig. 11b) [23].

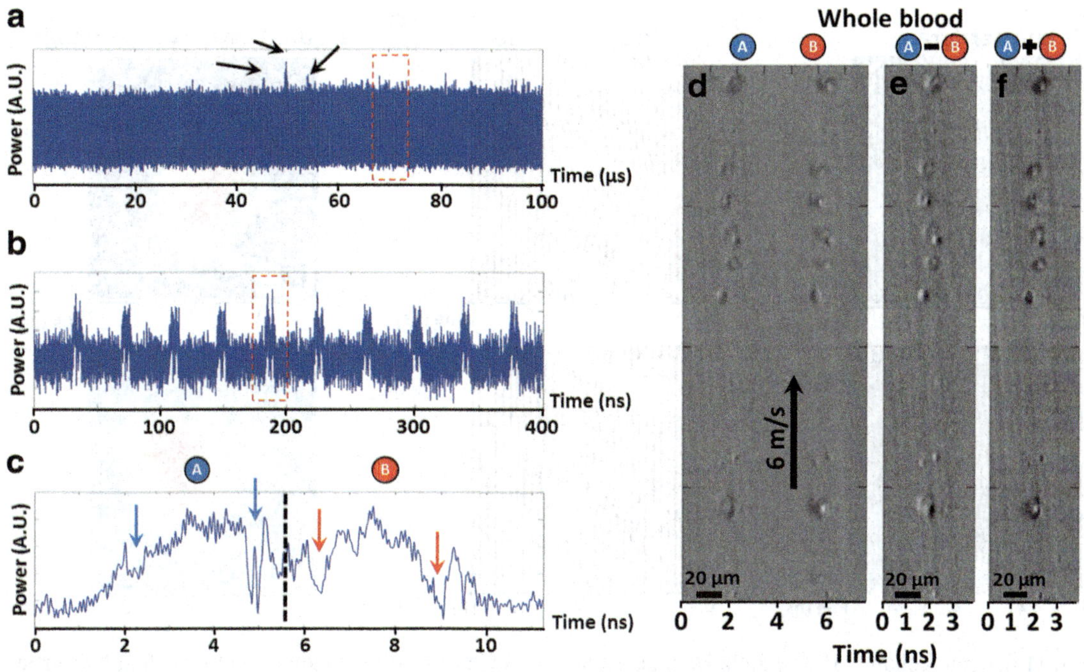

Fig. 10 (**a**) Raw data of ATOM. Arrows indicates the detected individual cells. (**b**) Zoomed-in view of the time-stretch pulses. Each pulse represents a single line in a 2D time-stretch image. Each line is digitally stacked to form the 2D image, as shown in (**e**). (**c**) Further zoomed-in view of a single line-scan. Arrows indicate the positions of the two channel side-walls. (**d**) An ATOM image of red blood cells reconstructed after **step 4** (Subheading 3.7). (**e**, **f**) Dual-contrast ATOM images: differential phase-gradient contrast and absorption contrast images formed by arithmetic subtraction and summation of dual-angle ATOM images

3. Goldstein's algorithm is used to unwrap the phase in the wrapped phase [24].
4. The slanted phase profile is then removed by subtracting the average phase profile in order to obtain the final quantitative phase image (Fig. 11c).

4 Notes

4.1 System Constructions

4.1.1. The diffraction grating used in ATOM plays a critical role on optical resolution and field of view, exact configuration depends on different applications and situations. For more details, refer to ref. 25.

4.1.2. A vertically built setup is preferred to eliminate the effect of gravity on the flow rate.

4.1.3. A half-wave plate may be inserted in the system for the best performance of diffraction grating as well as optimal visibility of the interferogram formed.

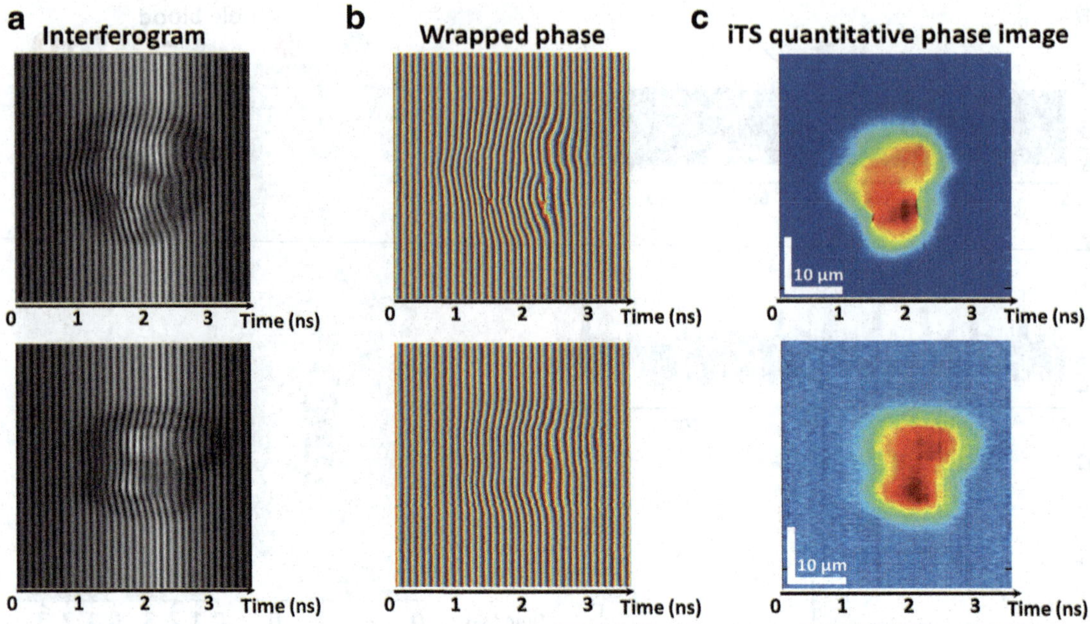

Fig. 11 Reconstruction of iTS quantitative phase image of HeLa cells. (**a**) Raw interferogram is first high-pass filtered to remove the DC-offset. (**b**) The wrapped phase is obtained from the complex analytical signal, which can be processed by Hilbert-transforming the filtered interferogram. (**c**) iTS quantitative phase images are then obtained by standard phase unwrapping algorithm, e.g., Goldstein's algorithm, and background subtraction

4.1.4. In time-stretch microscopy, multiple parameters can affect the imaging performance, such as the groove density of the diffraction grating, numerical aperture (NA) of the objective lenses, the specification of the laser, and the dispersive fiber. Below is an example of the configuration of the time-stretch microscopy. The laser source is a home-built ytterbium-doped mode-locked fiber laser (repetition rate = 26 MHz; center wavelength = 1064 nm, 3-dB bandwidth = ~10 nm, and a pulse width = ~4 ps). The diffraction grating used is a transmission holographic grating with a groove density of 1200 lines/mm. The NA of the objective lens is 0.66, such that a sub-cellular resolution can be retained. For the asymmetric detection and interferometer parts, there are no specific requirements for the components used. In the time-stretch module, the dispersive fiber has a GVD of 0.35 ns/nm to ensure the resolution of the system is limited by free-space components. Optical fiber-based amplifiers may be implemented to compensate the fiber loss and the off-axis fiber coupling loss, the optical gain achieved by the semiconductor optical amplifier (SOA) is ~500, such that the signal level of the time-stretched waveform is strong enough to be detected. Finally, a high-speed single-pixel photodetector with electrical bandwidth of 8 GHz and a real-time oscilloscope with sampling rate of 40 GS/s are used for data sampling and acquisition. With the above configuration, our current system achieves the diffraction-limited image resolution at ~1.2 μm.

4.1.5. In time-stretch microscopy, optical amplification plays a critical role to ensure high-sensitivity operation, i.e., to boost the time-stretch signals to a detectable signal level of the photodetector. This is an inherent advantage time-stretch imaging compared to the conventional imaging modalities, which capture images by CCD/CMOS cameras. Several types of optical amplifiers can be employed for signal amplification in time-stretch imaging, e.g., fiber Raman amplifiers, fiber optical parametric amplifiers (FOPA), SOA, and ytterbium-doped fiber amplifiers (YDFA) [6, 8, 13, 26]. The optimal choice of amplifiers is system and wavelength dependent. Here the general strategies of amplifier selection are listed:

(a) Optical gain should be set such that the signal is brought higher than the noise equivalent power of the photodetector. Noise figure of the amplifier is also another important metric to be taken into account [25].

(b) Gain bandwidth (optical bandwidth) of optical amplifiers should match with the operation wavelength band of time-stretch imaging.

(c) Optical amplifiers operated with optical pumps, e.g., Raman amplifiers, FOPA, and YDFA, are preferred.

(d) Multiple stages of amplification are also preferred for a higher optical gain, instead of a single stage of amplification [25].

(e) After amplification, the output signal spectrum should remain its input spectral shape.

4.1.6. In setting up the interferometer in the iTS microscope, once both the path lengths of the reference and sample arms have been roughly set the same, we should further fine-adjust the temporal overlap between the two laser pulses by monitoring them using an oscilloscope—to make sure the reference arm delay matches the sample arm length. Carefully tune the translation stage in reference arm to find out an optimal fringe frequency, such that the diffraction-limited resolution can be preserved in the interferogram. This can be done by verifying if the fringe separation in wavelength matches with or even smaller than the diffraction-limited resolution. It can be estimated by the knowledge of the mapping scales among spatial coordinates, wavelengths, and time, *see* refs. 13, 27.

4.1.7. Make sure that the spectral shape does not change. Any spectral change could be an indicator of an out-of-focus situation. The spectral shape should be cross-checked with the input spectrum into the time-stretch module.

4.1.8. Locating this optimal position can be facilitated by installing another additional white-light illumination and a standard camera in the spectrally-encoded imaging module.

4.1.9. Make sure that there is no air bubbles in the channel. Otherwise it could result in channel leakage or cracks because of the high pressure.

4.1.10. Source requirements: A broadband pulsed laser is essential for time-stretch imaging, since the field of view of a time-stretch microscope is directly related to the bandwidth [25], and is preferably to be an optical fiber-based laser. In addition to the bandwidth requirement, the pulse-to-pulse stability and spectral stability should also be carefully engineered, in order to have a high signal-to-noise ratio in the reconstructed time-stretch image. For a compact system design, optical fiber based lasers, e.g., mode-locked fiber laser, or supercontinuum (SC) fiber laser, should be used. In order to retain the image resolution in time-stretch module, the pulse width of a laser pulse should also be carefully considered such that spectral components do not overlap after time-stretch process. Therefore, a femtosecond or picosecond laser source is preferred. **Note 4.1.4** shows an example of the laser source.

4.1.11. The fiber delay line used to time-multiplex two single-angle ATOM images should be shorter than the repetition time of the laser, such that the effective imaging line-scan rate of ATOM is not compromised.

4.2 Sample Preparations

4.2.1. Pipette the cell suspension up and down before loading it into the microfluidic channel and imaging.

4.2.2. Cell straining is preferred before use.

4.3 Fabrications of Microfluidic Channel

4.3.1. Clean each surface before every bonding process.

4.3.2. Fabrication in a clean room environment is recommended.

4.3.3. A plastic mold should be used to make sure the chip height is thinner than the total working distance of the objective lens. This region should also be large enough to allow in-plane positioning without contacting the objective lenses.

4.3.4. A large output channel junction is designed right after the narrowest section of the channel in order to effectively reduce the buildup of high flow pressure near the imaging region, facilitating ultrahigh flow speed.

4.3.5. Channel design considerations: The microfluidic platform consists of two parts: a focusing section followed by an

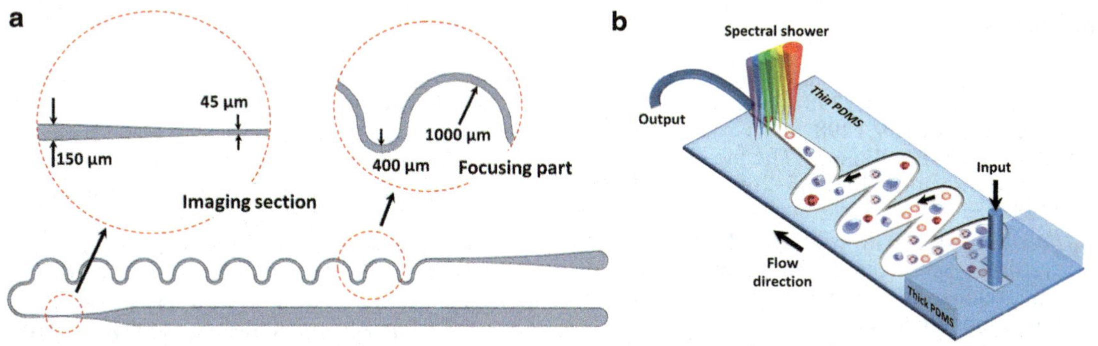

Fig. 12 Schematic of the microfluidic channel platform. (**a**) Detailed channel structure drawn to scale. The channel is divided into a focusing part and an imaging section. The focusing part helps focus the cells in the channel along the same streamline to facilitate subsequent imaging. The positions of the individual cells in the channel can be manipulated by adjusting the balance between the inertial lift force and the viscous drag force through changing the fluid flow rate. (**b**) The focused cells flow into the imaging section of the device, which consists of a microfluidic channel in a thinner PDMS slab (thick PDMS: ~4 mm, thin PDMS: ~1.5 mm). The spectral shower is illuminated in the imaging section, with the spectral encoding direction perpendicular to the microfluidic flow direction. The flow speed in the microfluidic channel is as high as ~10 m/s, which is limited only by the pressure that the channel can withstand. In the imaging of cells, this flow speed corresponds to an imaging throughput up to ~100,000 cells/s

imaging section (Fig. 12a). The focusing section consists of multiple pairs of connected curved channels with radii of curvature 400 µm and 1000 µm, respectively. There are 16 turns in total. (Fig. 12a). The width (150 µm) and height (50 µm) of the channel were chosen such that the channel is suitable for focusing cells with a size ranging from ~5 to 30 µm. In the imaging section in which the spectral shower is illuminated onto the channel, the channel width is narrowed to 45 µm to further boost the flow speed. Note that laminar flow condition is still satisfied at such an ultrafast flow. The Reynolds number of our current microfluidic channel design is at 600, which is far below the limit of 2000, beyond which the turbulence flow occurs [22]. The thicknesses of the top and bottom channel walls have been minimized to accommodate the high NA objective lens, which typically has a working distance of less than 1 mm (Fig. 12b).

Acknowledgements

This work was partially supported by grant from the Research Grants Council of the Hong Kong SAR, China (Project No. HKU 717510E, HKU 7172/12E, 717911E, 720112E, and 17207714), and University Development Fund of HKU.

References

1. May M (2014) Big biological impacts from big data. Science 344:1298–1300. doi:10.1126/science.344.6189.1298
2. Marx V (2013) Biology: The big challenges of big data. Nature 498:255–260. doi:10.1038/498255a
3. Perfetto SP, Chattopadhyay PK, Roederer M (2004) Seventeen-colour flow cytometry: unravelling the immune system. Nat Rev Immunol 4:648–655
4. Basiji DA, Ortyn WE, Liang L, Venkatachalam V, Morrissey P (2007) Cellular Image Analysis and Imaging by Flow Cytometry. Clin Lab Med 27:653–670
5. Baker M (2011) Faster frames, clearer pictures. Nat Methods 8:1005–1009. doi:10.1038/nmeth.1777
6. Goda K, Tsia KK, Jalali B (2009) Serial time-encoded amplified imaging for real-time observation of fast dynamic phenomena. Nature 458:1145–1149. doi:10.1038/nature07980
7. Goda K, Ayazi A, Gossett DR, Sadasivam J, Lonappan CK, Sollier E, Fard AM, Hur SC, Adam J, Murray C, Wang C, Brackbill N, Carlo DD, Jalali B (2012) High-throughput single-microparticle imaging flow analyzer. Proc Natl Acad Sci U S A 109:11630–11635. doi:10.1073/pnas.1204718109
8. Wong TT, Lau AK, Ho KK, Tang MT, Robles JD, Wei X, Chan AC, Tang AH, Lam EY, Wong KK, Chan GC, Shum HC, Tsia KK (2014) Asymmetric-detection time-stretch optical microscopy (ATOM) for ultrafast high-contrast cellular imaging in flow. Sci Rep 4:3656. doi:10.1038/srep03656
9. Goda K, Jalali B (2013) Dispersive Fourier transformation for fast continuous single-shot measurements. Nat Photon 7:102–112. doi:10.1038/nphoton.2012.359
10. Wong TT, Lau AK, Wong KK, Tsia KK (2012) Optical time-stretch confocal microscopy at 1 μm. Opt Lett 37:3330–3332. doi:10.1364/OL.37.003330
11. Tearney GJ, Webb RH, Bouma BE (1998) Spectrally encoded confocal microscopy. Opt Lett 23:1152–1154
12. Kang D, Yelin D, Bouma BE, Tearney GJ (2009) Spectrally-encoded color imaging. Opt Express 17:15239–15247
13. Lau AK, Wong TT, Ho KK, Tang MT, Chan AC, Lam EY, Shum HC, Wong KK, Tsia KK (2014) Interferometric time-stretch microscopy for ultrafast quantitative cellular and tissue imaging at 1 μm. J Biomed Opt 19(7):076001. doi:10.1117/1.JBO.19.7.076001
14. Popescu G (2011) Quantitative Phase Imaging of Cells and Tissues. McGraw Hill, New York
15. Mahjoubfar A, Chen C, Niazi KR, Rabizadeh S, Jalali B (2013) Label-free high-throughput cell screening in flow. Biomed Opt Express 4:1618–1625. doi:10.1364/BOE.4.001618
16. Park YK, Diez-Silva M, Fu D, Popescu G, Choi W, Barman I, Suresh S, Feld MS (2010) Static and dynamic light scattering of healthy and malaria-parasite invaded red blood. J Biomed Opt 15:020506. doi:10.1117/1.3369966
17. Mourant JR, Freyer JP, Hielscher AH, Eick AA, Shen D, Johnson TM (1998) Mechanisms of light scattering from biological cells relevant to noninvasive optical-tissue diagnostics. Appl Opt 37:3586–3593
18. Alexandrov SA, Hillman TR, Sampson DD (2005) Spatially resolved Fourier holographic light scattering angular spectroscopy. Opt Lett 30:3305–3307
19. Haldar JP, Wang Z, Popescu G, Zhi-Pei L (2010) Label-free high-resolution imaging of live cells with deconvolved spatial light interference microscopy. Conf Proc IEEE Eng Med Biol Soc 2010:3382–3385. doi:10.1109/IEMBS.2010.5627917
20. Wang Z, Tangella K, Balla A, Popescu G (2011) Tissue refractive index as marker of disease. J Biomed Opt 16:116017. doi:10.1117/1.3656732
21. Park Y, Diez-Silva M, Popescu G, Lykotrafitis G, Choi W, Feld MS, Suresh S (2008) Refractive Index Maps and Membrane Dynamics of Human Red Blood Cells Parasitized by *Plasmodium falciparum*. Proc Natl Acad Sci U S A 105:13730–13735. doi:10.1073/pnas.0806100105
22. Di Carlo D (2009) Inertial microfluidics. Lab Chip 9:3038–3046. doi:10.1039/b912547g
23. Ikeda T, Popescu G, Dasari RR, Feld MS (2005) Hilbert phase microscopy for investigating fast dynamics in transparent systems. Opt Lett 30:1165–1167
24. Ghiglia DC, Pritt MD (1998) Two-dimensional phase unwrapping: Theory, Algorithms, and Software. Wiley, New York
25. Tsia KK, Goda K, Capewell D, Jalali B (2010) Performance of serial time-encoded amplified microscope. Opt Express 18:10016–10028. doi:10.1364/OE.18.010016

26. Wei X, Lau AK, Xu Y, Zhang C, Mussot A, Kudlinski A, Tsia KK, Wong KK (2014) Broadband fiber-optical parametric amplification for ultrafast time-stretch imaging at 1.0 μm. Opt Lett 39:5989–5992. doi:10.1364/OL.39.005989

27. Bhaduri B, Edwards C, Pham H, Zhou R, Nguyen TH, Goddard LL, Popescu G (2014) Diffraction phase microscopy: principles and applications in materials and life sciences. Adv Opt Photon 6:57–119. doi:10.1364/AOP.6.000057

Chapter 4

Applications of Imaging Flow Cytometry for Microalgae

Mark Hildebrand, Aubrey Davis, Raffaela Abbriano, Haley R. Pugsley, Jesse C. Traller, Sarah R. Smith, Roshan P. Shrestha, Orna Cook, Eva L. Sánchez-Alvarez, Kalpana Manandhar-Shrestha, and Benjamin Alderete

Abstract

The ability to image large numbers of cells at high resolution enhances flow cytometric analysis of cells and cell populations. In particular, the ability to image intracellular features adds a unique aspect to analyses, and can enable correlation between molecular phenomena resulting in alterations in cellular phenotype. Unicellular microalgae are amenable to high-throughput analysis to capture the diversity of cell types in natural samples, or diverse cellular responses in clonal populations, especially using imaging cytometry. Using examples from our laboratory, we review applications of imaging cytometry, specifically using an Amnis® ImageStream®X instrument, to characterize photosynthetic microalgae. Some of these examples highlight advantages of imaging flow cytometry for certain research objectives, but we also include examples that would not necessarily require imaging and could be performed on a conventional cytometer to demonstrate other concepts in cytometric evaluation of microalgae. We demonstrate the value of these approaches for (1) analysis of populations, (2) documentation of cellular features, and (3) analysis of gene expression.

Key words Imaging cytometry, Microalgae, Phytoplankton, Intracellular processes, Population heterogeneity

1 Introduction

Unicellular microalgae are a highly diverse group of organisms, having been present during most of the evolutionary history of the planet. They encompass prokaryotes and eukaryotes, and include distinct and multiple endosymbiotic events [1], leading to substantial morphological and metabolic diversity. Microalgae also display diverse responses to environmental changes. Because of their diversity and abundance, microalgae are ideal subjects for high-throughput multiparametric cytometric interrogation. Analysis of large numbers of individual cells enables characterization at the population level as well as variation within populations.

Natasha S. Barteneva and Ivan A. Vorobjev (eds.), *Imaging Flow Cytometry: Methods and Protocols*, Methods in Molecular Biology, vol. 1389, DOI 10.1007/978-1-4939-3302-0_4, © Springer Science+Business Media New York 2016

Standard flow cytometry is a well-developed tool for microalgal analysis, especially from natural populations [2]. Flow cytometers were originally developed for biomedical applications, but have contributed to many important oceanographic discoveries, such as the first detection of the ubiquitous marine cyanobacteria *Prochlorococcus* [3]. In a typical analysis, various wavelength or scatter parameters are varied to optimize the separation of distinct populations of cells. In a flow cytometer equipped with a cell sorter, these cells can then be identified using microscopy or molecular techniques. The inability to image cells directly in a cell population can impose limitations on sorting flow cytometry. These include (1) the possible need to sort large numbers of subpopulations to identify all constituents, (2) the lack of precise size and morphological information, and (3) the inability to determine the spatial arrangement of intracellular components. Size estimates by standard flow cytometry are relative, being based on forward scatter as a proxy for spherical diameter, which restricts the amount of information that could be gained as compared with direct microscopic imaging. Intracellular imaging is critical for the understanding of many cellular responses, because it can provide a bridge between macroscale cellular processes that result from molecular-scale events.

Several imaging cytometry devices have been developed that can be applied to microalgae. In addition to generating scatterplot analyses of typical flow cytometers, every digital representation of a cell as a "dot" in the scatterplot is derived from and connected to a corresponding set of images. One instrument is the FlowCam™ (Fluid Imaging Technologies), which images and performs particle analysis for the identification and classification of aquatic microorganisms, to monitor environmental populations, assess water quality, or detect invasive species. This instrument relies on a flash LED to back light the sample while the camera is triggered synchronously with the flash to provide a "frozen" image. It also has the ability to add fluorescence values to further discriminate cell types. The FlowCam can operate on natural water samples without sheath fluid, and can accommodate a much wider range of particle sizes (from 2 μm up to 2 mm) than other cytometers. Analysis software identifies individual cells within an image and can then perform up to 26 different measurements, in terms of morphological features (diameter, length, width, perimeter, circularity, etc.), grayscale features (intensity, transparency, color information, etc.), and spectral features (peak area and width of fluorescence channels). The analysis can assign "weights" to various parameters to enable automated identification and classification. Another instrument is the Imaging FlowCytobot (McLane Research Laboratories) which is a submersible imaging-in-flow system used for species identification and monitoring in the environment. The instrument is capable of taking images and measuring chlorophyll fluorescence [4]. Seawater samples are injected into reusable sheath fluid that passes through a flow cell illuminated by a red diode laser. Particles either scatter light or emit chlorophyll fluorescence, which trigger a xenon flash lamp that provides Kohler illumination for imaging [4]. Automated image analysis software applies a 22-category classifier that enables taxonomic identification [5]. A third instrument, the ImageStream®X (Amnis® part of EMD Millipore), enables multimode imaging of cells simultaneously in bright field, dark field (analogous to flow cytometry measured side scatter laser light, SSC), and a broad range of fluorescence wavelengths using a time delay integration CCD camera that integrates images passing through the field, generating a summed image with high resolution. Three objectives (20×, 40×, and 60×) are featured, and up to seven lasers are possible. Rather than using specific cutoff filters, fluorescent signals are separated into spectral "channels" using long-pass filters. Although the wavelength range of an individual channel can be broad (up to 80 nm), compensation can be done to eliminate cross talk between channels. Up to 5000 cells per second can be imaged at resolutions rivaling dedicated fluorescence microscopes, with the caveat that optical sectioning cannot be done. Input of samples in a 96-well format is possible. Image analysis is done using IDEAS® software that allows measurement of nearly 1000 photometric and morphometric features. Feature classifiers can be customized and combined in a proportional manner, customized image masks can be defined, and image analysis algorithms generated using Boolean logic.

The FlowCam has been primarily used to assess plankton community and size structure in natural water populations [6–8]. The Imaging FlowCytobot has similar applications, but has also been used for long term measurement of phytoplankton abundance and bloom development [5, 9, 10]. The ImageStream®X was not specifically designed for in situ marine applications; however it features high-resolution cellular imaging in bright-field and multiple fluorescence channels not possible with the other instruments. Although natural samples can be analyzed with this instrument, the caveat is that these may need to be concentrated to obtain data at a reasonable rate. Our laboratory has extensively used an ImageStream®X, which provide the data in this chapter. Our lab-based work has focused on chlorophytic algae (*Chlorella*), and diatoms—which are unicellular microalgae with silica-based cell walls, but other classes of microalgae have been amenable to examination.

In this introduction we include examples of various applications of ImageStream®X analysis from our lab. Some of these examples highlight advantages of imaging flow cytometry for certain research objectives, but we also include examples that would not necessarily require imaging and could be performed on a conventional cytometer to demonstrate other concepts in cytometric evaluation of phytoplankton. We present examples from three general categories to which we have applied the ImageStream®: (1) analysis of populations, (2) documentation of cellular features, and

(3) analysis of gene expression. Interspersed within these categories is the ability to document epigenetic or life cycle stage variations.

Heterogeneity can occur in a sample containing mixed species or with clonal constituents. An example of diversity within a mixed sample is shown in Fig. 1, which depicts an analysis of seawater from the Scripps Institution of Oceanography pier. A unique ability of imaging relative to conventional cytometry is the ability to plot the aspect ratio (measurement of width over height) of objects. If we consider area in an imaging cytometer as a proxy for forward scatter in a conventional cytometer, a plot of area vs. chlorophyll shows an unresolved grouping (Fig. 1a). However, a plot of aspect ratio vs. chlorophyll of the seawater sample resolves the sample into two general groups, especially resolving populations R1 and R2 (Fig. 1b), which are not resolved in area vs. chlorophyll cytogram (Fig. 1a). There are still unresolved subpopulations in the aspect ratio vs. chlorophyll plot; however, the ability to evaluate image information enables resolution of subpopulations ranging from filamentous cyanobacteria to larger or smaller eukaryotic plankton cells with distinct shapes and sizes (Fig. 1b). By comparing individual images near the borders of gates, consistent definitions can be made, even when populations blend into each other. Even in gates containing mixed constituents, the ability to analyze individual cells enables their manual categorization, and more accurate evaluation of percentages of the different types of cells.

The ability to image allows documentation of variation between features of individual cells in a clonal population. Clonal populations under controlled culture conditions may be thought of as being homogeneous, however a variety of non-genetically based variations are possible [11–14]. Heterogeneity in clonal populations can be quantified by manifestation of an observable phenotype, either directly through morphological changes or through the use of dyes that serve as proxies for physiological state. The breadth of variation cannot be directly documented with bulk measurements.

Examples of heterogeneity in clonal populations demonstrate several important concepts. One example in microalgae is during accumulation of the neutral nonpolar lipid class triacylglycerol (TAG). TAG can be monitored by specific lipophilic dyes such as Nile Red [15, 16] and BODIPY [17]. In Fig. 2, we show the analysis of TAG accumulation after glucose addition to the culture medium in the chlorophytic algae *Chlorella protothecoides*, monitored by fluorescence of Nile Red. *C. protothecoides* reproduces by forming autospores, in which multiple daughter cells are contained within a single mother cell [18]. Imaging of cells enables a clear distinction between mother and daughter cells based on size and number of chloroplasts visible per cell (Fig. 2). We defined four categories (Fig. 2) based on cell type and lipid content and monitored TAG accumulation after 24 h of glucose treatment. Single

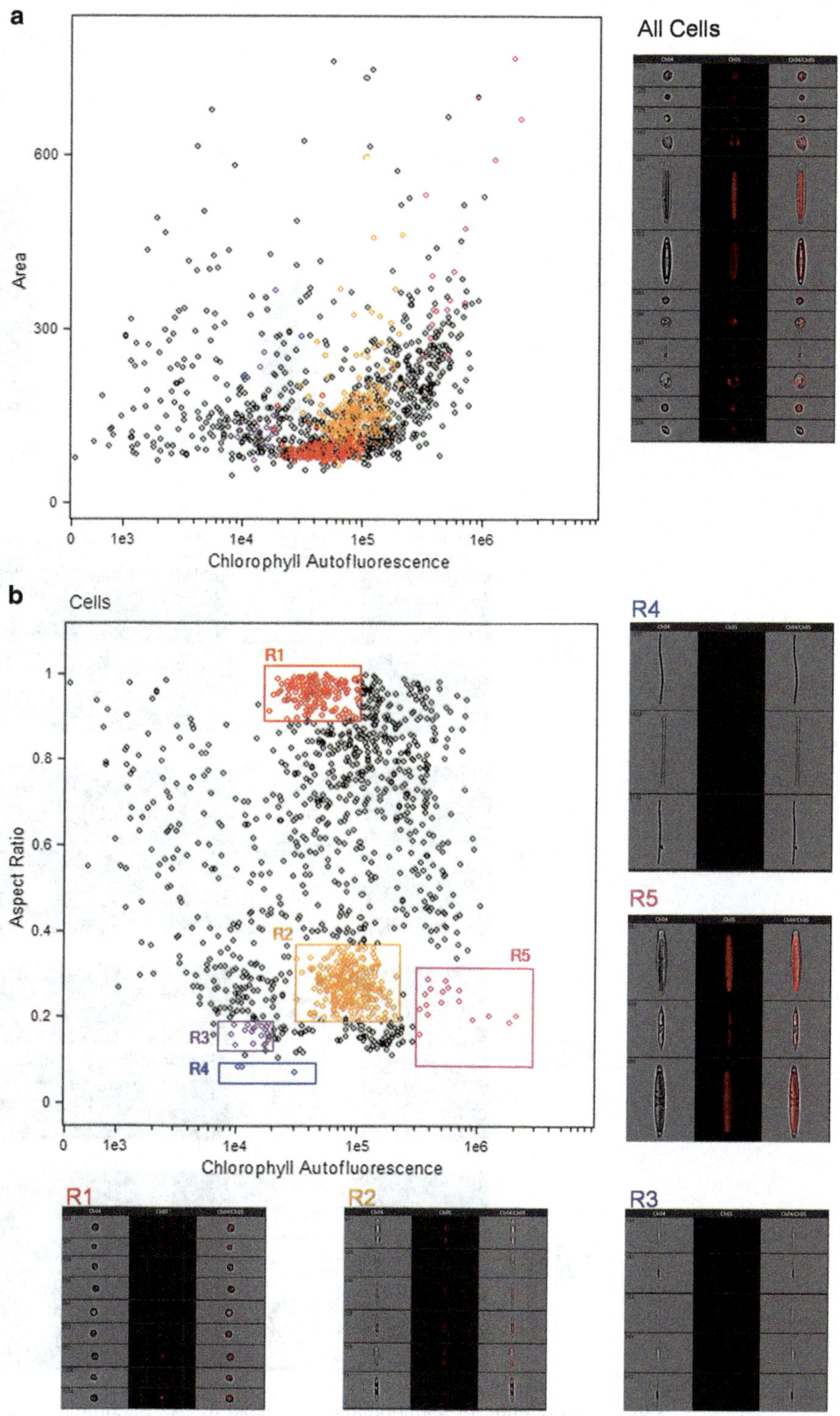

Fig. 1 Analysis of a natural seawater population. (**a**) Scatterplot of area vs. chlorophyll fluorescence as an example of a conventional cytometer output. Subpopulations are not readily distinguishable. (**b**) An aspect ratio vs. chlorophyll scatterplot, typical of imaging cytometry. Distinct subpopulations (R1 and R2) are visible on the plot, and examination of images identified additional subpopulations (R3–5). Colors of gates correspond to sets of images displayed at the periphery of the plot. *Colored dots* in (**a**) are the populations defined in (**b**), demonstrating the lack of resolution of area vs. chlorophyll

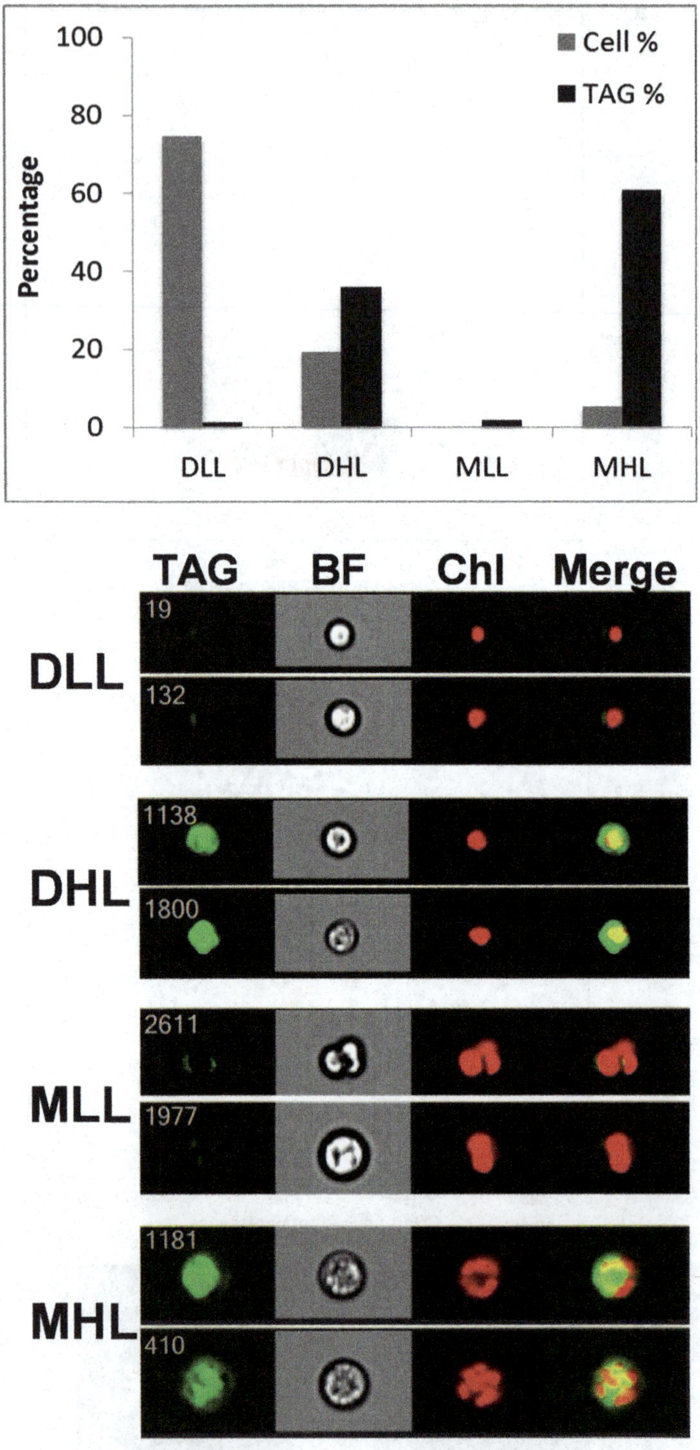

Fig. 2 Contribution of subpopulations of cells to TAG accumulation in *Chlorella protothecoides*. Four classes of cells were identified, Daughters Low Lipid (DLL), Daughters High Lipid (DHL), Mothers Low Lipid (MLL), and Mothers High Lipid (MHL), two examples of each class are shown at bottom. Images are TAG, bright field (BF), chlorophyll (Chl), and merged TAG and chlorophyll (Merge). *Top*, the percentage of cells in each class and their percentage contribution to total TAG in the culture are plotted

daughter cells comprise 94 % of the population. A subpopulation of the daughter cells consisting of 20 % of the total cell number accumulated 36 % of the total TAG in the culture, but the remainder of the daughter cells (75 % of the total in the population) did not accumulate TAG (Fig. 2). Approximately 6 % of the total cell number was mother cells that accounted for 61 % of the total TAG accumulated in the culture, and a small fraction of the mother cells did not accumulate TAG (Fig. 2). This analysis highlights the ability of imaging cytometry to define different life stages of cells, particularly by enumerating the number of chloroplasts, and shows that the response of individual cells within a given life stage can respond dramatically differently from each other.

During the course of TAG accumulation, fluorescent staining can transition from diffuse (where TAG is dispersed within the cell), to concentrated (TAG-containing lipid droplets accumulate). Droplet sizes can also vary in distinct ways, from small droplets to large aggregates that constitute a majority of the cell volume. This is illustrated in Fig. 3, which monitors the accumulation of TAG in the diatom *Cyclotella cryptica* during a time course of silicon starvation [19]. Six distinct subpopulations were identified based on BODIPY and chlorophyll fluorescence (Fig. 3b). These different populations represent distinct phases in TAG accumulation and can be considered as phenotypes [19]. A useful feature of the IDEAS® software is that displayed images can be selected as those with the true average value of fluorescence of a particular population, as is shown in Fig. 3b. Population shifts based on cellular phenotypic differences can be used to make informed choices for other analyses. A bulk analysis of the experiment in Fig. 3 indicates a biphasic process of TAG accumulation with a slower phase between 0 and 48 h, followed by a much more rapid phase between 48 and 72 h (Fig. 3c). This suggests that the hyperaccumulating high chlorophyll subpopulation, which becomes dominant during this time (Fig. 3d), is undergoing a fundamentally different process of TAG accumulation than occurred with earlier subpopulations. Sampling time points between which substantial shifts in subpopulations occur would be highly informative in terms of understanding the mechanisms underlying the metabolic shifts. The additional phenotypic information provided by imaging makes this approach more informative than a bulk measurement or by conventional cytometry, where identifying the phenotype would be more cumbersome.

Growth conditions or selection processes can generate subpopulations with overall morphology or size characteristics relevant to life cycle or growth processes; such information can be readily captured by imaging cytometry. Because diatoms have rigid silica-based cell walls that come in a wide variety of species-specific shapes, they are especially amenable to size and morphological analysis. With some diatom species that colonize surfaces, individual

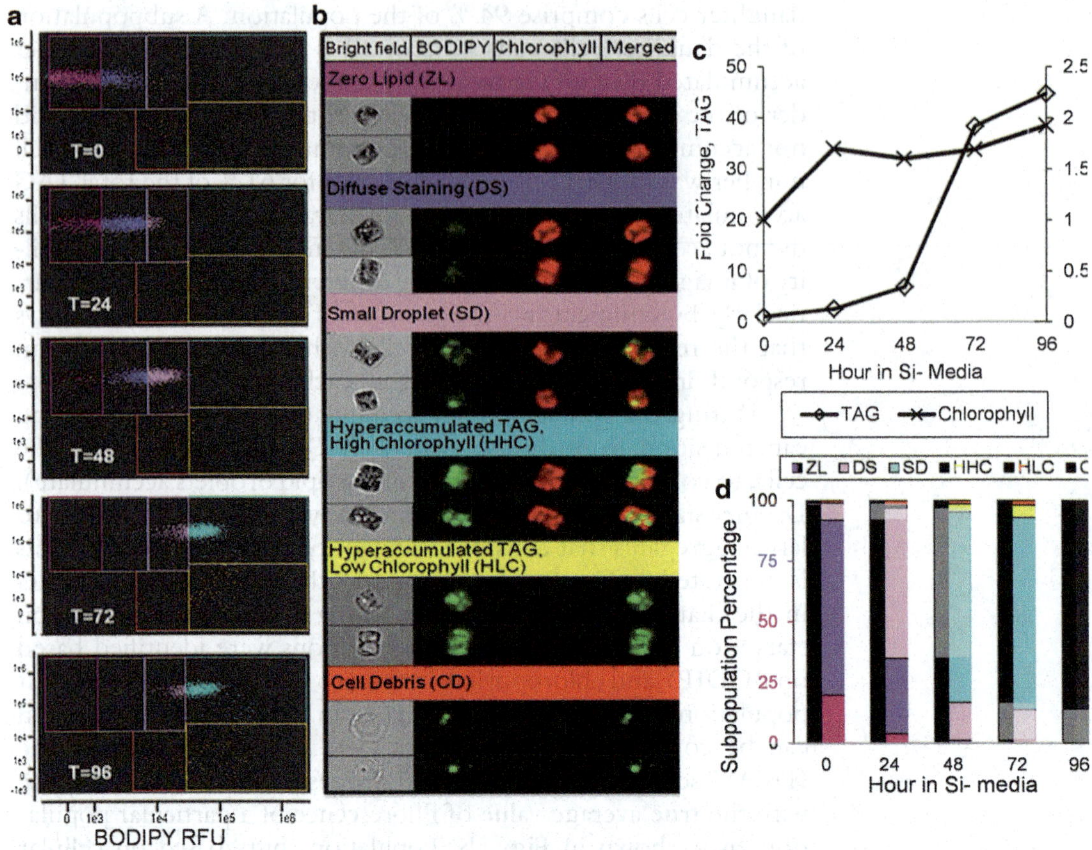

Fig. 3 Imaging cytometry analysis of TAG accumulation during silicon starvation of *Cyclotella cryptica*. A scatter plot time course analysis of changes in chlorophyll and BODIPY fluorescence (**a**) revealed six sub-populations of cells (**b**). (**b**) shows representative cells that are the actual mean for the subpopulation, and the color of the heading corresponds to a particular gated subpopulation in (**a**). Each cell has four images: bright field, BODIPY fluorescence (*falsely colored green*), chlorophyll autofluorescence (*falsely colored red*), and a merged image of BODIPY and chlorophyll. (**c**) Bulk measurement of the mean fold change in TAG and chlorophyll content per cell during the silicon starvation time course. (**d**) Relative percentage of cells within the different subpopulations during the time course. Colors of each bar correspond to the color of each gate in (**a**). Changes in the rate of TAG accumulation correspond to shifts into the HHC subpopulation. Figures are from Traller and Hildebrand [19]

cells cannot be easily separated due to aggregation resulting from secretion of adhesive material. This prevents accurate morphological measurements of individual cells. Taking advantage of the silica cell wall, which can be cleaned free of organic material by acid treatment, we developed a method to image subpopulations of these cell types [20]. After acid cleaning [21] diatom silica can be fluorescently stained with APS-FITC [22, 23], and silica structures can subsequently be analyzed by imaging cytometry (Fig. 4). In Fig. 4, we identify two different sized subpopulations of the surface-associated diatom *Craspedostauros australis* by first identifying structures called valves in a scatterplot (Fig. 4a), and then plotting

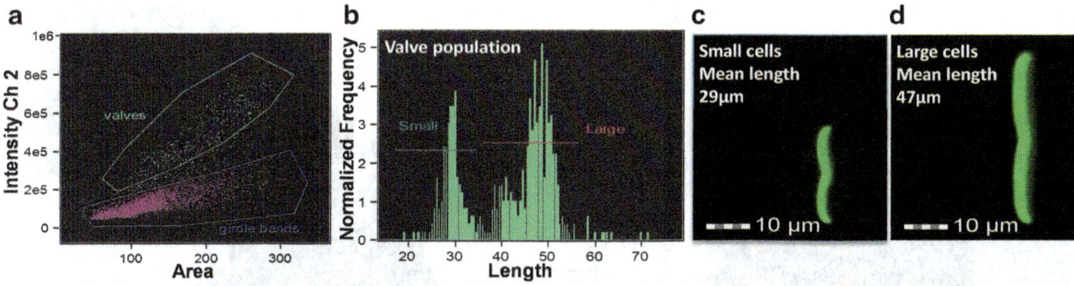

Fig. 4 Variation in *C. australis* cell wall morphology analyzed after acid-cleaning and staining with APS-FITC. (**a**) Scatterplot showing cell wall substructures called valves and girdle bands gated into separate populations. (**b**) Histogram of the length of the valves demonstrating two populations of different sizes. (**c**) Image of a representative cell from the bin with the mean value of the small population. (**d**) Image of a representative cell from the bin with the mean value of the large population

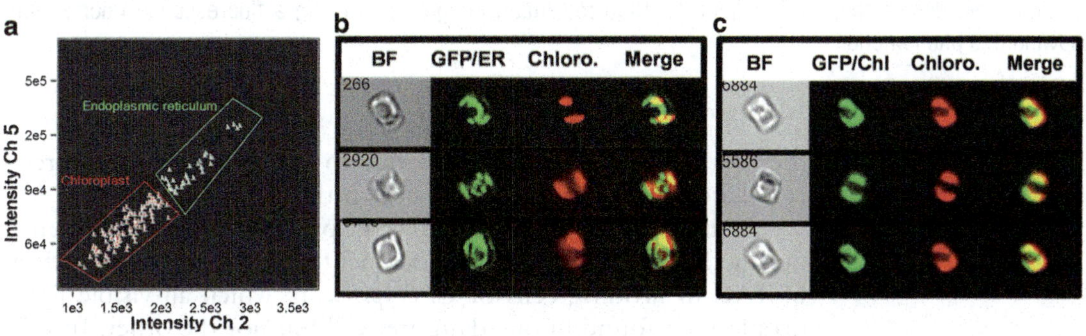

Fig. 5 Distinction of intracellular location of a GFP-tagged protein using the IDEAS software Co-localization wizard. (**a**) Scatterplot showing subpopulations identified as having endoplasmic reticulum (ER) or chloroplast localization. (**b**) Cell images with ER localization of the protein in bright field (BF), GFP, chloroplast (Chloro.), and merged channels. (**c**) Cell images with chloroplast localization of the protein with the same channels

the valve length in a histogram (Fig. 4b). These clearly separate into large and small subpopulations (Fig. 4b–d).

Another valuable application of imaging cytometry is in monitoring cellular processes by examining changes in cellular features. The location of intracellular components is an important aspect of understanding their function. Conventional cytometry lacks the ability to localize intracellular components, and conventional microscopy lacks high-throughput capability to generate robust statistics, as well as being subject to unintentional biases. In Fig. 5, we demonstrate the ability of IDEAS® software to distinguish the intracellular location of a GFP-tagged protein which, depending on conditions, is located in the chloroplast or in the endoplasmic reticulum (ER). A factor that complicates this analysis is that in diatoms the ER also surrounds the chloroplast [24]. The "co-localization" wizard in IDEAS® is able to distinguish between

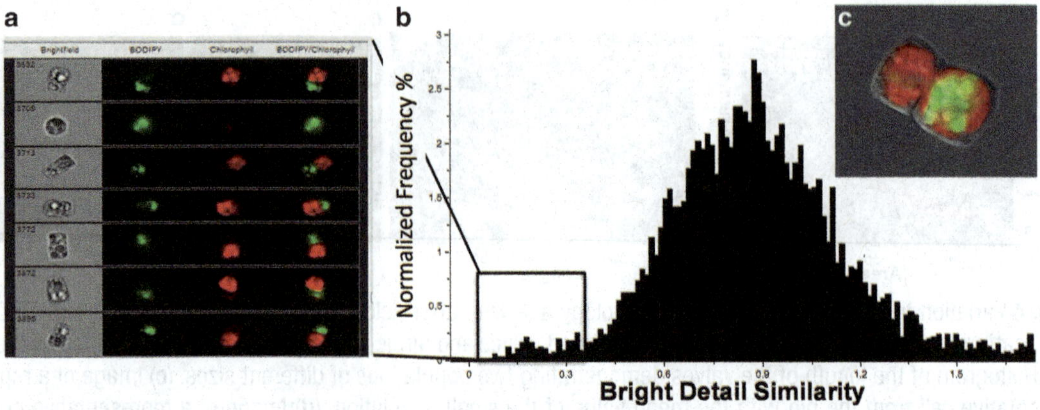

Fig. 6 Differential partitioning of chloroplasts and lipid droplets during cell division as shown by BODIPY and chlorophyll fluorescence. (**a**) Images of subpopulation of cells identified using the bright detail similarity feature of the two channels of interest to localize chlorophyll (*red*) and lipid droplets (*green*) within one cell. Numbers in bright field denote the cell identifier. (**c**) High-resolution image taken using a fluorescence microscope showing this phenomenon

the two possibilities, most likely due to the distinct GFP fluorescence around the nucleus in the cells with ER localization (Fig. 5).

An important cellular process is the partitioning of components during division. In Fig. 6, we document an unequal accumulation of TAG in dividing cells of *C. cryptica*, in which all visible lipid droplets are found in one daughter cell but not the other. In this analysis, a bright detail similarity feature for chlorophyll and BODIPY channels enables determination of the relative localization of probes within single cells. A histogram plot easily identifies those cells in which the bright detail of chlorophyll and BODIPY fluorescence are least similar (i.e., overlap the least).

Another analysis of cellular components is monitoring organelle abundance and features. In Fig. 7, we applied imaging flow cytometry to characterize chloroplasts and their replication in diatoms. By defining a custom mask on chlorophyll fluorescence and utilizing the spot count feature in IDEAS®, a variety of aspects of chloroplasts can be characterized. These include the total number of chloroplasts per cell, the average area within a cell that is occupied by a chloroplast (a proxy for chloroplast size), and the fluorescence intensity per chloroplast (a proxy for the extent of chloroplast pigmentation). Using imaging flow cytometry during an experimental time course of silicon starvation in the diatom *Thalassiosira pseudonana*, chloroplast replication along with changes to chloroplast size and pigmentation could be documented (Fig. 7). We were able to quantify what fraction of the increase in bulk chlorophyll fluorescence was attributed to increases in the number, size, and pigmentation of chloroplasts.

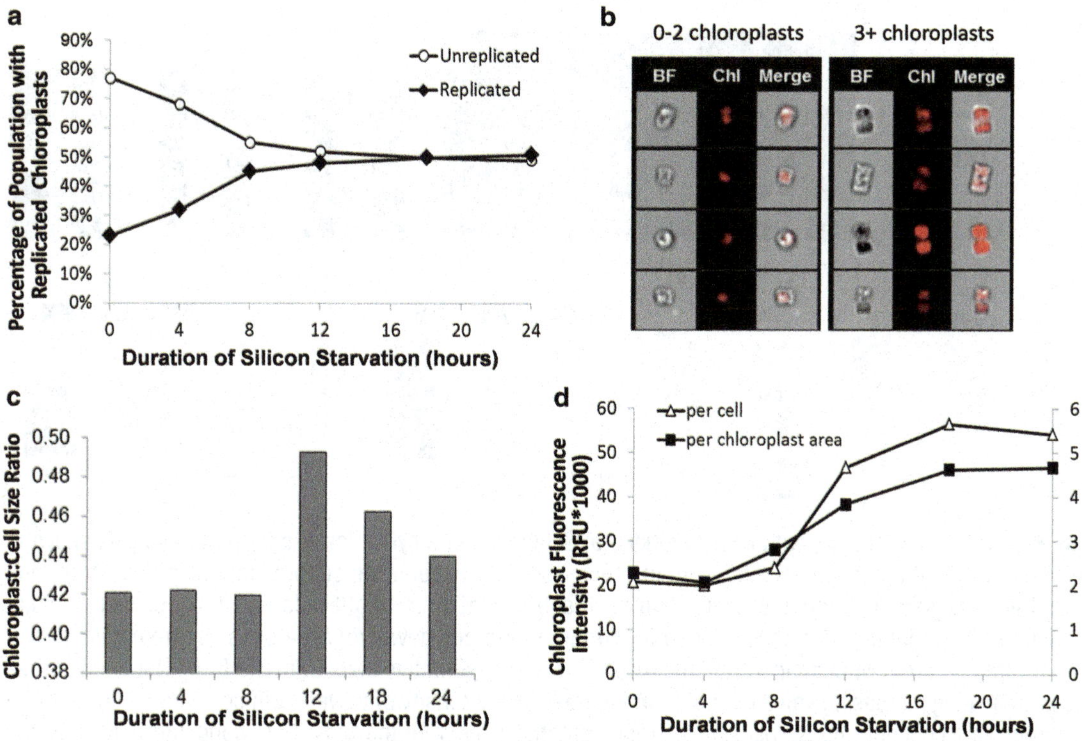

Fig. 7 Changes to chloroplast number, size, and fluorescence intensity in silicon-starved *T. pseudonana* determined by imaging flow cytometry. (**a**) Percentage of population with replicated chloroplasts. (**b**) Example images of cells with (n) chloroplasts. (**c**) Ratio of the average chloroplast size (μm^2) to average cell size (μm^2). (**d**) Average chloroplast fluorescence intensity normalized on a per cell basis and per chloroplast area

In some cases, the angle of imaging is important in analysis, especially with non-spherical cells. In Fig. 8, we show the analysis of silica incorporation into the cell walls of *T. pseudonana*, a cylindrically shaped diatom. Silica incorporation is monitored using the fluorescent dye PDMPO, which quantitatively incorporates into forming silica [25]. Different possible viewing angles of the cell are shown in Fig. 8a; the major features to consider are structures called valves and girdle bands. In this example, specific classifiers were developed to identify cells in different orientations. A custom classifier (the "Round" classifier) combined eight individual features in a proportional manner, as well as customized image masks and image algorithms using Boolean logic to distinguish diatoms in a valve vs. girdle band view (*see* Subheading 3). Even when cells are cultivated in the presence of excess dye, it is likely that a portion of the population will lack dye incorporation in both valves. These cells can only be detected when imaged in the girdle plane, and will have much lower mean fluorescence intensity than fully labeled cells. Cells imaged in valve view have slightly higher mean fluorescence intensity than cells imaged in girdle view (Fig. 8d). This is likely

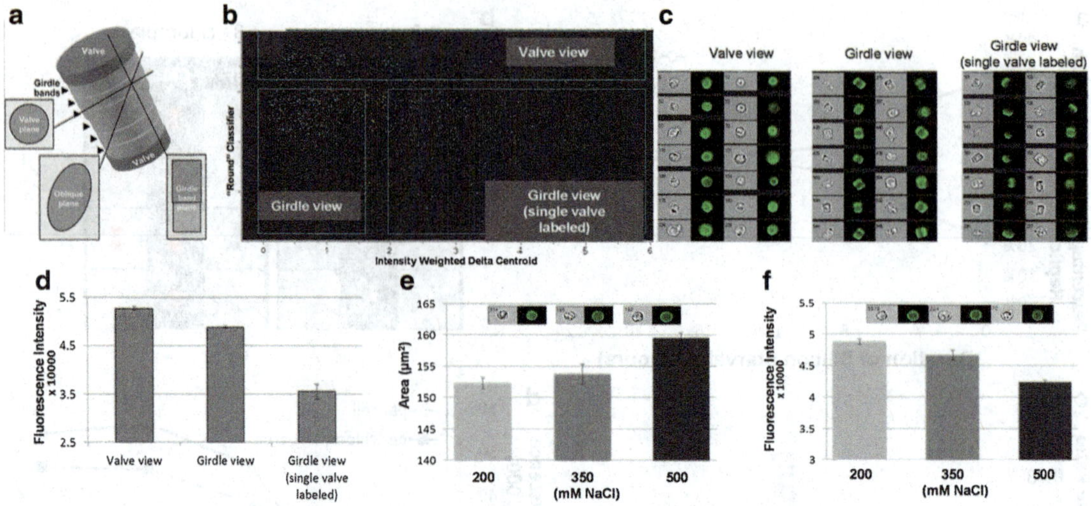

Fig. 8 Orientation effects on data interpretation. (**a**) Schematic of a cylindrically-shaped diatom cell, showing views at different angles. (**b**) Scatter plot of processed data to separate subpopulations of cells from different view angles. (**c**) Images of cells from each subpopulation in bright field and PDMPO fluorescence. (**d**) Fluorescence intensity of cells imaged in the three different views. (**e**) Area of valves from cells grown under three different NaCl concentrations. *Inset* shows representative cells from the three NaCl concentrations, indicating increasing valve diameter. (**f**) Fluorescence intensity of valves from cells grown under three different NaCl concentrations. *Inset* shows representative cells from the three NaCl concentrations, indicating decreasing PDMPO fluorescence

due to an additive effect of imaging fluorescence intensity of both valves and the girdle bands stacked on top of one another. The fluorescently labeled valves and girdle bands are viewed independently when cells are imaged in the girdle plane and display slightly lower mean fluorescence intensity (Fig. 8d).

In Fig. 8e, we demonstrate slight, but statistically significant, changes in valve diameter during growth under different salinities. Interestingly, PDMPO fluorescence decreases with increasing salinity (Fig. 8f), indicating that valves are becoming either thinner or less dense under these conditions, which is consistent with previous findings based on other, more involved, methods [26, 27]. Without the robust statistics enabled by the custom classifier and large number of images, reaching this conclusion would not have been possible.

We have also applied ImageStream® analysis to the study of gene expression and epigenetic effects related to that. In Fig. 9, we document variation in cellular responses in a clonal population of a promoter deletion series by monitoring GFP expression. The data show a conserved subpopulation with similar GFP fluorescence in the intact control and with all three regions of deletion. In addition, in all constructs there is a subpopulation of cells with higher GFP fluorescence, but the percentage of cells in this subpopulation in del 3 is 12- to 85-fold higher than the others. This suggests the

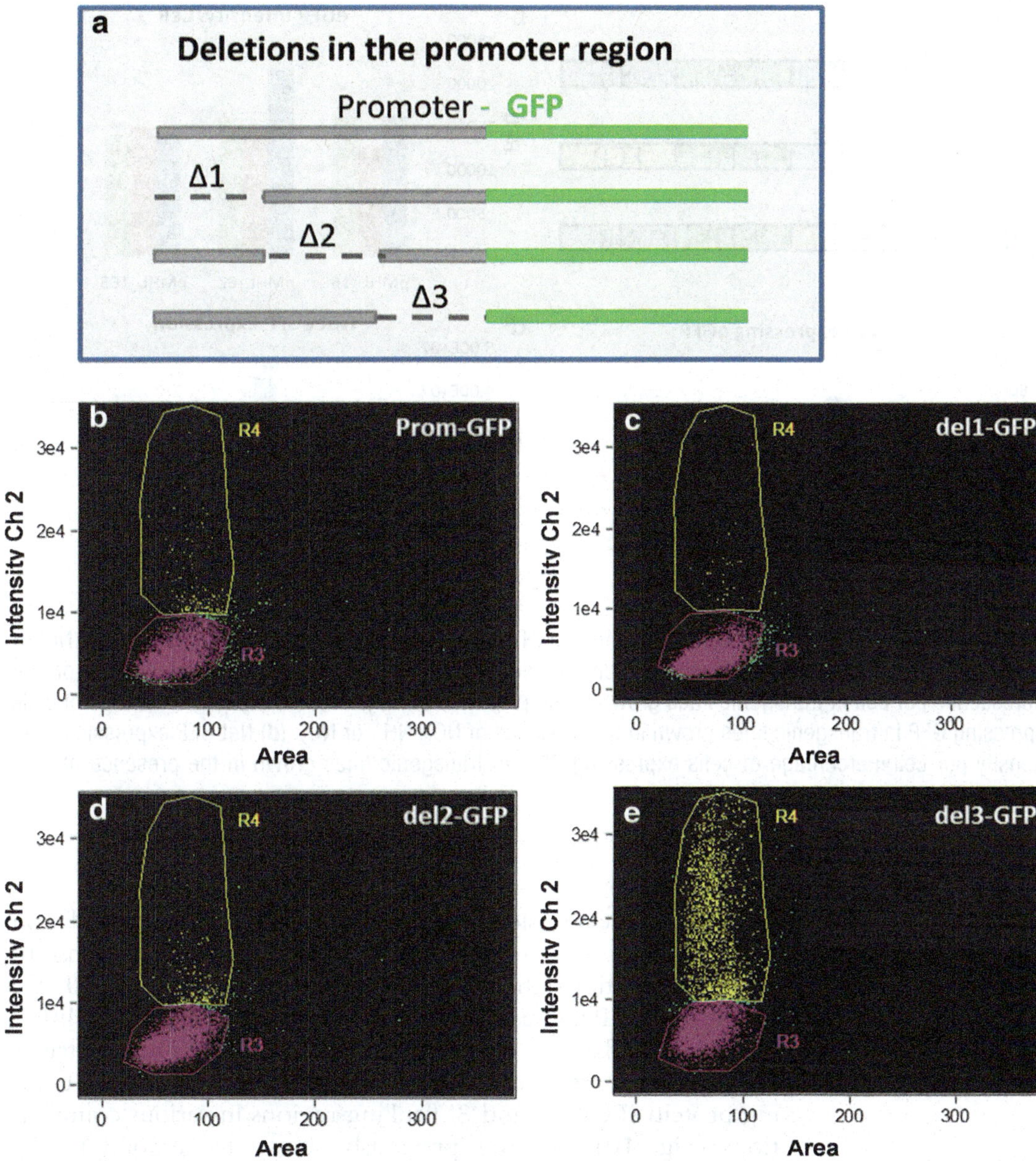

Fig. 9 GFP expression from a full-length promoter and deletion constructs. Full length promoter of a diatom gene and constructs with deletions in three different sections of the promoter were fused to GFP coding sequence (**a**). GFP expression in transgenic diatom lines that contain the full-length promoter-GFP construct (**b**) and constructs with deletions 1, 2, and 3 (**c**, **d**, **e**, respectively) was measured. Single cells were grouped into two subpopulations with different GFP expression level, R3 and R4, based on their fluorescence intensity

presence of a partially leaky repressor element in the del 3 region, which is functional only in a percentage of the population under the conditions tested. Substantial cell-to-cell variation in expression has been documented in other unicellular organisms or within the population of a particular cell type [11–13]. The ability to monitor

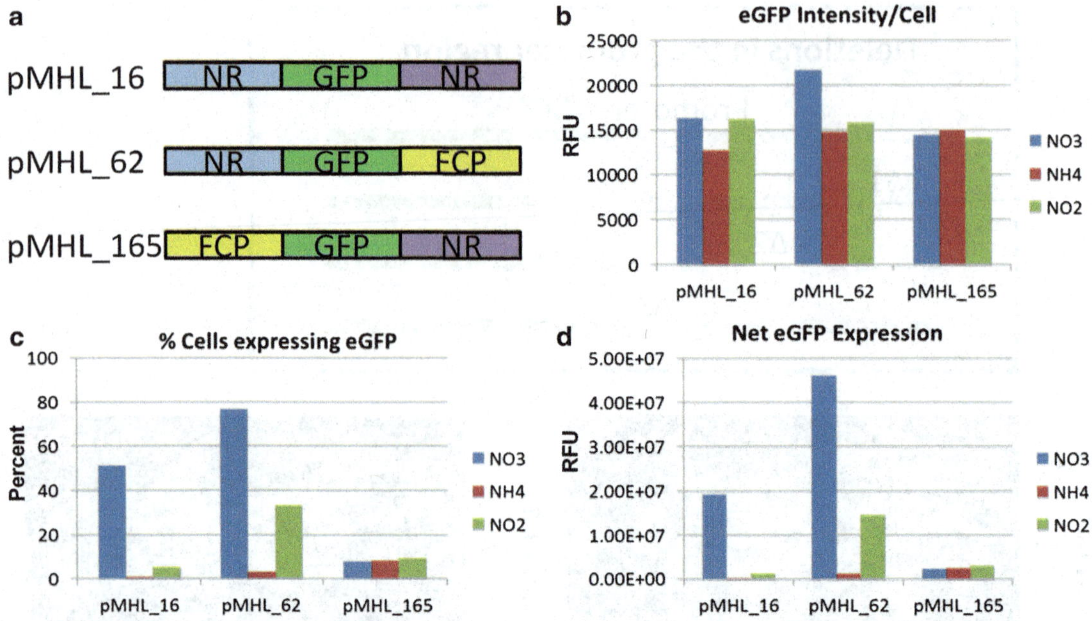

Fig. 10 Mapping of expression control elements. (**a**) Diagram of constructs using nitrate reductase (NR) or fucoxanthin chlorophyll binding protein (FCP) up- and downstream regions fused to GFP. (**b**) Intensity of GFP fluorescence per cell in transgenic lines grown in the presence of NO_3, NH_4, or NO_2. (**c**) Percentage of cells expressing GFP in transgenic lines grown in the presence of NO_3, NH_4, or NO_2. (**d**) Net GFP expression (GFP intensity per cell•Percentage of cells expressing GFP) in transgenic lines grown in the presence of NO_3, NH_4, or NO_2

levels of gene expression in individual cells within the population provides more accurate description regarding cellular response to various conditions than a bulk measurement. In Fig. 10, we monitored GFP fluorescence to evaluate the contribution of different 5′ and 3′ flanking regions to expression levels. The constructs included nitrate reductase (NR) and fucoxanthin chlorophyll binding protein (FCP) 5′ and 3′ flanking regions in various combinations (Fig. 10). NR was previously shown to respond to the presence or absence of nitrate, and to be repressed by ammonium [28], but it is unclear where the control elements mediating this response are located. In cells expressing GFP, the mean fluorescence was similar with the different constructs and conditions (Fig. 10a), indicating that the extent of the response was tightly constrained; that is, there was little tailoring of the level of response. The major difference was in the percentage of cells expressing GFP; the NR5′ flanking region was essential for expression in the majority of cells (Fig. 10b). However, even with NR 5′ and 3′ flanking regions only 50 % of the cells expressed GFP in the presence of nitrate, possibly suggesting that nitrogen status differs between cells. Control is exerted by both the 5′ and 3′ flanking region of NR, the presence of the 5′ flanking region enables expression in the majority of cells, but this is improved by replacing the 3′ flanking region with that of FCP, suggesting that a repression element may be present in this portion of the NR 3′ flanking region or a stimulating element is present in the 3′ flanking region of FCP.

2 Materials

2.1 Growth Medium

1. ASW medium: NaCl (400 mM), $CaCl_2$ (7.48 mM), glycylglycine (5.08 mM), $MgSO_4 \bullet 7H_2O$ (39.7 mM), $MgCl_2 \bullet 6H_2O$ (40.4 mM), H_3BO_3 (37 μM), NaEDTA • $2H_2O$ (32.3 μM), KCl (10 mM), KNO_3 (3 mM), K_2HPO_4 (200 μM), Sodium silicate (400 μM), $ZnCl_2$ (4.6 nM), $CuCl_2 \bullet 2H_2O$ (1.54 nM), $NaMoO_4 \bullet 2H_2O$ (1.15 nM), $CoCl_2 \bullet 6\ H_2O$ (1.76 nM), $FeSO_4 \bullet 7H_2O$ (8.96 nM), sodium tartrate (7.7 nM), Na_2SeO_3 (0.1 pM), thiamine HCl (0.6 nM), vitamin B12 (2.95 pM), biotin (8.2 pM). ASW is made as follows. Per liter of water NaCl (23.6 g), $CaCl_2 \bullet 2H_2O$ (1.1 g), and glycylglycine (0.67 g) is added and dissolved. 33.3 mL of Solution I (consisting of 147 g $MgSO_4 \bullet 7\ H_2O$ and 123 g $MgCl_2 \bullet 6\ H_2O$ per liter of water), and 16.7 mL of Solution II (consisting of 0.137 g H_3BO_3, 0.72 g NaEDTA • $2H_2O$) are added and mixed. 1 mL of trace elements [consisting of 0.568 g H_3BO_3, $ZnCl_2$ (0.628 g), $CuCl_2 \bullet 2H_2O$ (0.268 g), $NaMoO_4 \bullet 2H_2O$ (0.252 g), $CoCl_2 \bullet 6\ H_2O$ (0.42 g), $FeSO_4 \bullet 7H_2O$ (2.49 g), sodium tartrate (1.77 g), Na_2SeO_3 (17.3 μg)] in 1 L of water is added. 1.2 mL of a 10 % solution of $Na2SiO_3 \bullet 9H_2O$ is added, and the pH adjusted to 8.0 with a 10 % NaOH solution. After autoclaving and cooling, 2 mL of a sterile 0.1 M K_2HPO_4 stock, and 1 mL of a sterile vitamin stock (consisting of 200 mg thiamine-HCl, 4 mg vitamin B12, 2 mg biotin per liter of water) are added.
2. Proteose medium supplemented with glucose: Per 1 L, add $NaNO_3$ (2.94 mM), $CaCl_2 \cdot 2H_2O$ (0.17 mM), $MgSO_4 \cdot 7H_2O$ (0.3 mM), K_2HPO_4 (0.43 mM), KH_2PO_4 (1.29 mM), NaCl (0.43 mM), glucose (5.5 mM), 8 mL of vitamin solution (stock consisting of 200 mg vitamin B1, 4 mg vitamin B12, and 2 mg vitamin H per liter stored at 4 °C), and 6 mL of PIV metal solution (stock consisting of 2 mM $Na_2EDTA \cdot 2H_2O$, 0.36 mM $FeCl_3 \cdot 6H_2O$, 0.21 mM $MnCl_2 \cdot 4H_2O$, 37 nM $ZnCl_2$, 8.4 nM $CoCl_2 \cdot 6H_2O$, and 17 nM $Na_2MoO_4 \cdot 2H_2O$ in water).

2.2 Dyes

1. Frozen cell pellets are resuspended in 2.3 % NaCl (for marine species) or water to fluorescently stain TAG. A Nile Red (9-diethylamino-5-benzo[α]phenoxazinone, Sigma) stock is dissolved in acetone at a 250 μg mL^{-1}. All samples from an experiment are analyzed at once by resuspending each pellet in 1 mL

of 2.3 % NaCl or water, and adding Nile Red to a concentration of 1.56 μg mL^{-1} (6.25 μL of the stock). Samples are incubated at room temp for 30 min for diatoms and 15 min for Chlorella. Staining time can be important (*see* **Note 1**). For BODIPY 493/503 (4, 4-difluoro-4-bora-3a, 4a-diaza-s-indacene, Molecular Probes), cells are resuspended in 500 μL of 2.3 % NaCl or water and then stained by adding 1.3 μL of a 1 mg mL^{-1} BODIPY stock dissolved in DMSO and incubated in the dark on ice for a minimum of 10 min. Cells are processed within 2 h of staining while the dye characteristics remained stable.

2. Silica in living diatoms can be stained by growing cultures in the presence of either of two dyes. For rhodamine 123 (6-amino-9-(2-methoxycarbonylphenyl)xanthen-3-ylidene] azanium chloride), a stock solution is prepared in water at 1 mg/mL^{-1} and then diluted in cultures to a concentration of 2 μg/mL^{-1}. For PDMPO [2-(4-pyridyl)-5-((4-(2-dimethylaminoethyl-amino carbamoyl)methoxy) phenyl) oxazole, Molecular Probes] the stock solution from the manufacturer is added to cultures to 100 ng/μL^{-1}.
3. FITC (fluorescein-5-isothiocyanate) is used in conjunction with APS to fluorescently stain cleaned diatom silica. A 10× stock solution was prepared at 10 mM in ethanol and then used in the procedure detailed in Subheading 3.

3 Methods

3.1 Algal Cultivation Conditions

Algae are cultivated under continuous light (150 $\mu mol \bullet m^{-2} \bullet s^{-1}$) under different conditions depending on the culture volume. Small cultures are cultivated in 50 mL in a 125 mL Erlenmeyer flask with a sponge stopper, under continuous shaking (100–200 rpm). Larger cultures (1–1.5 L) are grown in 2 L Erlenmeyer flasks with stirring provided by a stir bar and with aeration by sterile air bubbled through a tube into the bottom of the flask. Diatoms in this study are cultivated at 18–22 °C, and *C. protothecoides* at 25 °C.

3.2 Acid Treatment to Remove Organic Material from Diatom Frustules

Acid treatment is used to remove organic material from diatom frustules. Diatoms are harvested by centrifugation at 3000 × *g* for 5 min to obtain a wet pellet volume of no more than 100 μL in a microcentrifuge tube. Cells are suspended in 400 μL concentrated sulfuric acid and placed in a boiling water bath for 10 min, cooled, then 20 mg KNO_3 is added and followed by boiling for an additional 10 min (*see* **Note 2**). Samples are cooled and the acid diluted by the slow addition of 700 μL of ultrapure water. Diatom silica is then pellet at 12,000 × *g* for 10 min (*see* **Note 3**). If the pelleted diatom silica appears white, then it is washed three times with ultrapure water to remove residual acid. If the pelleted material retains color, then the procedure is repeated until all of the organics (based on the presence of color) have been removed. Samples can be stored in 95 % ethanol at 4 °C.

3.3 APS-FITC Staining of Diatom Silica

Acid cleaned diatom silica structures are stained with (3-amino-propyl) trimethoxysilane—fluorescein-5-isothiocyanate (APS-FITC) to enable their clear visualization during imaging cytometry. After diluting the FITC stock to 1×(1 mM), APS is added in excess as a 1:200 molar ratio (FITC/APS) and mixed for 2 h at room temperature. The solution is stable for several months if protected from light. One microliter of the APS-FITC solution is added to 100 μL of silica sample and allowed to stain for 5 min. To reduce background fluorescence during imaging, it is best to pellet the silica and wash the sample once with ultrapure water to remove excess dye. The excitation and emission of APS-FITC is 485 nm and 535 nm, respectively.

3.4 Chlorophyll Fluorescence

Photosynthetic phytoplankton contain chlorophyll and accessory pigments that enable highly efficient light capture under a wide wavelength range, and which also fluoresce brightly. Chlorophyll fluorescence can be saturating at the lowest laser power setting on the ImageStream®. There are two solutions to this problem (*see* **Note 4**).

3.5 ImageStream® Data Acquisition

Sample volumes of 50–100 μL are sufficient for data acquisition. A concentration of $1–2 \times 10^7$ cells • mL^{-1} provides maximal event rates without coincident events of two or more cells in a segmented image. Typical collection of 10,000 cell events is realized in minutes. Lower concentrations will require longer collection times (*see* **Note 5**). Since the CCD camera is a static detector (unlike photomultipliers that are voltage-tuned for best signal-to-noise on traditional flow cytometers), the laser energy is adjusted to ensure sufficient fluorescence energy without saturation.

3.6 ImageStream® Data Analysis

The analysis software IDEAS® corrects for spectral cross talk between channels by creating a compensation matrix using single color control samples of chlorophyll and fluorochromes. The compensation matrix is then applied to the image files in order to isolate probed images into single imaging channels. It is important to note that the single fluorochrome samples should not have chlorophyll. This may require the labeling of another cell type that does not contain chlorophyll with the fluorochrome (*see* **Note 6**).

The compensated image files are analyzed using image-based algorithms in the IDEAS® software. The analysis wizards in IDEAS® guide the investigator through image analysis of focus, single cells, and fluorescent-positive cells. Gates are user-assigned populations selected in software that create distinct divisible subpopulations based on rational or exploratory views of the collected data. Using

IDEAS® software, users can define cell images in optimal focus (histogram of gradient RMS scores, a definition of sharp image), define single-cell images (bivariate display of an image Area and image Aspect Ratio) and further define multiple subpopulations using image metrics that could be ascribed to size, intensity of signal, shape, texture, or any of 85 pre-programmed algorithms or custom derived morphometric "features" and user-defined image masks. Besides the base masks and features, IDEAS® allows the user to customize image masks and image features and allows for the combining of multiple images masks using Boolean logic to create even more custom features.

The mask is the set of pixels containing a region of interest, and defines a specific area of an image to use for feature-value calculation. The IDEAS® application contains a Mask Manager for viewing existing masks and creating new ones. The default masks in IDEAS® are quite generous as they contain all pixels that are detected as different from the background. In many cases masks may need to be adjusted or new masks created that only include the specific area of interest in the cell. The Mask Manager in IDEAS® has two ways to make new masks; function masks are created with user input or combined masks are created using Boolean logic.

IDEAS® contains the Feature Finder wizard to expedite identifying the maximally discriminating image algorithms with user-selected image sets. IDEAS® ranks features using Fisher's discriminant ratio (RD) as the metric to quantify the amount of separation between the two populations. In addition to picking just one feature, classifiers can be created that linearly combine top-ranked features with weight proportional to their RD values until maximum separation is achieved. This approach was used to separate diatoms in different orientations as they flowed through the flow cell (Fig. 8).

There are additional software wizards for some of the fundamental features analyzed in IDEAS®. Nuclear Localization measures the nuclear localization of a probe by comparing the probe of interest to an image of the fluorescently stained nucleus. Co-localization measures the Bright Detail Similarity between two probes with punctate staining in a cell. Cell cycle and mitosis use nuclear and bright-field images to identify the three interphase stages and the dividing stage mitosis. Apoptosis uses changes in bright-field and nuclear morphology to identify apoptotic events. Shape change measures the relative circular morphology of a cell, and deviations from that which occur. Internalization measures if a probe is on the outside of a cell or inside the cell by defining the cell periphery using a mask that can be customized. Spot count enables separation of user-defined populations with different numbers of fluorescent spots per cell.

Since the ImageStream®X collects and stores large libraries of high-resolution TIF images for each file, the resultant files can be gigabytes in size (*see* **Note** 7).

An attractive feature of IDEAS® software is the ability to do batch processing of multiple data files. We typically adjust image features, define gates and display desired plots on a representative set of data, and then save those settings as a template that can then be applied to all other sets of data from a particular experiment or experiments.

4 Notes

1. Nile Red staining is nonlinear with regard to staining time and cell concentration also needs to be adjusted for specific cell types to obtain a linear response [10, 11].
2. It is advisable to use microcentrifuge tube cap locks or a special boiling rack (e.g., Beaker Buddy boiling rack, USA Scientific) to prevent caps from opening during boiling steps.
3. It is important that the acid be sufficiently dilute during this step otherwise the high viscosity prevents diatom silica from pelleting.
4. There are two solutions to chlorophyll spillover. One is to include an attenuation filter to reduce the laser light intensity. This is useful, but also reduces fluorescence from desired fluors; for example attenuation of the 488 nm laser is typically required to reduce chlorophyll autofluorescence, but it also reduces fluorescence of fluors such as GFP. In a practical sense, we have found that other fluors are still visible, but there is a sensitivity issue, and weaker fluorescence makes high resolution imaging challenging. The alternative approach is to put a chlorophyll blocking filter (Chroma Tech 642 Notch Filter) into the filter wheel of the instrument. Even though this eliminates chlorophyll fluorescence from the channel where is it strongest, it does not eliminate spill over from chlorophyll into other channels. Compensation must be done using chlorophyll containing cells in which the blocking filter was used. This works well, but eliminates some spatial information relative to chlorophyll that may be obtained.
5. A practical limit for the time required to acquire data from a sample in our hands is 20 min; however, if numerous samples are being run, 5 min is reasonable. If necessary, cells should be concentrated to decrease the acquisition time. Noticeable particulates in the sample could cause clogging. If necessary, let particulates settle, and transfer supernatant to new microfuge tube, or filter through an appropriate sized nylon mesh filter.

6. Any cell type can be used to generate compensation data, for example GFP expressed in *E. coli* will serve as a control for GFP expressed in phytoplankton.
7. To avoid file sizes that are onerous even for advanced computational systems, a workable strategy is to divide very large file collection into multiple file replicates (i.e., 10 collection files of 100,000 events each). These individual files can be analyzed by identical means and selected populations exported and concatenated into a single file in the analysis software.

References

1. Keeling PJ (2013) The number, speed, and impact of plastid endosymbiosis in eukaryotic evolution. Annu Rev Plant Biol 64:583–607. doi:10.1146/annurev-arplant-050312-120144
2. Dubelaar GBJ, Casotti R, Tarran GA, Biegala IC (2007) Phytoplankton and their analysis by flow cytometry. In: Doležel J, Greilhuber J, Sudam J (eds) Flow cytometry with plant cells: analysis of genes, chromosomes, and genomes. Wiley-VCH Verlag, Weinheim, pp 287–322
3. Chisholm SW, Olson RJ, Zettler ER, Goericke R, Waterbury JB, Welschmeyer NA (1988) A novel free-living prochlorophyte abundant in the oceanic euphotic zone. Nature 334:340–343. doi:10.1038/334340a0
4. Olsen RJ, Sosik HM (2007) A submersible imaging-in-flow instrument to analyze nano- and microplankton: Imaging FlowCytobot. Limnol Oceanogr Methods 5:195–203. doi:10.4319/lom.2007.5.195
5. Sosik HM, Olsen RJ (2007) Automated taxonomic classification of phytoplankton sampled with imaging-in-flow cytometry. Limnol Oceanogr Methods 5:204–216. doi:10.4319/lom.2007.5.204
6. Buskey EJ, Hyatt CJ (2006) Use of the FlowCAM for semi-automated recognition, and enumeration of red tide cells (Karenia brevis) in natural plankton samples. Harmful Algae 5:685–692. doi:10.1016/j.hal.2006.02.003
7. Alvarez E, Lopez-Urrutia A, Nogueira E, Fraga S (2011) How to effectively sample the plankton size spectrum? A case study using FlowCAM. J Plankton Res 23:1119–1133. doi:10.1093/plankt/fbr012
8. Chang C-W, Miki T, Shiah F-K, Kao SJ, Wu J, Sastri AR, Hsieh CH (2014) Linking secondary structure of individual size distribution with nonlinear size-trophic level relationship in food webs. Ecology 95:897–909
9. Campbell L, Olson RJ, Sosik HM, Abraham A, Henrichs DW, Hyatt CJ, Buskey EJ (2010) First harmful *Dinophysis* (*Dinophyceae, Dinophysiales*) bloom in the US is revealed by automated imaging flow cytometry. J Phycol 46:66–75. doi:10.1111/j.1529-8817.2009.00791.x
10. Brosnahan ML, Farzan S, Keafer BA, Sosik HM, Olson RJ, Anderson DM (2014) Complexities of bloom dynamics in the toxic dinoflagellate *Alexandrium fundyense* revealed through DNA measurements by imaging flow cytometry coupled with species-specific rRNA probes. Deep Sea Res Part 2 Top Stud Oceanogr 103:185–198. doi:10.1016/j.dsr2.2013.05.034
11. Novick A, Weiner M (1957) Enzyme induction as an all-or-none phenomenon. Proc Natl Acad Sci U S A 43:553–566
12. Ko MSH, Nakauchi H, Takahashi N (1990) The dose dependence of glucocorticoid-inducible gene expression results from changes in the number of transcriptionally active templates. EMBO J 9:2835–2842
13. Newman JRS, Ghaemmaghami S, Ihmels J, Breslow DK, Noble M, DeRisi JL, Weissman JS (2006) Single-cell proteomic analysis of *S. cerevisiae* reveals the architecture of biological noise. Nature 441:840–846. doi:10.1038/nature04785
14. Meyer B, Wulf M, Hakansson H (2001) Phenotypic variation of life-cycle stages in clones of three similar *Cyclotella* species after induced auxospore production. Diatom Res 16:343–361. doi:10.1080/0269249X.2001.9705525
15. Cooksey K, Guckert J, Williams S (1987) Fluorometric determination of the neutral lipid content of microalgal cells using Nile Red. J Microbiol Method 6:333–345

16. Chen W, Zhang C, Song L, Sommerfeld M (2009) A high throughput Nile Red method for quantitative measurement of neutral lipids in microalgae. J Microbiol Methods 77:41–47. doi:10.1016/j.mimet.2009.01.001
17. Govender T, Ramanna L, Rawat I, Bux F (2012) BODIPY staining, an alternative to the Nile Red fluorescence method for the evaluation of intracellular lipids in microalgae. Bioresour Technol 114:507–511. doi:10.1016/j.biortech.2012.03.024
18. Yamamoto M, Fujishita M, Hirata A, Kawano S (2004) Regeneration and maturation of daughter cell walls in the autospore-forming green alga *Chlorella vulgaris* (Chlorophyta, Trebouxiophyceae). J Plant Res 117:257–264
19. Traller JC, Hildebrand M (2013) Application of high throughput imaging to the diatom *Cyclotella cryptica* demonstrates substantial intrapopulation heterogeneity in the rate and extent of triacylglycerol accumulation. Algal Res 2:244–252. doi:10.1016/j.algal.2013.03.003
20. Kopanska KS, Tesson B, Lin H, Meredith JC, Hildebrand M, Davis A (2014) Morphological features involved in adhesion of acid-cleaned diatom silica. Silicon 6:95–107. doi:10.1007/s12633-014-9178-2
21. Tesson B, Hildebrand M (2013) Characterization and localization of insoluble organic matrices associated with diatom cell walls: Insight into their roles during cell wall formation. PLoS One 8, e61675. doi:10.1371/journal.pone.0061675
22. Van Blaaderen A, Vrij A, Blaaderen AV (1992) Synthesis and characterization of colloidal dispersions of fluorescent, monodisperse silica spheres. Langmuir 81:2921–2931
23. Desclés J, Vartanian M, El Harrak A, Quinet M, Bremond N, Sapriel G, Bibette J, Lopez PJ (2008) New tools for labeling silica in living diatoms. New Phytol 177:822–829. doi:10.1111/j.1469-8137.2007.02303.x
24. Gibbs SP (1979) The route of entry of cytoplasmically synthesized proteins into chloroplasts of algae possessing chloroplast ER. J Cell Sci 35:253–266
25. Shimizu K, Del Amo Y, Brzezinski MA, Stucky GD, Morse DE (2001) A novel fluorescent silica tracer for biological silicification studies. Chem Biol 8:1051–1060
26. Vrieling EG, Beelen TPM, van Santen RA, Gieskes WWC (2000) Nanoscale uniformity of pore architecture in diatomaceous silica: A combined small and wide angle X-ray scattering study. J Phycol 36:146–159
27. Vrieling EG, Sun Q, Tian M, Kooyman PJ, Gieskes WW, van Santen RA, Sommerdijk NA (2007) Salinity-dependent diatom biosilicification implies an important role of external ionic strength. Proc Natl Acad Sci U S A 104:10441–10446. doi:10.1073/pnas.0608980104
28. Poulsen N, Kroger N (2005) A new molecular tool for transgenic diatoms—Control of mRNA and protein biosynthesis by an inducible promoter-terminator cassette. FEBS J 272:3413–3423.doi:10.1111/j.1742-4658.2005.04760.x

Part II

IFC in Detection of Nuclei and Nucleic Acids

Chapter 5

The Analysis of Cell Cycle, Proliferation, and Asymmetric Cell Division by Imaging Flow Cytometry

Andrew Filby, William Day, Sukhveer Purewal, and Nuria Martinez-Martin

Abstract

Measuring cellular DNA content by conventional flow cytometry (CFC) and fluorescent DNA-binding dyes is a highly robust method for analysing cell cycle distributions within heterogeneous populations. However, any conclusions drawn from single-parameter DNA analysis alone can often be confounded by the asynchronous nature of cell proliferation. We have shown that by combining fluorescent DNA stains with proliferation tracking dyes and antigenic staining for mitotic cells one can elucidate the division history and cell cycle position of any cell within an asynchronously dividing population. Furthermore if one applies this panel to an imaging flow cytometry (IFC) system then the spatial information allows resolution of the four main mitotic phases and the ability to study molecular distributions within these populations. We have employed such an approach to study the prevalence of asymmetric cell division (ACD) within activated immune cells by measuring the distribution of key fate determining molecules across the plane of cytokinesis in a high-throughput, objective, and internally controlled manner. Moreover the ability to perform high-resolution, temporal dissection of the cell division process lends itself perfectly to investigating the influence chemotherapeutic agents exert on the proliferative capacity of transformed cell lines. Here we describe the method in detail and its application to both ACD and general cell cycle analysis.

Key words Cell cycle, Asymmetric cell division, Cell proliferation, CellTrace Violet, Dye dilution, Mitosis

1 Introduction

In all higher organisms, genetic fidelity and tissue homeostasis is maintained by exerting tight molecular controls over cell cycle commitment, progression, and exit. If a single cell accumulates multiple loss or gain of function mutations that affect these controls then this can lead to oncogenesis [1, 2]. In the majority of cases, successful completion of the cell cycle generates two quantitatively and qualitatively matched daughter progeny because any pre-existing macromolecules (DNA, RNA, protein) or pre-formed structures (endosomes) are apportioned equally across the plane of

Natasha S. Barteneva and Ivan A. Vorobjev (eds.), *Imaging Flow Cytometry: Methods and Protocols*, Methods in Molecular Biology, vol. 1389, DOI 10.1007/978-1-4939-3302-0_5, © Springer Science+Business Media New York 2016

cytokinesis. However in certain biological systems these key elements are not distributed in a symmetrical fashion leading to what is described as an asymmetric cell division (ACD). ACD has been shown to play a key role in cellular differentiation in developmental and stem cell biology [3–5]. It has also been suggested to play a role in the adaptive immune response to direct the development of effector and memory populations [6–10] although this remains highly controversial and unsupported by various studies that have elegantly tracked the immunological fate of single cells [11–15]. Any technique designed to determine the prevalence and role of ACD must firstly be capable of providing multi-parameter fluorescence-based imagery that can be measured on a quantifiable (relative) scale. Secondly it should be capable of high-throughput sample acquisition in order to identify statistically relevant numbers of rare, short-lived mitotic phase cells without the need for any enrichment using chemical inhibitors. Thirdly there should be a temporal component to the method that is able to cope with the asynchronous nature of cell proliferation, even within seemingly homogenous lymphocyte populations. Lastly there should be an objective and controlled analysis framework whereby asymmetry is identified/scored based on some kind of known internal control limit for symmetrical inheritance. To this end we recently described a novel method for reporting the cell cycle position and division history of asynchronously dividing cells using CellTrace Violet (CTV) fluorescence dye dilution, propidium iodide (PI) for DNA content, and co-staining for mitotic phase antigens such as MPM2 [16]. When used in conjunction with an imaging flow cytometry (IFC)-based instrument platform, it is able to provide the spatial information to subdivide mitosis into prophase, metaphase, anaphase, and telophase to then measure the distribution of various key fate determinants across the cytokinetic plane. This has been successfully employed to study ACD in both T and B cells responding to antigenic stimulation [12, 17]. As a secondary but equally powerful function, our method can also be used to study the role of chemotherapeutic agents on transformed cell lines to determine where and how they influence cell cycle progression. The method is relatively simple to set up using virtually any cell type and in this chapter we describe the protocol in detail so that it can be reproduced with additional notes covering the major issues and pitfalls.

2 Materials

2.1. *Essential cytometry hardware.* The minimum IFC instrument (Amnis, Seattle) requirements are one from the following (*see* Fig. 1):

- ImageStream 100 (IS100): 10-bit CCD camera, dynamic range 0–1024, 6 camera channels, and up to 5

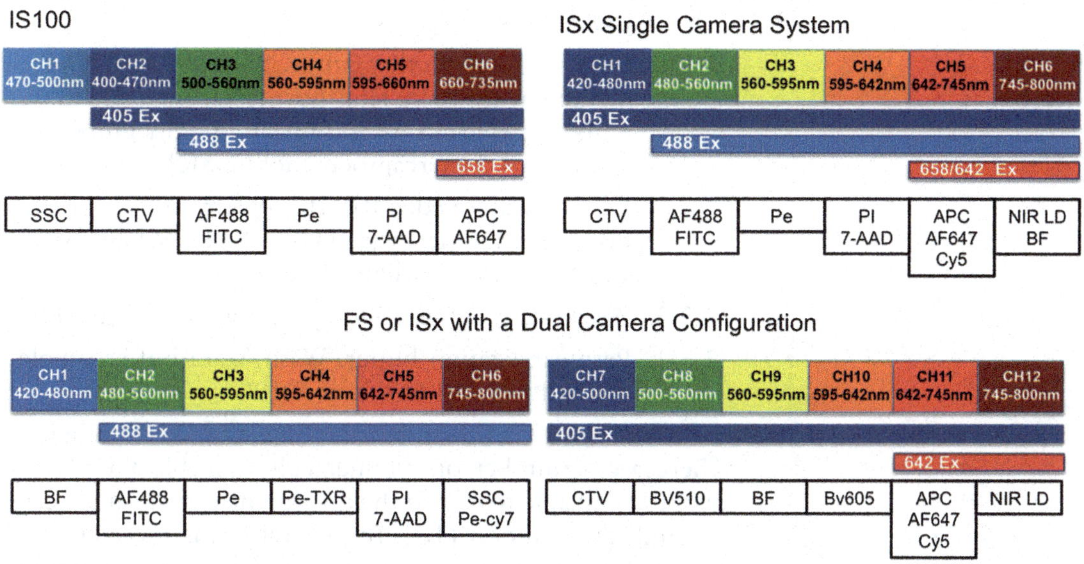

Fig. 1 The spectral layout of four different IFC systems we have successfully used to implement versions of our method. In all cases, the excitation lasers and approximate emission channel widths are shown with major fluorochrome examples to aid panel design. Although the IS100 offers limited flexibility, we have successfully tracked division with CTV, identified mitotic cells with PI/PH3 and measured one fate-determining signal within these cells [17]. The newer generation systems offer greater experimental flexibility and data content

of these can be fluorescent stains plus side scatter, fixed at 40× magnification.

- FlowSight (FS), 12-bit CCD camera, dynamic range 0–4096, 12 camera channels up to 10 of which can be used for fluorescent stains, fixed at 20× magnification.
- ImageStreamX MKI or MKII (ISXMKI/II): 12-bit CCD camera, dynamic range 0–4096, 6 or 12 camera channels of which 5/10 of which can be used for fluorescent stains depending on configuration, 20, 40, or 60× magnification with multi-mag upgrade otherwise fixed at 40×.

- All systems should possess 488 nm, 405 nm, and 654/642 nm excitation lasers, with a 561 nm laser optional (not available for the IS100).
- All systems should be fully ASISST calibrated with additionally quality control performed using a multi-level fluorescent microsphere set. Particular attention should be paid to the resolution of peaks 2 and 3 (dim 1 and dim 2) at laser powers used to excite experimental samples.

2.2. *A conventional flow cytometer (CFC) system* with an analogous excitation laser and detector configuration to the IFC systems listed above (*see* **Note 1**).

2.3. Media and buffers:

2.3.1. Primary T cell/Jurkat culture media: RPMI supplemented with 10 % fetal bovine serum (FBS), 100 units of penicillin/streptomycin, 2 mM glutamine, and 50 μM 2-mercaptoethanol (2-Me).

2.3.2. B cell culture media: RPMI supplemented with 10 % FBS, 50 μM 2-Me, 25 mM HEPES, 2 mM GlutaMAX, and 10 U/ml penicillin/streptomycin.

2.3.3. Wash/stain buffer: PBS supplemented with 2 % FBS.

2.3.4. Permeabilization buffer: Wash/stain buffer supplemented with 0.1 % Triton-X 100.

2.4. *CellTrace Violet*™ (Cat No A10198: Life Technologies): There are a number of commercially available fluorescent dyes that are able to record the division history of single cells through dye dilution [18]. In general terms there are two categories of reagents, succinimidyl esters and lipophilic dyes (*see* **Notes 2–4**). CTV is a succinimidyl ester.

2.4.1. CTV is supplied as a lyophilized powder (stored at −20 °C till the manufacturer's expiration date) that is reconstituted in 20 μl of DMSO to a stock concentration of 5 mM. Working concentrations range between 1 and 10 μM and should be determined empirically for each cell type. Aliquots of reconstituted dye can be stored at −20 °C and freeze-thawed once. It is excited by the 405 nm laser and emits at 450 nm (Fig. 1, lower panel).

2.4.2. The exact *activation reagents* required depend on the immune cell type being studied for ACD. All culture, activation and staining/washing steps are done in 96 well round bottom plate or 24-well flat-bottomed plates (various suppliers).

2.4.3. F5 transgenic T cell activation reagents:

- NP68 peptide (synthesized in house): The stimulatory peptide for the F5 receptor.
- Anti-CD3 antibody (clone 145-2C11): A generic T cell activator.
- Anti-CD28 antibody: A co-stimulatory signal.
- Recombinant ICAM-1: Thought to be essential for driving ACD.

2.4.4. B cell activation reagents:

- CD40 ligand. Mimics effect of a helper T cell signal.

- Anti-IgM antibody: Stimulated through the B cell receptor.
- Interleukin-4: T cell-derived cytokine.

2.5. *Antibodies against intracellular and extracellular targets* should be chosen depending on the molecules/cell cycle phases under investigation. For T cell studies CD69 or Ki67 are used to identify cells that have been triggered to divide. This allows one to determine the best harvest window in terms of the greatest number of cycling cells [12]. Antibodies should be directly conjugated when possible with fluorochromes compatible with the overall panel and spectral properties of the IFC instrument. Species, clones, and cross-reactivity should also be considered (*see* **Note 5**). For particular markers/antigens we strongly recommend using the indicated fluorochromes (shown in bold type). For markers where no particular fluorochrome is recommended, one can be selected that fits into a free channel on the IFC system used (*see* Fig. 1). Examples of some marker/fluorochrome combinations we have used are:

2.5.1. Anti-Ki67: Used for measuring cell cycle entry.

2.5.2. Anti-CD69 (clone H1.2F3): Denotes commitment to proliferation.

2.5.3. Anti-MPM2 (clone MPM2), for example FITC or Cy5 conjugated (*see* **Note 5**): Identifies all stages of mitotic cells.

2.5.4. Anti-PKC zeta goat anti rabbit (cat no SC-216-G, Santa Cruz, USA): Strongly implicated as an asymmetrically inherited fate marker.

2.5.5. Donkey anti-goat PE (CH3) or AlexaFluor-647, (CH11): To detect the unconjugated PKC zeta goat antibody (*see* **Note 5**).

2.5.6. Anti-CD8 (clone 53-6.7): A possible marker for T cell ACD.

2.5.7. Anti-CD25 (clone PC61.5): A possible marker for T and B cell ACD.

2.5.8. Anti-CD86 (clone GL1): A possible marker for B cell ACD.

2.5.9. Anti-IA/IE (clone M5/114.15.2): A possible marker for B cell ACD.

2.6. FC-block reagent (anti CD16/32 clone 2.4G2): Used to block nonspecific binding of antibodies via their Fc regions.

2.7. *Near-infrared live/dead (NIRLD) fixable dye* (CH12): Note that this dye is not compatible with an IS100, but can be used on a single-camera ISx system in CH6. BF illumination can also feature in this channel as NIRLD signal from dead cells will cause "white out" in the BF image and dead cells can be gated out using the gradient RMS feature for focus along with defocused cells.

2.8. Fixatives: Will depend on the overall antibody panel used and purpose of experiment (*see* **Note 6**).

   2.8.1. 70 % Ethanol (EtOH): Made up in deionized water (DI H_2O) and stored at 4 °C.

   2.8.2. 4 % Formaldehyde (FA) stock: Diluted in PBS (*see* **Note 6**).

2.9. *Propidium iodide/7-AAD:* Both are available from several commercial sources. Stocks are made up at 50 μg/ml in DI-H_2O.

2.10. *Siliconized 1.5 ml microfuge tubes* or protein "low-bind" version.

2.11. *Copies of IDEAS (Amnis), FlowJo (Treestar Inc.) or equivalent FCS-based software analysis package*: Image J (open-source image analysis package, NIH, USA). This protocol assumes a good working knowledge of these software packages. For IDEAS, users should understand compensation, masking, and feature creation. Users should also know how to export feature values and raw images for analysis in third party software. More information can be found in the respective user guides.

3 Methods

3.1 CTV Labeling

3.1.1 Labeling Primary T and B Cells with CTV

Cells are washed once in complete culture media and resuspended for cell counting. Centrifugation conditions are 1200 rpm/300 × *g* for 3–5 min followed by the removal of all supernatant (hereby referred to in this protocol as "washing cells"). Cell viability should also be determined using a dye exclusion method such as trypan blue.

CTV staining solution is made up in protein-free sterile PBS and pre-warmed at 37 °C prior to use. Concentrations for primary T and B cells should be determined empirically based on viability and uniformity of the CTV signal [19]. Labelling solutions should be made up at 2× concentrations. Typical final concentration used for T and B cell labeling is 2.5 μM.

Cells should be spun down and the supernatant carefully aspirated away as any residual protein can compete out the intracellular CTV labeling. The cell pellet should be resuspended in one volume of protein-free PBS so that once the 2× labeling solution is added at an equal volume, the final cell concentration should range between 1 and 3×10^6/ml. The cells should then be incubated at 37 °C for 10 for 5 min (B cells) or 10 min (T cells).

After incubation, neat FBS should be added to the labeling solution to a final concentration of 10 % and incubated for a minimum of 5 min at room temperature. This will terminate intracellular labeling by binding any remaining free dye from the solution.

Cells should then be washed once in full media and resuspended for counting and viability assessment. An aliquot of cells should also be analyzed by CFC in order to determine the intensity and uniformity of CTV labeling as this affects division peak resolution (*see* **Notes 2** and 7). Cells can then be adjusted to the appropriate density for subsequent culture and activation.

3.1.2 Labeling Cell Lines

Cell lines are labeled with CTV as described above with no further modifications.

In our hands, CTV successfully labels a number of cell lines (e.g., Jurkat/A549); however the labeled width always exceeds the limits for division peak resolution due to cell-intrinsic heterogeneity (*see* **Note** 7). This requires cells to be sorted prior to culture in order to narrow the input width sufficiently to resolve division peaks. This method is beyond the scope of this protocol and is described in full detail elsewhere [19].

3.2 Culture and Activation Conditions

3.2.1 T Cells

An example of the in vitro T cell activation protocol from our lab has been described elsewhere [12]. However, briefly CTV-labeled T cells are seeded in 24-well plates at a density of 1×10^6/ml in the presence of various activators, including peptide loaded dendritic cells or anti-CD3 (145-2C11, plate bound at 10 μg/ml or soluble at 0.5 μg/ml) with or without plate bound CD28 and/or recombinant ICAM-1 (both at 10 μg/ml). The cells are harvested at the point of cell cycle commitment based on Ki-67/CD69 expression in the CTV-undivided population (*see* **Note 8**). Typically this tends to be between 32 and 36 h. for MHCI-restricted F5 CD8 T cells [12].

3.2.2 B Cells

CTV-labeled B cells are seeded in 24 or 96 well-flat bottom plates at a density of 1×10^6 cells/ml. To stimulate B cell proliferation we add CD40L at concentrations ranging from 1 to 0.001 μg/ml or anti-IgM at concentrations ranging from 10 to 0.001 μg/ml. In all cases, B cell culture media must be supplemented with 5 ng/ml IL-4. In our system, B cell proliferation begins 48 h after stimulation with an optimal harvest time of 72 h.

3.2.3 Culturing Cell Lines for Chemotherapeutic Drug Assays

CTV-labeled cell lines, pre-sorted to optimal input width for division peak resolution, are placed back into culture at a density of 0.5×10^6 cells/ml in complete media with a total minimum cell number of 1×10^6 cells (2 ml) per well per analysis time point.

An aliquot of post-sorted cells (1×10^6 cells) are fixed and stained to determine the input cell cycle distribution. Cells are allowed to recover post-sorting for a minimum of 24 h before addition of any cell cycle modulating compounds.

At the point of drug addition, a time zero sample is taken (1×10^6 cells) and fixed under appropriate conditions as described later in the protocol. As a control, cells are also incubated in the absence of drug for the full culture period. Cells can then be cultured for defined periods in the presence or absence of compound.

To model the effects of drug metabolism and removal, one can also investigate the effects of washing the drug away and re-culturing the cells to determine long-term irreversible effects on proliferation to model the effects of drug clearance. The ability to track cells over multiple cell divisions using CTV (~6 rounds) makes such long-term tracking a viable option compared to other temporal methods. As a control at this stage, a sample of cells is taken and fixed at the point of drug removal. A further control is set up where the presence of drug is maintained throughout the culture period.

3.3 Live/Dead Fixable Dyes (NIRLD)

Harvest cells from culture and wash ×1 in wash buffer. Add 2 μl of LD-NIR per ml and incubate at RT for 10 min. Wash cells again then proceed to extracellular staining.

3.4 Extracellular Antibody Labeling

1. This stage is only required if surface markers are to be investigated. Furthermore, if the surface epitopes are unaffected by fixative treatment, then all antibodies (extra- and intracellular) can be added post-fixation/permeabilization. All antibodies and dyes should be optimally titrated for use on an IFC system paying particular attention to the relative brightness of all fluorochromes excited by the same laser (*see* **Note 9**).
2. Make up the cocktail of extracellular antibodies at the pre-optimized concentrations in stain/wash buffer. Add 100 μl per well to stain 1×10^6 cells. If you are using tandem dyes, be sure to check that they remain stable after fixation. Also pay attention to the fact that tandem dyes will be prone to cross laser excitation (*see* **Note 9**).
3. For staining, cells should be seeded in a round-bottomed 96-well plate at a density of 1×10^6 cells/ml. Often we stain 2–3 million cells in total per sample. This will require setting up multiple wells per sample to be pooled at the end of staining.
4. If required, incubate cells with anti-FC block made up in stain/wash buffer for 10 min, then wash. After removing the supernatant, briefly press plate bottom onto a vortex mixer to break up the cell pellets. This can be done after every wash step.
5. Add 100 μl of antibody cocktail per well and incubate cells in the dark at RT for at least 1 h. Wash cells twice in stain/wash buffer and incubate with secondary labelled antibodies if required.

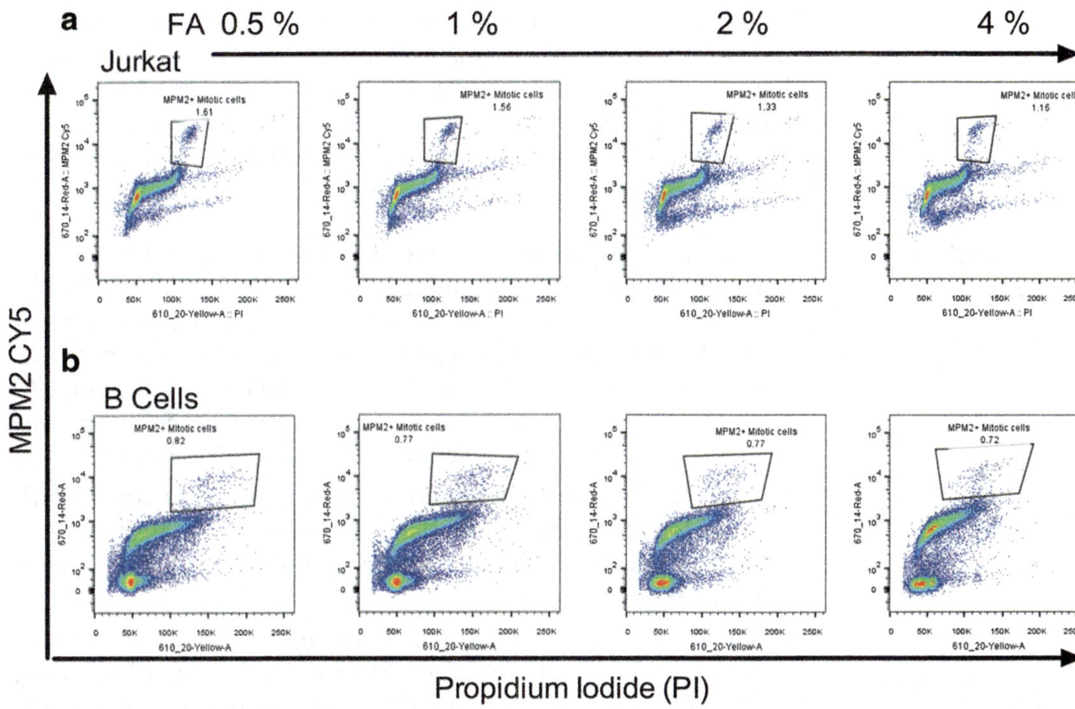

Fig. 2 Identifying mitotic cells by MPM2 staining. (**a**) Jurkat cells and (**b**) primary mouse B cells were fixed with the indicated concentrations of FA as outline in Subheading 3 and stained with PI (*x*-axis) and MPM2 (*y*-axis)

6. Single stained controls for each fluorochrome in the panel should be prepared and processed exactly as experimental samples. Allow for the inclusion of single stains against intracellular targets by preparing these wells in advance. Also set up further control wells containing unlabelled cells at exactly the same cell density as experimental samples. These will serve as the PI compensation single cell control when the dye is added to the samples at the end of the staining process.

3.5 Fixation (See Note 6)

3.5.1 70 % EtOH Fixation

If you are only measuring cell cycle distribution without extensive antigenic staining then 70 % EtOH fixation is optimal as it gives the best DNA profiles. It is also compatible with MPM2 staining for mitotic phases [19].

For fixation, ice-cold 70 % EtOH is added to the pellet and then samples are incubated for a minimum of 1 h at RT. EtOH fixed cells can be kept in fixative for 24–48 h without detrimental effects on the CTV peak resolution.

3.5.2 Formaldehyde Fixation

ACD analysis will require staining for antigens in addition to MPM2; therefore EtOH will likely destroy the epitopes necessitating a switch to FA. We have found that a lower concentration of formaldehyde (0.5 %) work best for DNA and MPM2 staining resolution in cell lines (Fig. 2a) and primary murine B cells (Fig. 2b).

The cell pellet is first resuspended in one volume of PBS to eliminate clumping. An equal volume of 2× FA solution is then added to a final desired fixative concentration (0.5 %). Incubation should be done at RT for a minimum of 1 h. Cells can be washed out of fixative and stored in stain/wash buffer for up to 24–48 h without loss of CTV peak resolution.

3.6 Intracellular Antibody Labeling

1. Cells are first permeabilized in 100 μl of permeabilization buffer for a maximum of 5 min, after which 100 μl of wash buffer is added prior to centrifugation. Next add the cocktail of intracellular antibodies made up at appropriate dilutions in wash/stain buffer and incubate at RT in the dark for minimum 1 h. At the same time, stain for any single-stain compensation samples that require antibodies against intracellular antigens.
2. Wash cells a minimum of two times. If required, incubate with fluorescently labeled secondary antibodies made up in wash/stain buffer for a minimum of 45 min in the dark at RT, after which wash cells once.
3. The final volume for sample resuspension should not exceed 60 μl (*see* **Note 10**). Pool multiple wells if needed by adding 60 μl of stain/wash buffer to the first well and transfer to subsequent wells until there is approximately $2–3 \times 10^6$ cells in the 60 μl volume. Transfer to a 1.5 ml siliconized, low-bind microfuge tubes.

3.7 DNA Dye Staining

1. PI concentration should be pre-titrated to avoid camera saturation at a given laser power (*see* **Note 9**) while still giving well-defined DNA profile based on G1 CVs (Fig. 3a).
2. Add PI to a final concentration of 1 μg/ml per 2–3 million cells and incubate samples for 10–20 min at RT to ensure saturation of the DNA. If cells are fixed in FA, then we use the same concentration of PI solution containing 0.1 % triton X-100 (*see* **Note 6**).

3.8 Sample Acquisition (12 Channel Systems, FS/ISx)

3.8.1 Setting Excitation Laser Powers

Collect bright-field (BF) illumination in CH1 and CH9. As side-scatter information is not required the 758 nm laser can be turned off.

Switch on all required excitation lasers at least 20 min before acquisition and set the powers based on the raw max pixel feature for each individual fluorescence channel. Aim to achieve the best possible total fluorescence signal (akin to pulse area on CFC instruments) while avoiding saturation of the camera based on Raw Max Pixel values (12-bit CCD, scale range from 0 to 4096, with 4096 denoting saturation akin to pulse height on a CFC system).

For primary cell acquisition on an ISx, select 60× magnification with the high-sensitivity fluidics mode. Cell lines can be analyzed effectively at 20× or 40×.

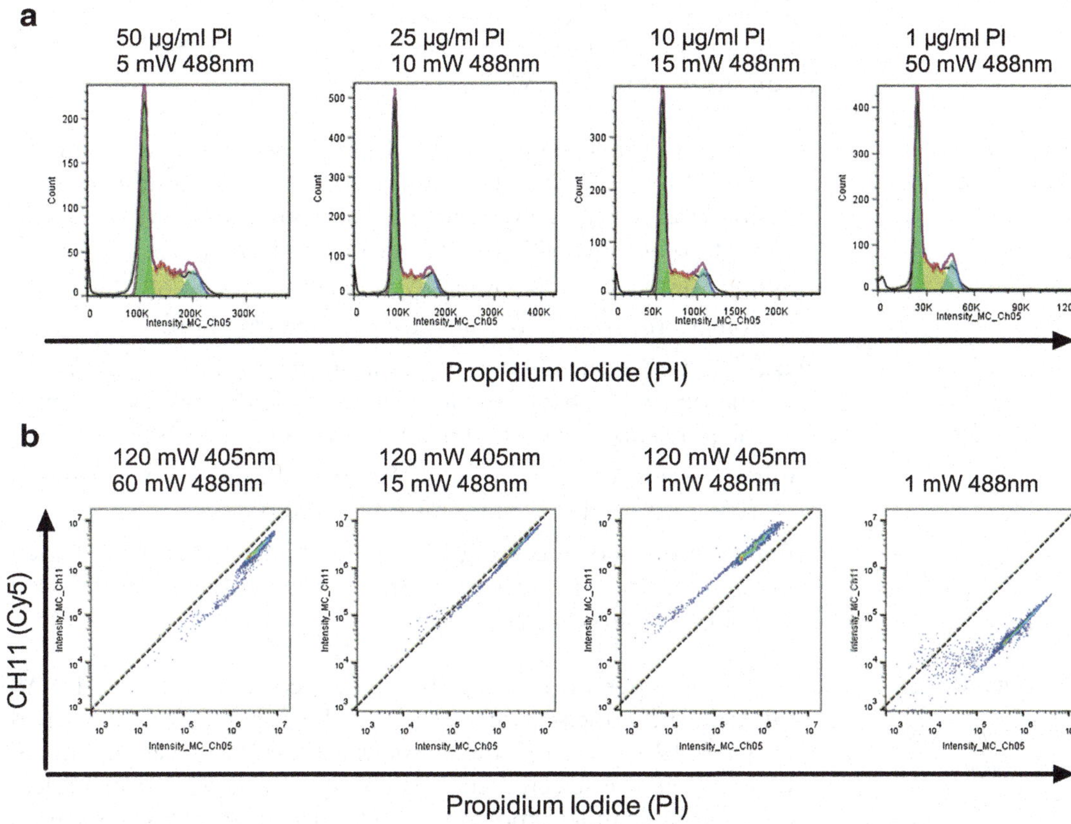

Fig. 3 Optimising the PI concentration for IFC analysis (**a**) Fixed Jurkat cells were stained with decreasing levels of PI and acquired at increasing 488 nm laser powers as indicated. The PI fluorescence was analyzed by exporting as an FSC file and using the modeling option in FlowJo [33]. In all cases the G1 CV was <9 and the % of cells modelled in each phase was comparable (data not shown). (**b**) Fixed Jurkat cells were labeled with 50 μg/ml PI and acquired on a 12-channel IFC system using the indicated 488/405 nm laser powers. The 45° lines drawn through the origin for each plot denote the limits of acceptable spill over whereby PI fluoresce becomes brighter in CH11 than CH5 and would therefore require over 100 % compensation (not possible by IFC)

As well as guarding against CCD channel saturation, look for issues caused by cross laser excitation of particular dyes in the panel. For example PI is excited maximally by the 488 nm but also 20 % by the 405 nm lasers (*see* **Note 9**). Staining with high PI concentrations will require setting a low 488 nm laser power to avoid signal saturation. If the 405 nm laser is set relatively high in order to excite CTV or other dyes, then it is possible PI emission will be maximal in CH11 rather than CH5 (Fig. 3b). If this happens, then the data is effectively ruined so this is why titration of PI is essential when working with a multicolor panel (Fig. 3a, b).

3.8.2 Setting Limits of Event Acquisition

Aim for a final analysis population of >100 cells and base the number to acquire on the frequency of these event within the total population. For example to obtain 200 telophasic cells from the undivided population at a target frequency of 0.001 %, aim to collect 200,000 live cells. Concatenate samples if needed post acquisition using IDEAS.

3.8.3 Sample Acquisition Order

First run a sample stained with all fluorochromes apart from the DNA dye. This sample should ideally have the brightest staining levels of all markers. Set the laser powers as outlined above and do not change them from this point onwards.

Acquire single-stained samples individually with bright-field illumination and the 758 nm scatter laser turned off.

Lastly, add PI to the unstained sample at the pre-optimized concentration/incubation time for the selected 488 nm laser power. Acquire this as the PI compensation control (*see* **Note 9**).

Add the same concentration of PI to the experimental samples (e.g., 1 μg/ml, 15 min) and collect the data.

If any samples not requiring PI need to be run after a PI-stained sample, clean the instrument by loading a tube of bleach-based solution followed by DI-H_2O.

3.9 Data Analysis

3.9.1 Multispectral Compensation

Perform compensation using the embedded wizard in the IDEAS software, the principles of which have been extensively described elsewhere [20, 16, 21]. Briefly, the .rif files for each single fluorochrome stain are selected and loaded into the wizard. Only select the imaging channels known to contain the peak emissions signals of the fluorochromes in your panel. IDEAS then calculates the coefficient of the fluorescence relationship for each dye in the specific channel versus each cross talk channel. Coefficients are presented in a 12 × 12 matrix format. Manually inspect each distribution to identify debris and autofluorescence that would skew the best fit line and affect compensation. If required, refine the compensation populations to eliminate any such skewing. Finally, validate the matrix values on the imagery for each fluorochrome channel.

3.9.2 Identifying Live, Focused, Single/Bilobular Cells

Do not use any of the embedded analysis wizards in IDEAS software. However the use of the building blocks option for predefined plots is very useful. Initial measurements and features are derived from the pre-calculated channel masks (e.g., defaults masks M0×, where × denotes channel number, or MC that denotes the sum of all masks):

First, identify live cells based on NIRLD total fluorescence levels (MC, CH12) and the area of the BF image (Fig. 4a).

Next use the area versus aspect ratio of BF image creating two gates, one for single cells and one for bilobular cells. Next, construct an "OR" Boolean population of single and bilobular cells for subsequent analysis so as to include telophasic cells (Fig. 4b)

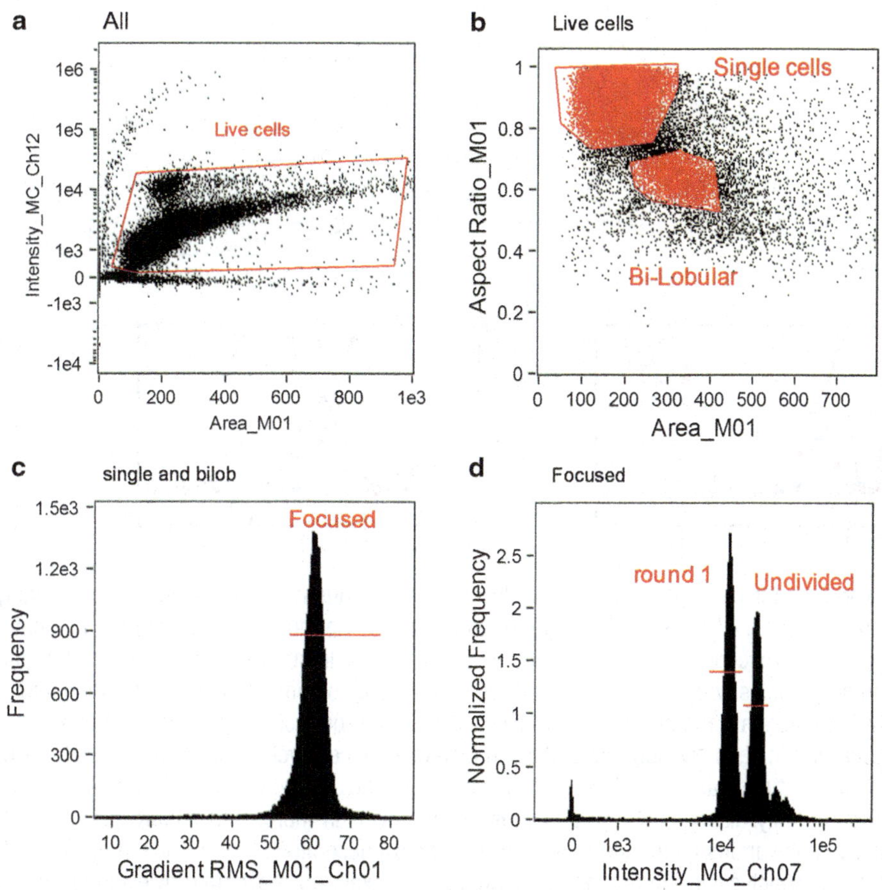

Fig. 4 Initial gating strategy for IFC-based cell cycle/ACD analysis. (**a**) Live cells are identified based on NIRLD fluorescence in CH12 (*y*-axis) and BF area (*x*-axis. (**b**) Single and bilobular cells are identified based on the area (*x*-axis) and aspect ratio (*y* axis) of the BF image. (**c**) Focused cells are selected by gating on the gradient RMS measurement of the BF image. Focused cells are gated as shown (>50). (D) CTV fluorescence in CH7 is plotted as a histogram and gated to identify different proliferative generations

Next identify focused cells based on the gradient RMS of the BF image (M01, CH1) and gate accordingly (Fig. 4c).

Manually inspect the CTV fluorescence histogram (MC, CH07) and set region markers to identify individual division rounds (Fig. 4d).

3.9.3 Analysis of Division History and Cell Cycle Distribution in the Presence of Chemotherapeutic Compounds (See Notes 11 and 12)

From the live, single/bilobular, focused population begin by using the CTV, PI and MPM2 distributions to gate and record the frequencies of the major cell cycle stages from within each division round (Fig. 5a). Adjust the frequency of cells within each CTV peak by 2^{n}, where n = the division round to account for the doubling effect every time an input cell has divided. It is then possible to derive the true frequency of the input population within each division round by dividing the adjusted frequencies by the sum of all adjusted frequencies (Fig. 5b, upper table).

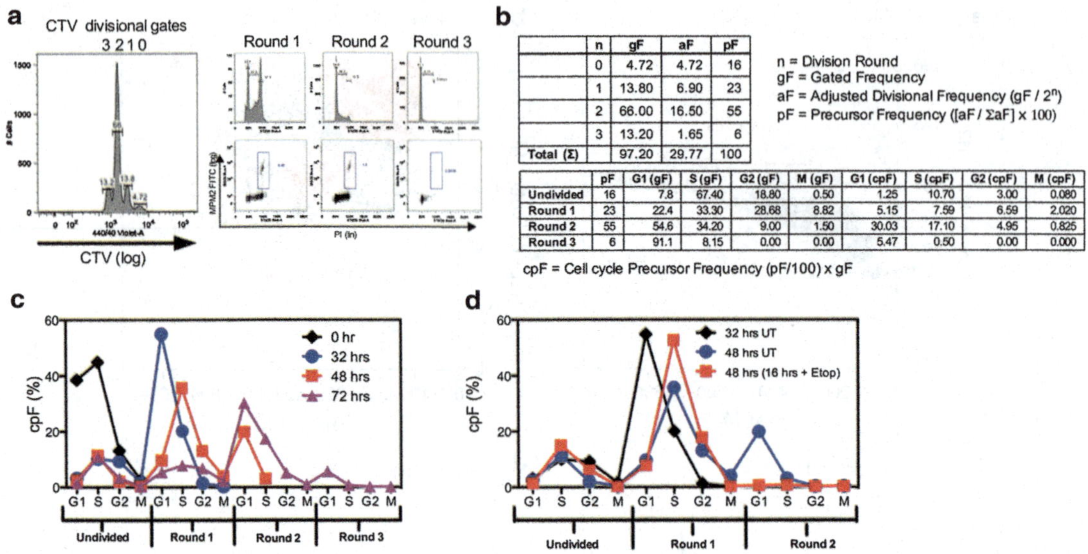

	n	gF	aF	pF
	0	4.72	4.72	16
	1	13.80	6.90	23
	2	66.00	16.50	55
	3	13.20	1.65	6
Total (Σ)		97.20	29.77	100

	pF	G1 (gF)	S (gF)	G2 (gF)	M (gF)	G1 (cpF)	S (cpF)	G2 (cpF)	M (cpF)
Undivided	16	7.8	67.40	18.80	0.50	1.25	10.70	3.00	0.080
Round 1	23	22.4	33.30	28.68	8.82	5.15	7.59	6.59	2.020
Round 2	55	54.6	34.20	9.00	1.50	30.03	17.10	4.95	0.825
Round 3	6	91.1	8.15	0.00	0.00	5.47	0.50	0.00	0.000

Fig. 5 The workflow for calculating the cell cycle precursor frequency (cPF) of an asynchronously dividing population stained with CTV, PI, and MPM2. (**a**) Region gates are set on the CTV histogram to identify division rounds. PI/MPM2 fluorescence is used to derive the major cell cycle stages within each peak through gating. (**b**) The gated frequencies within each CTV peak are corrected as outlined to account for the input population expanding by 2*n*, where n is the division round. These adjusted frequencies represent the true frequency of the input population within each division round. We can then use the cell cycle stage frequencies to determine the frequency of the input population within each phase (cPF). (**c**) The cPF values (*y*-axis) can be plotted as a function of each temporally distinct cell cycle phases (*x*-axis) over an increasing culture period. (**d**) Using this method, one can add various chemotherapeutic drugs (the s-phase blocker etoposide in this case) for a short time window and conclusively determine what effect they exert on the temporal evolution of the labeled population

Take the adjusted frequencies for each division round and break those down by cell cycle phase using the gated frequencies from the PI versus MPM2 plots (Fig. 5b, lower table).

This then allows one to plot the temporal evolution of the input population as it moves through and becomes distributed over multiple cell cycles (Fig. 5c).

If one harvests, stains and analyses cells prior to the addition of chemotherapeutic agents it is possible to establish how the cells are distributed prior to exposure. One can then incubate cells in the presence and absence of drug (Fig. 5d) and determine exactly where and when the drug effect occurs.

By then using imagery to subdivide the mitotic phase (*see* next section), one can incorporate the frequencies of prophase, metaphase, anaphase, and telophase cells into the analysis framework.

3.9.4 Identifying MPM2 Positive Mitotic Cells with Defined Division Histories

The most reproducible way to identify and subdivide mitotic cells is to stain for a unique mitotic antigen (MPM2) and then use custom masking and morphometric/intensity-based features. Relying solely on spatial parameters without unique fluorescence, as employed by

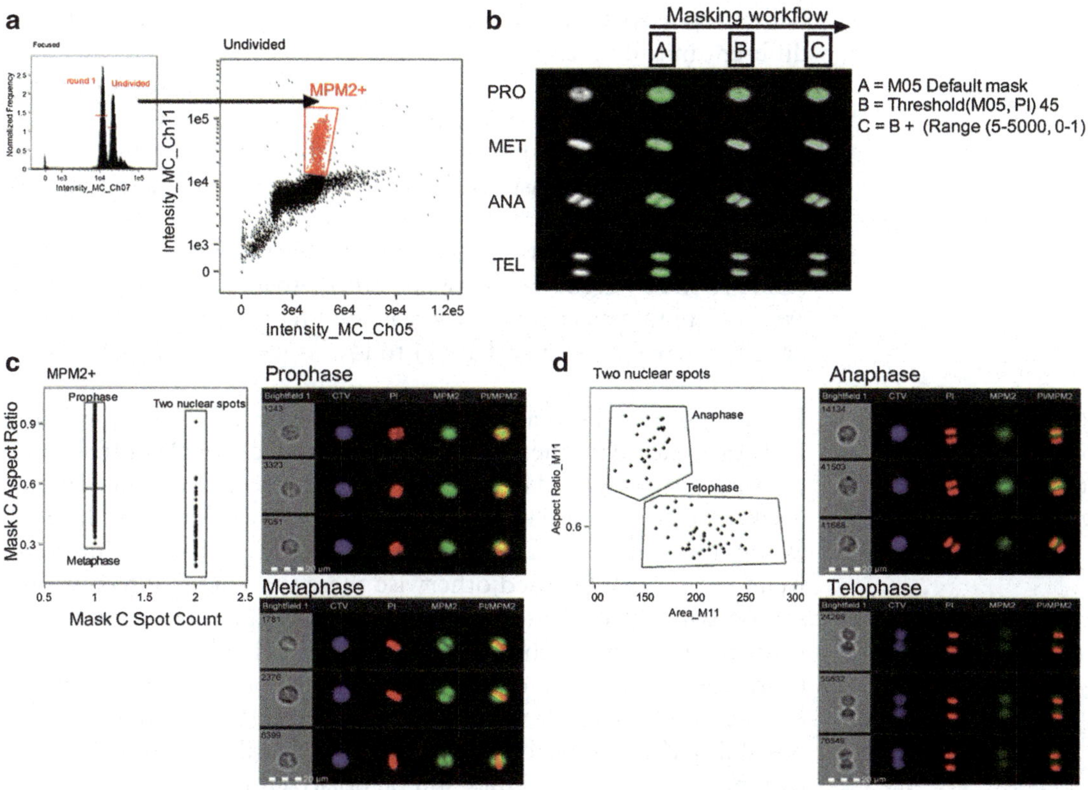

Fig. 6 Using IFC-derived image data to subdivide mitosis. (**a**) Cells undergoing first mitosis can be identified based on CTV (CH7), PI (CH5), and MPM2 (CH11) intensity as shown. (**b**) In order to best resolve the different mitotic phases from within the MPM2+ gate based on the PI-stained nuclear image, the default channel mask must be adapted as shown. Note that the default mask does not distinguish the two nuclear poles in anaphase or telophase cells without adaptation. The masking workflow is shown using the language of IDEAS for audit and reproduction purposes. (**c**) A bivariate plot of the spot count (*x*-axis) and aspect ratio (*y*-axis) of mask **c**. Gates are set to identify prophase and metaphase cells from within the 1-spot population. Multispectral images are shown as examples at 20× magnification. (**d**) The two spot population is further refined using a bi-variate plot of the area (*x*-axis) and aspect ratio (*y*-axis) of the MPM2 image (M11). Gates are set to distinguish anaphase from telophase cells. Multispectral images are shown as examples at 20× magnification

the IDEAS cell cycle wizard, leads to a high degree of misclassification (data not shown).

Firstly, create bivariate plots of PI total intensity (CH05, MC) versus MPM2 total intensity (CH11 Cy5, or CH2 FITC, MC) for each division round. Set a gate to include all MPM2$^+$ events that have a 4N DNA content based on PI intensity (Fig. 6a). The MPM2 phospho-epitope, unlike pH 3, is maintained in telophasic cells making it easier to resolve them from debris and doublets rather than relying on complex masking [16].

Use a simple set of masking adaptations based on the default PI channel mask (M05). The masking adaptation workflow is shown in Fig. 6b (*see* **Note 13**). The aim is to create a mask that better represents the morphological differences in the nuclei of different mitotic phases [21].

Calculate the aspect ratio and spot count features from the final mask. Create a bivariate plot of spot count (x) versus aspect ratio (y) and gate to subdivide 1 spot, high aspect ratio cells as prophase, 1 spot mid-aspect ratio events as metaphase, and all two spot events (Fig. 6c).

The two spot events are plotted on a bivariate graph of MPM2 (CH11, M11) area versus aspect ratio. Anaphase cells have low area and high aspect ratio, whereas telophasic cells (*see* **Note 14**) have an increased area and a mid to low aspect ratio (Fig. 6d).

3.9.5 Measuring Signal Symmetry/Asymmetry (Only Required for Analysis of ACD)

For prophase cells that have no clear plane of division we use IDEAS to calculate the delta between the BF channel centroid and the intensity-weighted centroid of the fluorescence channel of our marker we are investigating (Fig. 7a). In both cases we use the MC as the input mask for this feature. It is essential that the intensity weighted centroid is used otherwise there is a potential to misclassify polarity depending on how the overall signal intensity is distributed (Fig. 7b). In this case, fluorescently labeled antigen-coated beads can be distributed in several possible configurations and the feature used to calculate signal polarity must be robust to the possibility that the beads may be globally distributed, but the relative intensity of the foci may still be polarized [17].

In metaphase, anaphase, and telophase populations where the plane of cytokinesis can be clearly defined (Fig. 7c) we export the 16-bit raw tiff channel images for the PI (CH5), CTV (CH7), and fate determinant (e.g., CD8 and PKCζ) fluorescence channels. This is done from the final gated populations through IDEAS export option.

These images can then be analyzed rapidly by software such as Image J either manually or, if cell numbers are impractical, using an automated script. In all cases, the putative daughter poles are segmented across the plane of cytokinesis using the nuclear PI image as a guide (Fig. 7d).

One can use the fluorescent distribution of CTV and PI as an internal control for symmetry and polar focus. This has been described as a "Morphometrically Relevant Biological" (MRB) control (*see* **Notes 15** and **16**). In prophase cells, the MRB control is the intensity weighted delta centroid between CTV and BF images.

Fig. 7 (continued) the respective centroid values shown for BF vs. beads and BF vs. CTV signals. (**c**) A cartoon depicting the use of CTV fluorescence distribution across the plane of division in metaphase, anaphase and telophasic cells to set the limits of symmetrical inheritance. The green signal exceeds the 40–60 % limit so is considered asymmetric (**d**) An example of using ImageJ to segment putative daughter poles in anaphase and telophasic cells to then measure the pixel intensities within these areas for all the indicated fluorescent channels. Images are shown at 40× magnification

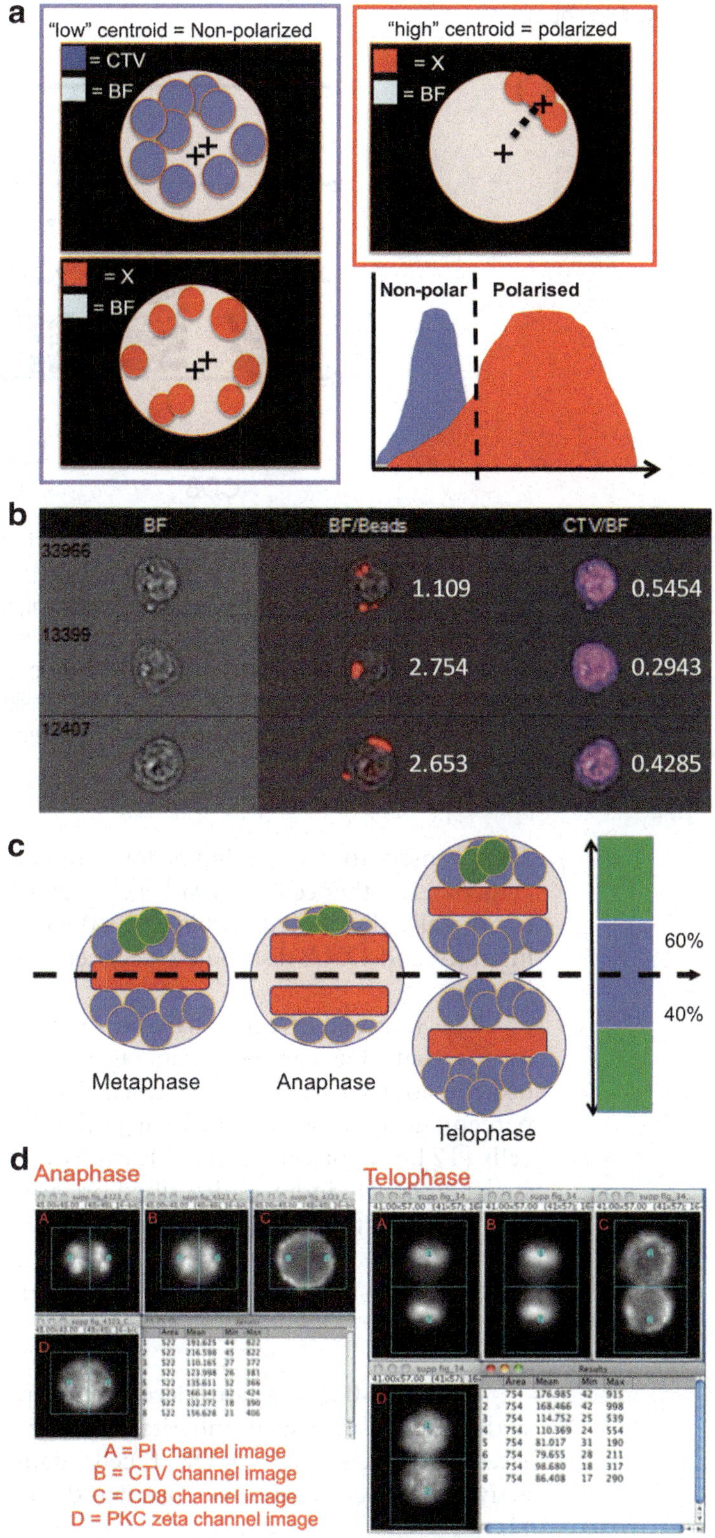

Fig. 7 Quantifying polarity and asymmetry using IFC-derived mitotic imagery. (**a**) A cartoon denoting how the intensity-weighted delta centroid measurement can be used to determine the limits of non-polarized and polarized fluorescence distributions. (**b**) An example of using these measurements on primary CTV-labeled B cells loaded with fluorescent antigen-labeled beads [17]. Multispectral images are shown at 40× magnification with

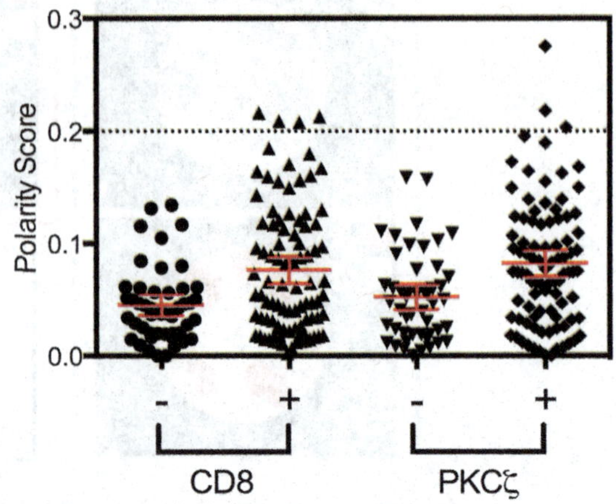

Fig. 8 Plotting the polarity index of CD8 and PKCζ in antigen-activated F5 CD8 T cells entering first division. (**a**) F5 CD8 T cells were labeled with CTV and activated with NP68 peptide under ACD-inducing conditions in the absence (–) or presence (+) of cytochalasin B for the last 4 h of culture. Late-stage mitotic cells from within the undivided CTV peak were then analyzed as described in Subheading 3 and the polarity index was calculated. Values >0.2 are considered to be polarized based on our MRB control. Data are adapted from Hawkins et al. [12]

This is used to set the limits for symmetrical signal distribution (Fig. 6a). In other cell types with a clear divisional plane, total protein is normally distributed within the 40–60 % limits based on CTV intensity and DNA distributed between 45 and 55 % based on PI intensity.

The number of cells analysed using our IFC-based method means that one can plot data as population distributions and use the appropriate statistical tests. Data can be expressed as the percentage polarized or median polarization score for prophase cells [17]. Or for cells with a clear plane of division one can construct a polarity index as described previously [12, 19]. The equation for polarity index calculation is shown below (Eq. 1):

$$\text{Polarity Index} = \frac{\sqrt{(\%\text{in pole1} - 50)^2}}{50} \tag{1}$$

A score >0.2 was considered to be polarized or asymmetric with respect to signal inheritance over the cytokinetic plane. Example data from F5 CD8 T cells stained with our panel and activated under conditions considered to drive ACD is shown in Fig. 8.

4 Notes

1. We use a Becton Dickinson LSRFortessa (BD, USA) system with 405 nm, 488 nm, 561 nm, and 633 nm laser lines and a standard filter configuration.
2. We have obtained the optimal division peak resolution using the succinimidyl ester CellTrace Violet™ (CTV). It is non-cytotoxic, retained within the cells, inherited in a symmetrical fashion by daughter cells during cytokinesis and fluorescently stable after fixation. The excitation and emission properties of the dye make it spectrally compatible with both CFC and IFC systems in combination with a range of other fluorochromes. Moreover it can be used to reliably define the limits of symmetrical protein inheritance as an internal control [19, 16, 12, 17].
3. Lipophilic dyes such as PKH26 tend to yield poorly resolved division peaks compared to most succinimidyl esterase dyes. This may be due to mechanisms such as membrane exchange [22] that influence the relationship between fluorescent intensity and how many times a cell has divided. If for example a cell that has divided twice takes up some dye from an adjacent cell, it may now have an intensity that makes it appear as though it has not divided at all. Conversely, a cell that has not divided may lose some dye and appear as though it has undergone division. For studies of ACD where the division history of cell must be accurately determined, such sources of error must be eliminated where possible.
4. Of the succinimidyl esterase-based dyes currently available, we have found eFluor proliferation dye 670™ highly unsuitable for division tracking as firstly it transfers heavily to bystander cells and secondly it labels a discrete number of cellular structures rather than all intracellular protein. This means that due to a stochastic process [23], the fluorescent signal is not always inherited within symmetrical limits rendering it unsuitable for accurately tracking cell proliferation [19].
5. As well as cross reactivity between primary and secondary antibodies, one should also check that mouse derived clones do not bind non-specifically to mouse cells (B and T). In experiments with B cells, special care must be taken regarding the usage of secondary antibodies because as B cells proliferate they start to express different immunoglobulin (Ig) isotypes on their surface, such as IgG1. To avoid issues with cross reactivity, it is strongly suggested to use directly conjugated antibodies, particularly for MPM2, which is a mouse IgG1. MPM2 can be obtained directly conjugated to both FITC and Cy5 (*see* Subheading 2). Another alternative is to select primary unconjugated antibodies that are not raised in mouse thereby avoiding the need for anti-mouse secondary labeling reagents.

6. The choice of fixative will depend on the application of our method. If it is purely for analyzing cell cycle using PI and MPM2 then 70 % EtOH works best as it dehydrates the cells allowing easy access to the nucleus for DNA-binding dyes without affecting the MPM2 epitopes. However, if more complex antigenic staining is required, often EtOH fixation will destroy binding epitopes meaning that FA must be used. FA fixed cells tend not to give such well resolved DNA histograms as the cross-linking affects dye access to the nucleus. This can be improved somewhat by including 0.1 % triton-X 100 when incubating with the dye. When making up FA fixative solutions, one should start from liquid state rather than paraformaldehyde powder. The reason for this is that the actual fixative is monomeric formaldehyde, or methyl-hydrate. Methyl hydrate formation is most rapid and stable when formaldehyde is diluted in neutral buffers such as PBS. As such we always make up our FA solution in PBS and leave it for at least 2 weeks at 4 °C prior to use. If cells are not properly fixed prior to permeabilization, then the cells will simply fall apart. Fixed cells should also have reduced forward and side scatter signals compared to unfixed cells. We routinely use these metrics to determine the activity of a fixative solution.
7. The ability to resolve division peaks using fluorescent dye dilution is determined by several interrelated factors. Firstly, cytometer performance must be optimal in terms of laser excitation and fluorescent detection so as to minimise the contribution of machine errors to the measurement. Secondly, labeling conditions must be optimised to ensure tight, uniform populations. However, heterogeneous cell types (cell lines, lymphocyte blasts) have variable protein content and will always label with a broad distribution even under fully optimized conditions. These cell types require cell sorting to narrow the input population below a certain threshold for peak resolution [19].
8. The harvest time is very important for the analysis of rare events such as telophasic cells. These targets can be less than 0.01 % of the total population. Typically we aim to collect between 100 and 500 cells for the final measurement as this allows us to use population statistics to describe the biological differences we may see. We want to avoid any inhibitors of cytokinesis or block and release protocols as we have shown that cytochalasin treatments may alter the natural cell state and induce artifactual asymmetry [12]. Therefore we need some idea of the window when cells are entering first division by using markers that denote entry into G1 such as Ki67 or CD69. We suggest that CTV-labelled primary cells are cultured under activating conditions and harvested at different time

points, fixed, and stained for Ki67/CD69 to denote activation/cell cycle commitment. This can be analysed by CFC and does not require an IFC system.

9. One of the biggest challenges with acquiring samples on an IFC system is the lack of control over individual camera channels. Unlike CFC systems where each PMT voltage can be controlled in order to maximise signal resolution, we have no such controls over individual camera channels. As such, we rely on altering laser powers to control signal. PI is maximally excited by a 488 nm laser line but is also excited by a 405 nm laser to ~20 % of maximum. If the PI intensity is very bright due to over-staining then the 488 nm laser output would need to be set comparatively low. If the CTV or other 405 nm excited fluorochromes require the 405 nm laser to be set near to maximum then there is a possibility that the 405 nm excitation of PI will exceed the 488 nm excitation meaning that PI is now maximally fluorescent in CH11. This is because the spectral widths of CH5 and CH11 are identical to one another (*see* Fig. 1). CH11 would not be usable now to detect AF-647 or equivalent dyes.
10. The sample run rate of any IFC system is slower compared to a CFC system. The overriding consideration for sample speed is image quality and the system ensures that the objects move past the camera at the appropriate rate. The CCD camera has no electronic dead time, therefore concentrating the sample into a density such as 20–30 × 10^6 cells/ml means that the sample run time is greatly reduced but there will be no detrimental effect on the data.
11. As IDEAS has no modelling options for either cell cycle or proliferation, one can export the data as an FCS file and use FlowJo for analysis. In this instance, IDEAS can be used to compensate data and to perform limited data reduction (single in focus cells for example) as well as any custom masks and feature extraction. We have found that FlowJo X handles IFC-derived FCS files particularly well due to the easy re-scaling feature within the software. However if the values of the exported features map between 0 and 1, or go into the negatives, there can be issues with displaying the data. FCS Express (DE Novo, USA) can also be used and does not suffer from this scaling issue.
12. CFC has had significant impact on the area of cell cycle analysis [24]. It is essential to consider that within an asynchronously growing population, not only will cells be in different stages of a given cell cycle; they may also temporally separated across multiple division rounds. As a result, the analysis of asynchronously growing cells always requires a multiparameter, temporal approach. One particularly powerful solution to this issue is

the bromodeoxyuridine (BrdU)-Hoechst quenching method whereby defined populations can be tracked from their origin at the point of drug addition as long as BrdU is added at the exact same moment [25]. However the technique limited by incompatibility with antigen staining, resolution of up to 2 two division rounds and a very complex analysis framework. In our hands, the use of CTV, PI, and MPM2 is a viable alternative. While our approach may lack the ability to track specific cohorts from defined cell cycle stages through progressive rounds, it does provide the temporal data required to elucidate the true impact a drug may exert intra- and inter-cycle while still being relatively simple to perform. By adopting an analysis strategy that is in effect a hybrid between the classical dye dilution calculation of precursory frequencies [26] and traditional cell cycle phase modeling, we can express the data as the frequency of the input population within a given cell cycle phase within each division round.

13. We have found that the overall masking modification string is similar ("threshold" followed by "range") whether images are collected at 20×, 40×, or 60× magnification. However the absolute threshold values for the adaptations will differ. For example images collected at 60× will have more pixel texture and information associated with them. As such, thresholding tends to form smaller satellite masks, especially with highly textured prophase nuclei. This requires a more stringent range mask adaptation based on the minimum masking area. The performance and appropriateness of a particular mask is also related to the staining quality in a given channel. We find that masks not only need to be adapted between magnifications but also between experiments due to differences in the signal to noise within the image. Therefore we suggest that the exact composition of the masking adaptations should be determined empirically.

14. One of the biggest challenges to analysing cells in telophase is misclassifying due to the presence of doublet cells. The fact that MPM2 is still able to identify telophasic cells above background levels of fluorescence is a massive improvement on our previous approach that relied on complex masking without a unique fluorescence signal [16]. There is still a possibility that two conjoined metaphase cells (MPM2 positive and with flat nuclei) could resemble a telophasic cell but as MPM2 is brighter on metaphase cells, one could use this feature to exclude such a possibility. Moreover an actual telophase cell would still have the same sum CTV intensity as a single undivided cell, simply distributed over two poles. The only remaining issue is if two cells from round 1 became attached they could look like a telophase cells about to divide for the first

time but if the population is analysed before cells enter the second mitosis, then these cells will not be present.

15. It has recently been suggested that all cellular protein is naturally inherited in an asymmetrical fashion within T cells and that this needs to be factored in to any dye-dilution-based calculations of cell division [27, 28]. We have specifically looked at CFSE and CTV inheritance across the plane of cytokinesis in telophasic cells and found no evidence of any asymmetrical inheritance [19] outside of what can be considered normal stochastic variation in the inheritance of protein [29]. We use this to define the "normal" limits of protein inheritance and to therefore define the boundary of asymmetry. We have termed this a "morphometrically relevant biological control" (MRB) [21] as like a fluorescence minus one control (FMO) in CFC is used to determine negative and positive thresholds based on gross fluorescence, it can be used to set gating thresholds on a spatial/morphometric parameter. The idea of an MRB control has been used successfully in other IFC-based analysis [30]. There has been some recent question over the validity of using a diffuse CTV signal to control for a more discrete signal [31]. However we have also stain for endosomes (discrete signal) and shown they are inherited in symmetrical fashion same as diffuse signal [16].

16. We have found that for telophasic cells, if one pole is imaged outside of the focal plane then this can affect the measurement of signal, tending to then lead to misclassifying a cell as asymmetric. However, the same effect is seen when one looks at the PI and CTV signals placing them outside the limits of symmetry. We therefore reject these cells from the analysis based on the MRB controls (PI/CTV). It has been suggested that the extended depth of field option (EDF) could be employed to overcome issues with defocus by brining all light back into the focal plane [32]. However as we have no control over the orientation of a telophasic cell as it flows, EDF would simply merge the fluorescent information from two poles imaged at an angle as they are brought in to the same focal plane by deconvolution. Therefore the defocus information gained by acquiring in non-EDF mode can be used to ones advantage in this regard to eliminate cells with one or more poles outside the focal plane.

Acknowledgements

AF, WD, and SP were funded by CRUK. AF also acknowledges support from the ISAC SRL Emerging Leaders program. NMM acknowledges support from the Maria Sklowdowska-Curie Fellowship.

References

1. Kastan MB, Bartek J (2004) Cell-cycle checkpoints and cancer. Nature 432:316–323. doi:10.1038/nature03097, nature03097 [pii]
2. Vermeulen K, Van Bockstaele DR, Berneman ZN (2003) The cell cycle: a review of regulation, deregulation and therapeutic targets in cancer. Cell Prolif 36:131–149, doi:266 [pii]
3. Cairns J (1975) Mutation selection and the natural history of cancer. Nature 255:197–200
4. Kusch J, Liakopoulos D, Barral Y (2003) Spindle asymmetry: a compass for the cell. Trends Cell Biol 13:562–569
5. Shen Q, Zhong W, Jan YN, Temple S (2002) Asymmetric Numb distribution is critical for asymmetric cell division of mouse cerebral cortical stem cells and neuroblasts. Development 129:4843–4853
6. Arsenio J, Kakaradov B, Metz PJ, Kim SH, Yeo GW, Chang JT (2014) Early specification of CD8+ T lymphocyte fates during adaptive immunity revealed by single-cell gene-expression analyses. Nat Immunol 15:365–372
7. Chang JT, Palanivel VR, Kinjyo I, Schambach F, Intlekofer AM, Banerjee A, Longworth SA, Vinup KE, Mrass P, Oliaro J, Killeen N, Orange JS, Russell SM, Weninger W, Reiner SL (2007) Asymmetric T lymphocyte division in the initiation of adaptive immune responses. Science 315:1687–1691
8. King CG, Koehli S, Hausmann B, Schmaler M, Zehn D, Palmer E (2012) T cell affinity regulates asymmetric division, effector cell differentiation, and tissue pathology. Immunity 37: 709–720
9. Ciocca ML, Barnett BE, Burkhardt JK, Chang JT, Reiner SL (2012) Cutting edge: Asymmetric memory T cell division in response to rechallenge. J Immunol 188:4145–4148
10. Barnett BE, Ciocca ML, Goenka R, Barnett LG, Wu J, Laufer TM, Burkhardt JK, Cancro MP, Reiner SL (2012) Asymmetric B cell division in the germinal center reaction. Science 335:342–344
11. Lemaitre F, Moreau HD, Vedele L, Bousso P (2013) Phenotypic CD8+ T cell diversification occurs before, during, and after the first T cell division. J Immunol 191:1578–1585
12. Hawkins ED, Oliaro J, Kallies A, Belz GT, Filby A, Hogan T, Haynes N, Ramsbottom KM, Van Ham V, Kinwell T, Seddon B, Davies D, Tarlinton D, Lew AM, Humbert PO, Russell SM (2013) Regulation of asymmetric cell division and polarity by Scribble is not required for humoral immunity. Nat Commun 4:1801. doi:10.1038/ncomms2796
13. Duffy KR, Wellard CJ, Markham JF, Zhou JH, Holmberg R, Hawkins ED, Hasbold J, Dowling MR, Hodgkin PD (2012) Activation-induced B cell fates are selected by intracellular stochastic competition. Science 335:338–341
14. Gerlach C, Rohr JC, Perie L, van Rooij N, van Heijst JW, Velds A, Urbanus J, Naik SH, Jacobs H, Beltman JB, de Boer RJ, Schumacher TN (2013) Heterogeneous differentiation patterns of individual CD8+ T cells. Science 340:635–639
15. Buchholz VR, Flossdorf M, Hensel I, Kretschmer L, Weissbrich B, Graf P, Verschoor A, Schiemann M, Hofer T, Busch DH (2013) Disparate individual fates compose robust CD8+ T cell immunity. Science 340:630–635
16. Filby A, Perucha E, Summers H, Rees P, Chana P, Heck S, Lord GM, Davies D (2011) An imaging flow cytometric method for measuring cell division history and molecular symmetry during mitosis. Cytometry A 79A:496–506
17. Thaunat O, Granja AG, Barral P, Filby A, Montaner B, Collinson L, Martinez-Martin N, Harwood NE, Bruckbauer A, Batista FD (2012) Asymmetric segregation of polarized antigen on B cell division shapes presentation capacity. Science 335:475–479
18. Quah BJ, Parish CR (2012) New and improved methods for measuring lymphocyte proliferation in vitro and in vivo using CFSE-like fluorescent dyes. J Immunol Methods 379:1–14
19. Begum J, Day W, Henderson C, Purewal S, Cerveira J, Summers H, Rees P, Davies D, Filby A (2013) A method for evaluating the use of fluorescent dyes to track proliferation in cell lines by dye dilution. Cytometry A 83: 1085–1095
20. Ortyn WE, Hall BE, George TC, Frost K, Basiji DA, Perry DJ, Zimmerman CA, Coder D, Morrissey PJ (2006) Sensitivity measurement and compensation in spectral imaging. Cytometry A 69A:852–862
21. Filby A, Davies D (2012) Reporting imaging flow cytometry data for publication: Why mask the detail? Cytometry A 81A:637–642
22. Lassailly F, Griessinger E, Bonnet D (2010) "Microenvironmental contaminations" induced by fluorescent lipophilic dyes used for noninvasive in vitro and in vivo cell tracking. Blood 115:5347–5354
23. Summers HD, Rees P, Holton MD, Rowan Brown M, Chappell SC, Smith PJ, Errington RJ (2011) Statistical analysis of nanoparticle dosing in a dynamic cellular system. Nat Nanotechnol. doi: nnano.2010.277 [pii] 10.1038/nnano.2010.277

24. Darzynkiewicz Z, Crissman H, Jacobberger JW (2004) Cytometry of the cell cycle: cycling through history. Cytometry A 58A:21–32
25. Ormerod MG, Kubbies M (1992) Cell cycle analysis of asynchronous cell populations by flow cytometry using bromodeoxyuridine label and Hoechst-propidium iodide stain. Cytometry 13:678–685
26. Filby A, Seddon B, Kleczkowska J, Salmond R, Tomlinson P, Smida M, Lindquist JA, Schraven B, Zamoyska R (2007) Fyn regulates the duration of TCR engagement needed for commitment to effector function. J Immunol 179: 4635–4644
27. Luzyanina T, Cupovic J, Ludewig B, Bocharov G (2013) Mathematical models for CFSE labelled lymphocyte dynamics: asymmetry and time-lag in division. J Math Biol. doi:10.1007/s00285-013-0741-z
28. Bocharov G, Luzyanina T, Cupovic J, Ludewig B (2013) Asymmetry of Cell Division in CFSE-Based Lymphocyte Proliferation Analysis. Front Immunol 4:264. doi:10.3389/fimmu.2013.00264
29. Darzynkiewicz Z, Crissman H, Traganos F, Steinkamp J (1982) Cell heterogeneity during the cell cycle. J Cell Physiol 113:465–474
30. Niswander LM, McGrath KE, Kennedy JC, Palis J (2014) Improved quantitative analysis of primary bone marrow megakaryocytes utilizing imaging flow cytometry. Cytometry A 85A:302–312
31. Shimoni R, Pham K, Yassin M, Ludford-Menting MJ, Gu M, Russell SM (2014) Normalized polarization ratios for the analysis of cell polarity. PLoS One 9, e99885. doi:10.1371/journal.pone.0099885
32. Ortyn WE, Perry DJ, Venkatachalam V, Liang L, Hall BE, Frost K, Basiji DA (2007) Extended depth of field imaging for high speed cell analysis. Cytometry A 71A:215–231
33. Watson JV, Chambers SH, Smith PJ (1987) A pragmatic approach to the analysis of DNA histograms with a definable G1 peak. Cytometry 8:1–8

Chapter 6

Quantitation of Chromosome Damage by Imaging Flow Cytometry

Lindsay A. Beaton-Green and Ruth C. Wilkins

Abstract

Biodosimetry is a method for measuring the dose of radiation to individuals using biological markers such as chromosome damage. Following mass casualty events, it is important to provide this information rapidly in order to assist with the medical management of potentially exposed casualties. Currently, the gold standard for biodosimetry is the dicentric chromosome assay, which accurately estimates the dose from the number of dicentric chromosomes in lymphocytes. To increase throughput of analysis following a large-scale mass casualty event, this assay has been adapted for use on the imaging flow cytometer. This chapter describes the methods for the identification and quantification of mono- and multicentric chromosomes using the imaging flow cytometer.

Key words Dicentric chromosome assay, Imaging flow cytometry, Lymphocytes, Chromosome damage, Biodosimetry

1 Introduction

The dicentric chromosome assay (DCA) converts the frequency of dicentric chromosomes scored in mitotic lymphocytes into a dose estimate of ionizing radiation exposure. The DCA is considered to be the gold standard for radiation biodosimetry [1–3], and can be useful in the case of an emergency involving radiological or nuclear material where physical dosimetry is absent. While the assay is sensitive, the time required to prepare slides for microscopy and to score them manually is a limiting factor particularly for high-throughput triage following a large-scale mass casualty event [4, 5].

Recent efforts to increase the sample throughput of the DCA for emergency triage have included the development of the Quickscan slide scoring protocol [6] and improvements in the automated analysis of slides using metaphase finders which both locate the metaphase spreads and identify dicentric chromosomes [7, 8]. Despite these efforts, sample throughput is still limited by

Natasha S. Barteneva and Ivan A. Vorobjev (eds.), *Imaging Flow Cytometry: Methods and Protocols*, Methods in Molecular Biology, vol. 1389, DOI 10.1007/978-1-4939-3302-0_6, © Springer Science+Business Media New York 2016

the time required for the preparation, scanning and scoring of microscope slides.

Previous efforts to develop the DCA method for analysis by flow cytometry have resulted in well-established sample preparation methods in which chromosomes and centromeres were successfully fluorescently labelled [9–12]. However with the sensitivity limitations of conventional flow cytometry, it remains difficult to reliably detect the difference between mono- and dicentric chromosomes or to distinguish dicentric chromosomes from chromosome aggregates or debris [9, 10]. Recently, with the development of new imaging flow cytometry technology, imaging capabilities of microscopy can be combined with the increased throughput of conventional flow cytometers. This has led to the adaptation of microscope-based DCA analysis to an automated cytometry method (flow-DCA) which can now identify and quantify damaged chromosomes [13].

Chapter 6 describes the methods to quantitate chromosome damage in four major sections: sample preparation and culture, chromosome isolation, hybridization and staining, and data acquisition and analysis.

2 Materials

2.1 Lab Equipment

1. Lithium-heparin Vacutainer® tubes (10 mL).
2. Histopaque®-1077 (Sigma-Aldrich Inc.).
3. AccuSpin™ tubes (Sigma-Aldrich Inc.).
4. Centrifugation tubes (15 mL).
5. Ventilated culture flasks (25 cm^2).
6. Safe-Lock tubes (1.5 mL).
7. Filters (0.2 μm).
8. Imagestream^X (EMD-Millipore) with 488 nm laser, 405 nm laser (*see* **Note 1**), 60× magnification and extended depth of field (EDF) option.

2.2 Reagents

1. Complete media: RPMI 1640 with 1 % 2 mM l-Glutamine–Penicillin–Streptomycin solution, 15 % v/v inactivated fetal bovine serum (FBS), 2 % phytohemagglutinin (PHA), and 1 % colcemid (10 μg/mL) (*see* **Note 2**).
2. Isotonic phosphate buffered saline (PBS) (sterile): 137 mM NaCl, 2.7 mM KCl, 4.3 mM Na_2HPO_4, 1.4 mM KH_2PO_4 (*see* **Note 3**).
3. Hypotonic solution: 55 mM KCl and 20 mM HEPES (*see* **Note 4**).
4. Chromosome isolation buffer: 10 % v/v each of 20 mM ethylenediaminetetraacetic acid (EDTA) in double-distilled water

(ddH_2O) (*see* **Note 5**), 5 mM ethylene glycol tetraacetic acid (EGTA) in ddH_2O (*see* **Note 6**) and 150 mM Tris–HCl, 800 mM KCl, and 200 mM NaCl in ddH_2O (*see* **Note 7**). 0.1 % v/v each 2-mercaptoethanol and Triton X-100. 0.05 % each of 0.4 M spermine in ddH_2O (*see* **Note 8**) and 1.0 M spermidine in ddH_2O (*see* **Notes 8** and **9**).

5. Hoechst 33258 stock solution: 100 μg/mL in ddH_2O (*see* **Note 10**).
6. PNA probe (F3006, PNA Bio, Thousand Oaks, CA): 5 μM stock in ddH_2O (*see* **Note 11**).
7. $MgCl_2$ buffer (20×): 82 mM Na_2HPO_4, 9 mM citric acid and 20 mM $MgCl_2$ in ddH_2O (*see* **Note 12**).
8. Blocking reagent stock solution (Perkin Elmer, 10×): 2.5 % w/v in ddH_2O (*see* **Note 13**).
9. Hybridization Buffer: 70 % formamide, 10 % (v/v) blocking reagent stock solution (0.25 % w/v. final concentration), 5 % (v/v) $MgCl_2$ buffer, 0.2–0.5 μg/mL fluorescently labeled PNA probe in ddH_2O (*see* **Note 14**).
10. Wash solution (for hybridization): 70 % formamide, 0.1 % (w/v) bovine serum albumin (BSA), and 5 % (v/v) $MgCl_2$ buffer in ddH_2O.

3 Methods

- Unless otherwise noted, all steps are conducted at room temperature (RT) which is assumed to range from 18 to 25 °C.
- Steps are conducted under normal lighting conditions unless otherwise noted.

3.1 Sample Preparation and Culture

1. Draw blood into lithium-heparin Vacutainers®.
2. Irradiate and/or challenge samples as required.
3. To separate lymphocytes bring Histopaque®-1077 to RT and pipet 3.0 mL Histopaque®-1077 into each upper chamber of the AccuSpin™ tubes.
4. Centrifuge at 800 × *g* for 30 s to place the Histopaque®-1077 below the frit.
5. Pipet 6.0 mL anticoagulated whole blood into each upper chamber of the AccuSpin™ tubes.
6. Centrifuge at 800 × *g* for 15 min at RT (*see* **Note 15**).
7. Carefully aspirate the plasma layer with a Pasteur pipette and discard.
8. Carefully transfer the mononuclear band with a Pasteur pipet to a 15 mL centrifuge tube (*see* **Note 16**).

9. Add 10 mL of isotonic PBS to the mononuclear cells and mix gently.
10. Centrifuge at 250 × *g* at RT for 10 min.
11. Aspirate the PBS and resuspend the pellet (*see* **Note 17**).
12. Repeat **steps 9–11** twice, each time resuspending pellet in 5 mL isotonic PBS.
13. Resuspend the pellet in complete media.
14. Divide cells in complete media, containing colcemid and PHA at an initial concentration of 1×10^6 cells/mL (*see* **Note 18**).
15. Incubate the cultures for 48 h at 37 °C and 5 % CO_2.

3.2 Chromosome Isolation

1. After incubation, transfer culture to 15 mL centrifuge tubes.
2. Spin down cultures at 350 × *g* for 5 min at RT.
3. Aspirate media from the tube, removing as much media as possible.
4. Disrupt the cell pellet by flicking the tube.
5. Resuspend cell pellet by slowly adding 10 mL hypotonic solution (*see* **Note 19**).
6. Incubate at RT for 20 min (*see* **Note 20**).
7. Spin down cells at 350 × *g* for 5 min at RT.
8. Gently aspirate the hypotonic solution from the tubes (*see* **Note 21**).
9. Disrupt the cell pellet by flicking the tube and resuspend the cell pellet by slowly adding 3 mL ice-cold chromosome isolation buffer.
10. Incubate on ice for 15 min.
11. Vortex vigorously for 75 s at the maximum vortex setting to break apart the cells and free the chromosomes into suspension (*see* **Note 22**).
12. Allow chromosome suspension to settle overnight at 4 °C (*see* **Note 23**).

3.3 Hybridization and Staining of Chromosomes in Suspension

1. Set water bath to 80 °C.
2. After chromosomes have been left to settle overnight, carefully remove excess buffer to concentrate the chromosomes (*see* **Note 24**).
3. Add 100 μL of pre-warmed hybridization buffer to the chromosomes, pipette to mix.
4. Denature the chromosome suspension by immersing in a water bath at 80 °C for 5 min (*see* **Note 25**).
5. Hybridize, in the dark, at RT for 60 min.
6. After hybridization, add 500 μL of pre-warmed wash solution to the chromosome suspensions.

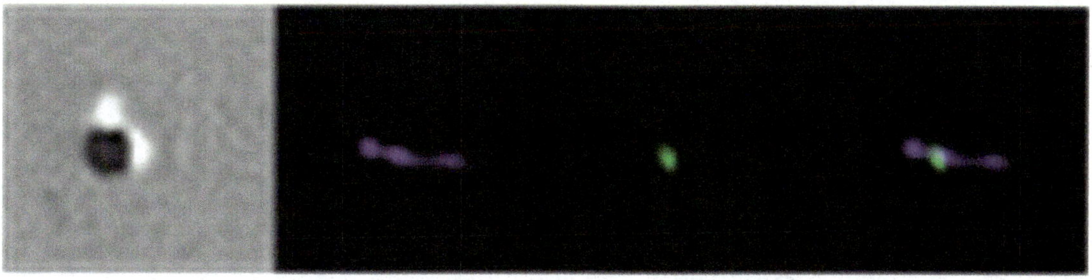

Fig. 1 Example of an object (#1605) in each of the collected channels (channel 5 bright field, channel 1 Hoechst 33258 and channel 2 PNA probe) as well as a composite view of channels 1 and 2 overlayed

7. Incubate for 5 min at 37 °C.
8. Spin down the chromosomes at 1675 ×*g* for 10 min at 4 °C.
9. Carefully remove the supernatant, leaving approximately 150 μL (*see* **Note 26**).
10. Add 2 μg/mL of Hoechst 33258 to each sample and let stand for 1 h at 4 °C (*see* **Note 27**).

3.4 Data Acquisition and Analysis

1. Transfer 75 μL of well-mixed sample to a 1.5 mL Safe-Lock tube, and load onto the Imagestream^X.
2. Set the Imagestream^X to 60× magnification and extended depth of field.
3. Set the laser power and cell classifiers to optimize collection of chromosomes while discarding intact cells and unstained debris (*see* **Note 28**).
4. Acquire samples (*see* **Note 29**).
5. Open the Imagestream analysis software package IDEAS.
6. Follow **steps 7–12** to identify the chromosome population; iterate as required to fine-tune the gating strategy.
7. Create a composite view by overlaying channel 1 (Hoechst 33258) and channel 2 (PNA probe) to allow for visual inspection of objects as in Fig. 1.
8. Create a plot of *Intensity* of channel 1 (Hoechst 33258) versus *Intensity* of channel 2 *(PNA)*, where the *Intensity* feature is the sum of the background subtracted pixel values within the masked area of the image. In this way you can select for two-colour positive populations (*see* Fig. 2). Note that cells can also be identified and rejected (identified as "cells" in Fig. 2).
9. Visually investigate identified populations to ensure that gates are including objects of interest without including too many unwanted objects (*see* Fig. 3).
10. Based on the two-color positive population, create a new scatterplot of the *Contrast* of channel 5 (bright field) versus the *Contrast* of channel 1 (Hoechst 33258), where *Contrast*

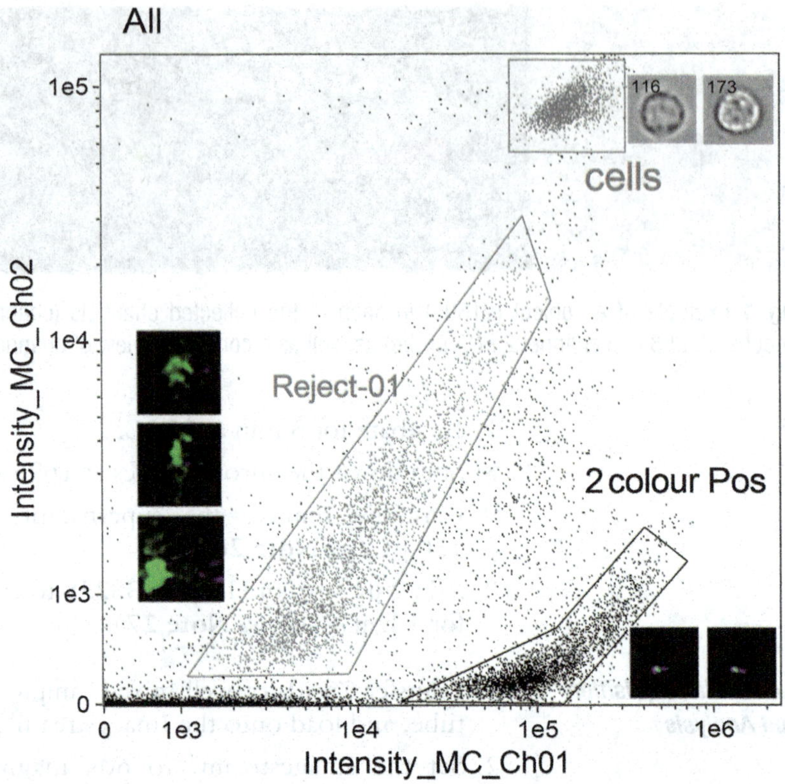

Fig. 2 Scatterplot of *Intensity* of channel 1 (Hoechst 33258) versus *Intensity* of channel 2 (PNA probe) to identify two-colour positive population "2colourPos". Noted on the plot is a population of "cells" which appear to be intact. Also shown is a population identified as "Reject-01" with some example objects to the *left*

Fig. 3 Selection of objects from the "2colourPos" population with a scale bar of 7 μm for reference. Note that some of the objects appear to be aggregates of multiple chromosomes (object #3674)

measures the sharpness quality of an image, to select a subselection of objects with good contrast (*see* Fig. 4, **Note 30**).

11. Again, visually inspect the populations to verify the stringency of the gating strategy. Figure 4 shows a selection of objects from each gate (Reject-02 and goodContrast) and illustrates the difference in image quality.

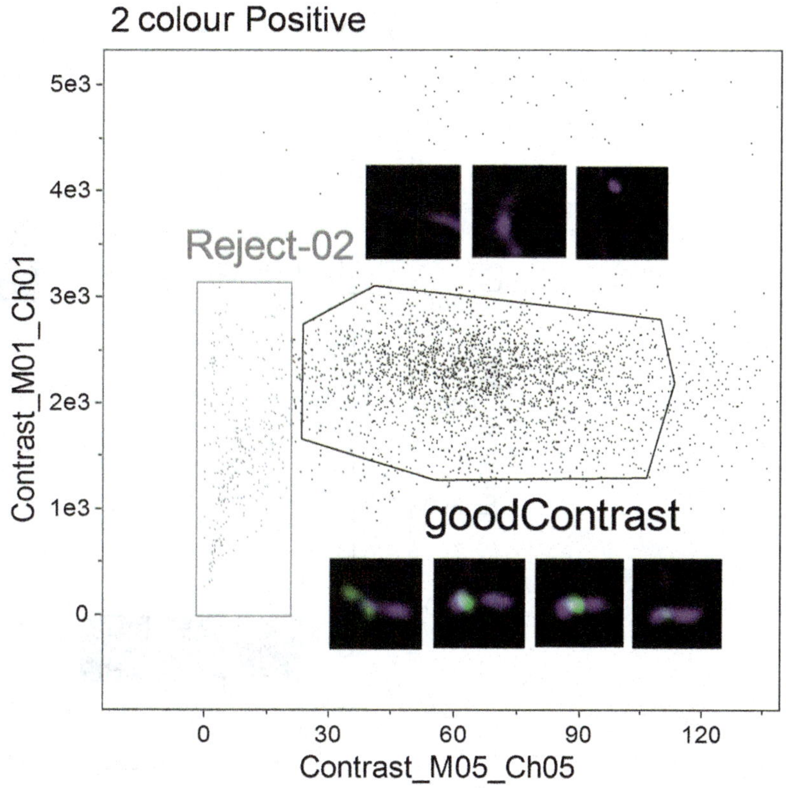

Fig. 4 Based on the two-colour positive chromosome population identified in Fig. 2, plot *Contrast* of channel 5 (bright field) versus the *Contrast* of channel 1 (Hoechst 33258) to identify objects with good contrast. Examples of objects in each of the "good Contrast" and "Reject-02" region are shown alongside the gates

12. From the goodContrast population, create a scatterplot of the *Area* of channel 2 (PNA probe), where *Area* is equal to the number of microns squared in a mask, versus the *Intensity* of channel 2 to remove objects that have too much PNA probe (*see* Fig. 5, **Note 31**).
13. From the centArea population created in **step 12**, create a scatterplot of the *Area* of channel 5 (bright field) versus the *Aspect Ratio Intensity*, defined to be the ratio between the intensity weighted minor axis of the image divided by the intensity weighted major axis of the image, of channel 6 (side scatter), to remove objects that appear to be doublets (Reject-04), leaving a population of apparently single chromosomes (Singles) (*see* Fig. 6).
14. To quantify the chromosome damage, follow **steps 15–17**; iterate as required to fine tune the masking strategy.
15. Create a spot mask based on the centromere signal as shown in Fig. 7 (*see* **Note 32**).

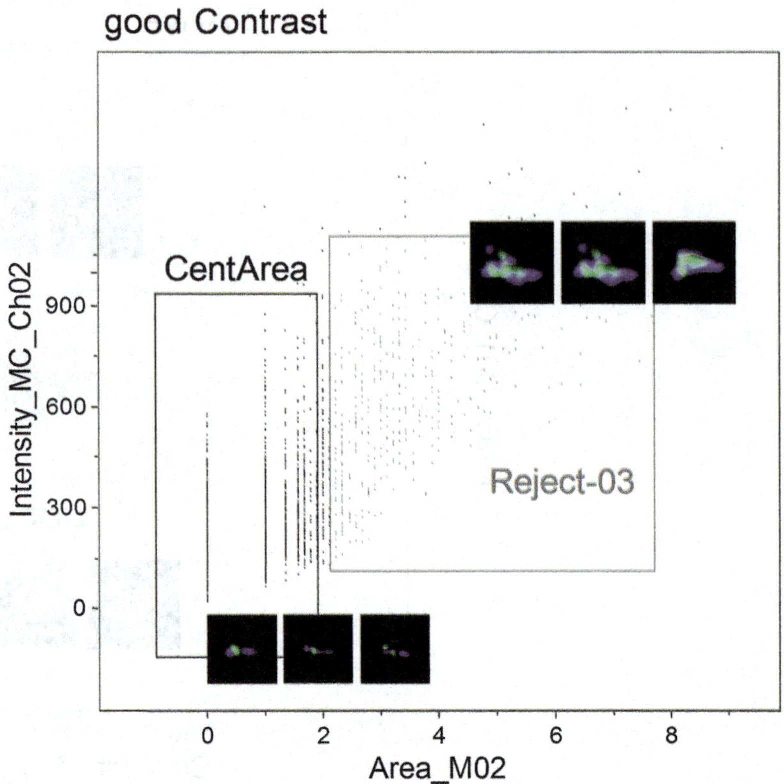

Fig. 5 Scatterplot of the *Area* of channel 2 (PNA probe) versus the *Intensity* of channel 2

16. Plot a histogram of spot counts based on the mask created in **step 15** (*see* Fig. 8).
17. Create gates on the histogram bins for chromosomes with each number of centromeres (spots). Check each of these gates to see if the included objects are good representations of chromosomes with the specified number of centromeres (*see* **Note 33**).

4 Notes

1. If a 405 nm laser is not available, the method can be adapted for use with a different laser and an alternative DNA stain, i.e., propidium iodide on the 658 nm laser.
2. Make fresh on the first day of the experiment. For a 70 mL final volume, start with 56.8 mL RPMI 1640 media and add 0.7 mL l-Glutamine-Penicillin-Streptomycin, 10.5 mL heat-inactivated FBS, 1.26 mL PHA, and 0.7 mL colcemid. Let warm in an incubator until needed.

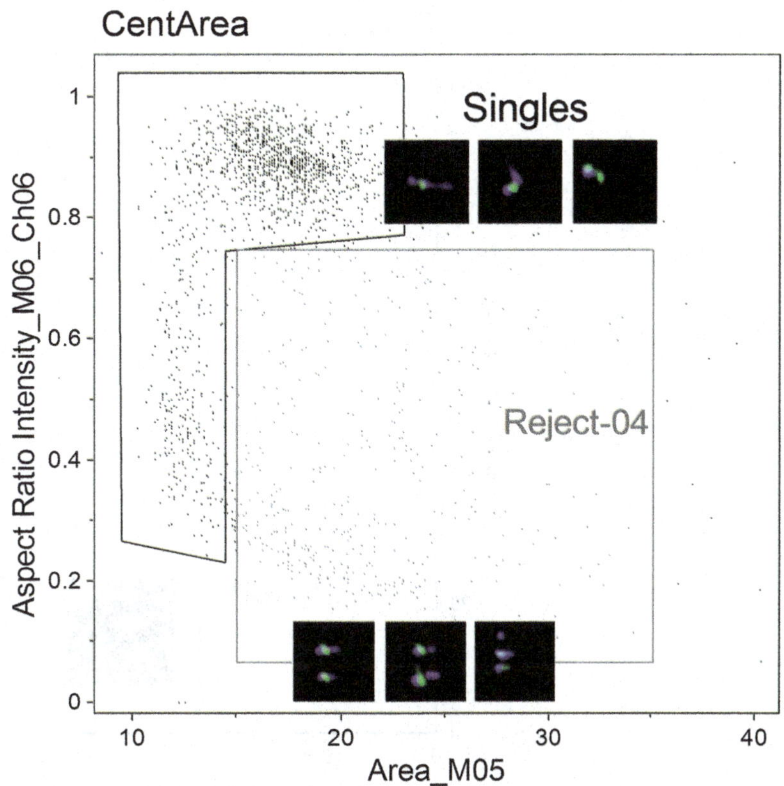

Fig. 6 Scatter plot of the *Area* of channel 5 (bright field) versus the *Aspect Ratio Intensity* of channel 6 (side scatter). Two populations are shown, one of objects considered to be single chromosomes "singles," and another of objects considered to have multiple chromosomes or aggregates "Reject-04." Examples of objects in each population are shown adjacent to the respective gates

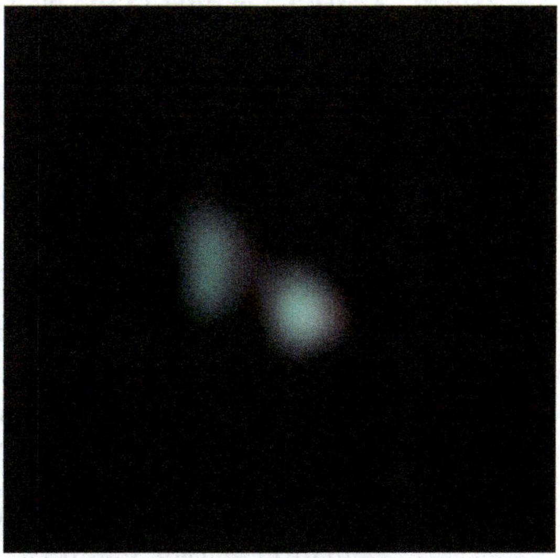

Fig. 7 Masking of PNA probe signal (*see* **Note 32**)

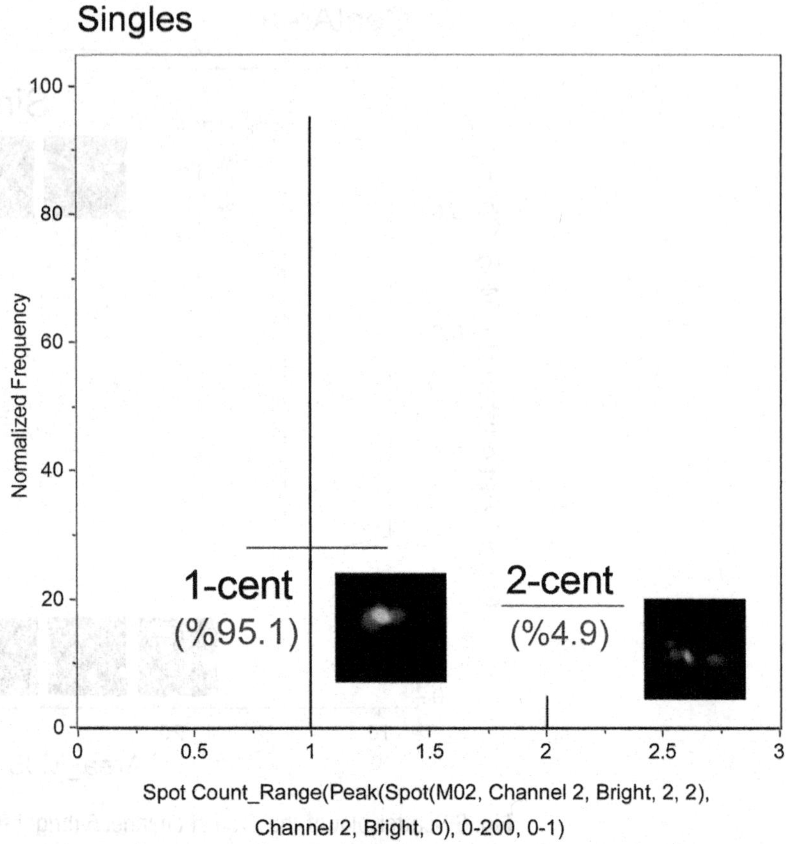

Fig. 8 Histogram of *Spot counts*. These gates represent monocentric (1-cent) and dicentric (2-centric) chromosomes

3. pH 7.4 in ddH_2O and filter sterile. Can be stored at RT for up to a year.
4. pH 7.4 in ddH_2O and filter (0.2 μm) and store at RT for up to several months.
5. Store at RT for up to several months.
6. Dissolve by adding 18 M NaOH dropwise. Store at RT for up to several months.
7. Store at RT for up to several months.
8. Filter (0.2 μm), aliquot, and store at −20 °C for up to several years.
9. Make fresh for each experiment. In a beaker, start with approximately 60 % of final volume ddH_2O; add 10 % each of 20 mM EDTA in ddH_2O, 5 mM EGTA in ddH_2O and 150 mM Tris–HCl, 800 mM KCl, and 200 mM NaCl in ddH_2O. Adjust the pH to 7.2 using 0.5 M NaOH. Add 0.1 % v/v each of 2-mercaptoethanol and Triton X-100. Cover and stir for 15 min.

Transfer to a volumetric flask and top up with ddH_2O. Filter sterile (0.2 μm). Add 0.05 % each of chromosome isolation buffer components 0.4 M spermine in ddH_2O and 1.0 M spermidine in ddH_2O. Keep on ice until required.

10. Aliquot and store at −20 °C. Protect from light.
11. Dissolve 5 nM in 1 mL ddH_2O to make 5 μM stock, aliquot and store frozen. Heat to 70 °C for 10 min before each use. Protect from light.
12. Aliquot and store at −20 °C for up to a year. If the buffer precipitates, warm at 37 °C to resuspend. Working solution can be kept at 4 °C for up to a month.
13. Dissolve by adding 18 M NaOH dropwise. Filter (0.2 μM), aliquot and store at −20 °C for up to a year. Working solution can be kept at 4 °C for up to a month.
14. For each condition, prepare 100 μl of hybridization mixture.
15. Centrifuge brakes can be used but if lymphocyte recovery is low, remove the brakes.
16. If lymphocyte recovery is low, harvest the plasma layer and repeat **steps 6–8** in Subheading 3.1.
17. To ensure a proper wash, aspirate as close to the pellet as possible.
18. A total volume of 10 mL of culture media is used in each vented 25 cm^2 culture flask. Note that including PHA stimulates the lymphocytes to enter metaphase, while the colcemid should prevent the cells from entering second metaphase. This is important as only the cells in first metaphase should be included. The addition of colcemid for 48 h results in smaller, tighter chromosomes, which works well for detection on the Imagestream^X. Time of incubation with colcemid can be varied (3–48 h), requiring only a few hours for some cell cultures.
19. The hypotonic solution will cause the cells to swell and become more fragile; therefore it is important to handle the tubes gently.
20. The incubation time for the hypotonic treatment varies for different cell types. Approximately 20 min works well, or a range between 17 and 22 min is also acceptable. Some testing might be required if using different cell types. *See* ref. [11] for details of other cell types.
21. Handle the tubes gently while aspirating as the cell pellet is more loose after hypotonic treatment.
22. Vortex time varies per cell type and vortex force. Test new cell types to determine vortex settings. Use caution: to prevent

chromosomes from uncoiling and clumping, do not vortex longer than required.

23. If need be, the chromosome suspensions can be stained on the same day (after 3–4 h at 4 °C) but for best results, let the chromosome suspensions settle and rest. Chromosome suspensions can be stored at 4 °C for up to a few weeks.
24. Typically approximately 0.5 mL of chromosomes are left in suspension; this step could result in chromosome loss if they have not settled well, and if the tubes are not handled carefully. Larger volumes can be pipetted with the use of a 1 mL pipette however for smaller volumes, a 200 μL pipette should be used. Unlike the method of Brind'Amour [11], RNase pretreatment was not used.
25. Denaturing requires very specific temperatures and timing—it is important to be as accurate as possible here.
26. Chromosomes do not pellet well in formamide and this step can lead to loss of chromosomes when removing the supernatant. As recommended by Brind'Amour [11], use a 1000 μL pipette to remove most of the volume, and a 200 μL pipette for finer control. Washing was performed only once, but **steps 6–9** in Subheading 3.3 can be repeated if problems are encountered with staining.
27. Stained chromosomes can be left at 4 °C for up to a week. A 405 nm laser is required to excite the Hoechst 33258; alternatively, propidium iodide can be used to stain the DNA and excited with a 658 nm laser.
28. The following acquisition settings were used in our laboratory: bright-field LED, a 488 nm laser at 80 mW and an unfiltered 405 nm laser at 2.5 mW. If, as is the case with most Imagestream^{X}, the 405 laser is filtered, the power will need to be increased. Images were collected from channel 1 (430–505 nm, DNA signal, Hoechst 33258), channel 2 (505–560 nm, centromere signal, PNA probe), channel 5 (660–745 nm, bright-field signal, LED) and channel 6 (side scatter). Note that bright field can be switched to a different unused channel if channel 5 is required for detection of another fluorophore. Cell classifiers were set to limit acquisition to images with a minimum of integrated intensity (less background) in channel 1 of 25 (arbitrary units [au]) and in channel 2 of 10 au, as well as a maximum area in channel 5 of 35 μm^2. Note that these settings will vary by machine, staining intensity and cell type, and will need to be optimized for different experiments.
29. Number of objects to acquire will vary on sample preparation and on how stringently the cell classifiers are set. It is recommended that smaller data files are collected at first to fine tune

which classifiers work best. Once the acquisition settings are optimized, then more objects of interest (and less debris) can be collected for good statistics without resulting in overly large data files.

30. Note that to optimize the potential of the IDEAS software, it is important to make use of the "Find the Best Feature" approach available to the user. Using the population tagger, the user can define populations of objects of interest (i.e., good chromosomes) as well as populations that the user would like to gate out (i.e., too much background, objects too large). Using the statistics functions, the user can compare the two populations and use the RD-mean statistic to find the features that best differentiate the two populations. As the user builds and applies more masks and features, there are more options to better separate the desired and undesired populations. These features, that best differentiate the populations, can be plotted on a scatterplot to create subsequent gated regions. Iterating through this approach leads to a better final population of desired objects, in this case, a chromosome population. Note that it is also very useful to show the tagged populations on the scatterplots to help create different regions.
31. If the area of the centromere signal is too large, it is difficult to distinguish if this is multiple centromeres.
32. This mask can be fine-tuned: the final mask used for this experiment was *Range*(*Peak*(*Spot*(M02, Channel 2, Bright, 2, 2), Channel 2, Bright, 0), 0–200, 0–1). The *Range* mask provides the ability to mask components in an image within a set area (i.e., 0–200) and set aspect ratio (i.e., 0–1). The *Peak* mask identifies intensity areas from an image with local maxima (i.e., bright peak) or local minima. Finally, the *Spot* mask (on the Bright option) obtains bright regions from the image by eroding the image leaving only bright areas. The spot to cell background ratio and radius thresholds are determined by the user (i.e., 2 and 2).
33. If the gates include too many chromosomes with the wrong number of centromeres, the masking may need to be adjusted. Repeat **steps 15–17** in Subheading 3.4 as needed to optimize the masking strategy. It is not unusual to have some false positive objects; to some extent, this can be taken into account in a calibration curve.
34. The sample preparation method described in the chapter uses a polyamine method to release the chromosomes into suspension. The KCl method has also been successful [13] but it was found that the polyamine method results in chromosomes with less aggregation and more stability than with the KCl method. The imaging flow cytometry method (data acquisition

and analysis) described in this chapter can also be used with chromosomes isolated using the KCl sample preparation method.

35. Samples were analyzed uncompensated after testing whether compensation improved the results. As little spectral overlap was observed, compensation was deemed unnecessary. This should be tested based on the conditions used and proper single-color controls to generate a compensation matrix should be collected.

Acknowledgements

The authors gratefully acknowledge the help of Catherine Ferrarotto for help with the sample preparation. This equipment was purchased through a Canadian Safety and Security Program Technology Acquisition Project.

References

1. International Organization for Standardization (2008) Radiation protection—performance criteria for laboratories performing cytogenetic triage for assessment of mass casualties in radiological or nuclear emergencies—general principles and application to dicentric assay. Geneva: ISO, ISO 21243
2. International Organization for Standardization (2004) Radiation protection—perfomance criteria for service laboratories performing biological dosimetry by cytogenetics. Geneva: ISO, ISO 19238
3. IAEA (2011) Cytogenetic dosimetry: applications in preparedness for and response to radiation emergencies. IAEA, EPR-Biodose, Vienna
4. Lloyd DC, Edwards AA, Moquet JE, Guerrero-Carbajal YC (2000) The role of cytogenetics in early triage of radiation casualties. Appl Radiat Isot 52:1107–1112
5. Voisin P, Benderitter M, Claraz M, Chambrette V, Sorokine-Durm I, Delbos M, Durand V, Leroy A, Paillole N (2001) The cytogenetic dosimetry of recent accidental overexposure. Cell Mol Biol 47:557–564
6. Flegal FN, Devantier Y, Marro L, Wilkins RC (2012) Validation of QuickScan dicentric chromosome analysis for high throughput radiation biological dosimetry. Health Phys 102:143–153. doi:10.1097/HP.0b013e3182307758
7. Vaurijoux A, Gruel G, Pouzoulet F, Gregoire E, Martin C, Roch-Lefevre S, Voisin P, Voisin P, Roy L (2009) Strategy for population triage based on dicentric analysis. Radiat Res 171: 541–548. doi:10.1667/RR1664.1
8. Vaurijoux A, Gregoire E, Roch-Lefevre S, Voisin P, Martin C, Voisin P, Roy L, Gruel G (2012) Detection of partial-body exposure to ionizing radiation by the automatic detection of dicentrics. Radiat Res 178:357–364. doi:10.1667/RR2728.1, http://dx.doi.org/
9. Fantes JA, Green DK, Elder JK, Malloy P, Evans HJ (1983) Detecting radiation damage to human chromosomes by flow cytometry. Mutat Res 119:161–168
10. Fantes JA, Green DK, Malloy P, Sumner AT (1989) Flow cytometry measurements of human chromosome kinetochore labeling. Cytometry 10:134–142
11. Brind'amour J, Lansdorp PM (2011) Analysis of repetitive DNA in chromosomes by flow cytometry. Nat Methods 8:484–486. doi:10.1038/nmeth.1601
12. Sillar R, Young BD (1981) A new method for the preparation of metaphase chromosomes for flow analysis. J Histochem Cytochem 29: 74–78
13. Beaton LA, Ferrarotto C, Kutzner BC, McNamee JP, Bellie PV, Wilkins RC (2013) Analysis of chromosome damage for biodosimetry using imaging flow cytometry. Mutat Res 756:192–195. doi:10.1016/j.mrgentox.2013.04.002

Chapter 7

Fluorescent In Situ Hybridization in Suspension by Imaging Flow Cytometry

Orla Maguire, Paul K. Wallace, and Hans Minderman

Abstract

The emergence of imaging flow cytometry (IFC) has brought novel applications exploiting its advantages over conventional flow cytometry and microscopy. One of the new applications is fluorescence in situ hybridization in suspension (FISH-IS). Conventional FISH is a slide-based approach in which the spotlike imagery resulting from hybridization with fluorescently tagged probes is evaluated by fluorescence microscopy. The FISH-IS approach evaluated by IFC enables the evaluation of tens to hundreds of thousands of cells in suspension and the analysis can be automated and standardized diminishing operator bias from the analysis. The high cell number throughput of FISH-IS improves the detection of rare events compared to conventional FISH. The applicability of FISH-IS is currently limited to detection of abnormal quantitative differences of hybridization targets such as occur in numerical chromosome abnormalities, deletions and amplifications.

Here, we describe a protocol for FISH-IS using chromosome enumeration probes as an example.

Key words FISH in suspension, ImageStream, Cytogenetic abnormalities, Acute myeloid leukemia

1 Introduction

Fluorescent in situ hybridization (FISH) is a slide-based assay used to identify a specific DNA or RNA sequence in a sample by hybridization with fluorescently labeled probes complementary to the targeted sequences. By designing and combining the use of specific probes, FISH can be used to detect numerical as well as structural chromosome defects. For example, numerical changes such as monosomies and structural changes resulting from amplification or deletions can be detected by the loss or gain of hybridization spots while translocations can be detected by the colocalization of hybridization spots when probes of two different colors are used that are specific to the translocation partners.

Diagnostic DNA FISH was initially utilized for the detection of chromosomal abnormalities in heritable disorders such as

Natasha S. Barteneva and Ivan A. Vorobjev (eds.), *Imaging Flow Cytometry: Methods and Protocols*, Methods in Molecular Biology, vol. 1389, DOI 10.1007/978-1-4939-3302-0_7,

Trisomy 21 in Down syndrome and Disomy X in males with Klinefelter syndrome [1]. Today, FISH is routinely used in the detection of chromosomal abnormalities important in the diagnosis, prognosis and response to therapy of patients suffering from a wide array of diseases including acute myeloid leukemia (AML). Approximately 55 % AML patients have one or more chromosomal abnormalities that will contribute to disease [2, 3]. The cytogenetic profile of an AML patient is important not only in the classification of the disease, and therefore the chemotherapy prescribed, but it is also important in the determination of minimal residual disease (MRD) at the time of clinical remission [4]. Clinical remission is designated as normal white cell counts, absence of blasts in the bone marrow, and the absence of the chromosomal abnormality [5, 6]. This third criterion, however, is not mandatory at remission and is not always utilized due to the lack of published studies showing its prognostic value. One of the main reasons for this is that the low numbers of abnormal cells present at remission are below the threshold for detection by the conventional FISH assay [7]. The lower threshold of sensitivity of conventional FISH is probe specific but is commonly at the 5–10 % level [8, 9]. A patient with just 2 % cells containing the abnormality would, therefore, be considered "negative." Relapse rates varying from 30 to 70 % depending on classification of the initial disease [10] indicate that approximately half of all patients are in fact false negatives.

The current protocol using imaging flow cytometry was developed in our lab to improve on the sensitivity of FISH. Its feasibility was previously described using Trisomy 8 in acute myeloblastic leukemia as a model [11]. This protocol can routinely analyze orders of magnitude higher numbers of cells than conventional FISH, and the analysis can quickly identify chromosomal abnormalities. The lower threshold of sensitivity of FISH-IS is at least 1 %, as previously described in ref. 11. This makes it ideal for the detection of abnormalities in rare cells, and therefore may improve the detection of MRD at clinical remission.

2 Materials

2.1 Buffers Used

1. Chloroform-free Carnoy's Fixative solution (3:1 methanol:acetic acid): 36 mL methanol and 12 mL glacial acetic acid carefully mixed in a fume hood in a 50 mL tube by inverting (*see* **Note 1**). Prepare a fresh stock of this solution for each experiment.
2. Wash buffer: 1× phosphate-buffered saline (PBS), 1 % bovine serum albumin (BSA). Weigh 0.5 g BSA and add 40 mL ice-cold 1× PBS (*see* **Note 2**). Mix until BSA has dissolved. Make up to 50 mL with ice-cold 1× PBS. Store at 4 °C (*see* **Note 2**).

3. SSC buffer I: 2× saline-sodium citrate (SSC), 0.1 % IGEPAL CA-630, pH 7.0. Add 4 mL 20× SSC to 26 mL 1× PBS. Add 40 μL IGEPAL CA-630 to solution and mix. Adjust pH to 7.0 with NaOH and make up to 40 mL with 1× PBS (*see* **Note 3**). This buffer should be freshly prepared each week (*see* **Note 4**).
4. SSC buffer II: 0.4× SSC, 0.3 % IGEPAL CA-630, pH 7.0. Add 800 μL 20× SSC to 26 mL 1× PBS. Add 120 μL IGEPAL CA-630 to solution and mix. Adjust pH to 7.0 with NaOH and make up to 40 mL with 1× PBS (*see* **Note 3**). This buffer should be freshly prepared each week (*see* **Note 4**).

2.2 Cell Preparation

1. Whole blood sodium heparin collection tubes (*see* **Note 5**).
2. Ficoll-Hypaque reagent (*see* **Note 6**).

2.3 DNA Hybridization Reagents

1. Vysis Chromosome Enumeration Probes (CEP) probes (Abbott Molecular, Des Plaines, IL) (*see* **Note 7**): These commercial probes are designed to target highly repetitive regions around the centromere of the chromosome. They are available in a number of fluorochromes. All these fluorochromes are available to study most chromosomal abnormalities. The fluorochrome used in this study is SpectrumGreen (Ex: 488 nm/Em: 520 nm). In this protocol, we use chromosome 8 as proof of principle. Store probes at −20 °C.
2. CEP Hybridization Buffer (Abbott Molecular, Des Plaines, IL): Buffer contains dextran sulfate, formamide, and SSC (*see* **Note 8**). Store at −20 °C.
3. RNase/DNase-treated water.
4. Hemocytometer (*see* **Note 9**).
5. Tabletop centrifuge.
6. Two dry heat blocks (*see* **Note 10**).
7. ImageStream^X (Amnis, part of EMD Millipore, Seattle, WA, USA) equipped with a 40× objective and extended depth of field (EDF) function (*see* **Note 11**).
8. IDEAS® analysis software (Amnis, WA, USA).

3 Methods

Carry out all procedures at room temperature unless otherwise specified.

3.1 Preparation of Peripheral Blood Mononuclear Cells

1. Collect approximately 10 mL whole blood from a donor/patient in a sodium heparin tube (*see* **Note 5**).
2. Isolate peripheral blood mononuclear cells (PBMCs) using Ficoll-Hypaque density centrifugation gradient according to the manufacturer's instructions (*see* **Note 6**).

3. Following isolation, wash cell pellet with 5 mL 1× PBS and centrifuge at 600 × *g* for 4 min.
4. Decant superatant. Resuspend cell pellet in 3 mL 1× PBS.
5. In a fume hood, add 12 mL chloroform-free Carnoy's Fixative solution slowly (adding approximately 1 ml/15 s) with gentle vortexing to prevent clumping of cells (*see* **Note 12**).
6. Incubate cell suspension at room temperature for 10 min to ensure complete fixation.
7. Store cell suspension at −20 °C (*see* **Note 13**).

3.2 Hybridization of DNA to Probe

1. Place 3 mL fixed cells into a 15 mL tube and wash with 5 mL ice-cold 1× PBS/1 % BSA (*see* **Note 14**).
2. Centrifuge at 600 × *g* for 4 min, and then decant supernatant.
3. Resuspend cells in 5 mL ice cold 1× PBS/1 % BSA again.
4. Centrifuge at 600 × *g* for 4 min, and then decant supernatant.
5. Resuspend cells in 2 mL SSC buffer I (*see* **Note 15**).
6. Count cells using a hemocytometer according to the manufacturer's instructions (*see* **Note 9**).
7. Place 1.5×10^6 cells into a fresh 15 mL tube and centrifuge at 600 × *g* for 4 min (*see* **Note 16**).
8. Decant supernatant. Resuspend cell pellet in 500 μL SSC buffer I and transfer cell suspension into a thin-walled 0.65 mL Eppendorf tube (*see* **Note 17**).
9. Centrifuge at 600 × *g* for 5 min at room temperature.
10. Carefully aspirate and discard supernatant (*see* **Note 18**).
11. Prepare hybridization buffer mix: 28 μL CEP hybridization buffer and 10 μL RNase/DNase-free water (*see* **Note 19**).
12. Add 38 μL hybridization buffer mix to cells.
13. Add 2 μL CEP probe to each appropriate tube (*see* **Note 20**). If adding more than one probe to a particular tube the volume of water should be reduced to ensure final volume is always 40 μL (*see* **Note 21**).
14. Place samples in a preheated dry heat block at 80 °C for 5 min (*see* **Note 10**).
15. Transfer samples into a preheated dry heat block at 42 °C for 14 h (*see* **Notes 10** and **22**).
16. Warm 200 μL SSC buffer II in a preheated dry heat block to 73 °C for 10 min (*see* **Note 23**).
17. Remove samples from 42 °C heat block and centrifuge cells at 6000 × *g* for 5 min at room temperature (*see* **Note 24**).

18. Aspirate and discard supernatant. Resuspend cell pellet in prewarmed 200 μL SSC buffer II.
19. Incubate cell suspension in the preheated dry heat block at 73 °C for 2 min to facilitate the removal of excess probe and probe that may been have nonspecifically bound to sequences partially homologous to the target sequence (*see* **Note 25**).
20. Add 200 μL ice-cold FBS to suspension to quickly drop the temperature and prevent clumping of cells.
21. Centrifuge cells at 6000 × *g* for 5 min at room temperature.
22. Aspirate and discard supernatant. Resuspend cell pellet in 100 μL ice-cold FBS (*see* **Note 26**).
23. The single color compensation control for SpectrumGreen can be prepared in one of two ways. A separate sample can be hybridized or a small aliquot can be removed before DAPI (nuclear dye) staining in **step 25**. We perform the latter and remove 10 μL cells into a fresh 1.5 mL Eppendorf tube before DAPI stain. Add 90 μL FBS to this to make solution up to 100 μL.
24. For the single-color control for DAPI, lyse red blood cells from a healthy donor whole blood sample. Fix the PBMCs with 4 % formaldehyde for 10 min at room temperature. Centrifuge at 600 × *g* for 4 min, and then decant supernatant. Wash cells with 5 mL 1× PBS and centrifuge at 600 × *g* for 4 min. Decant supernatant and resuspend cells in residual volume of 1× PBS, which should total approximately 100 μL.
25. Add 10 μL 5 μg/mL DAPI to test sample and DAPI compensation control.

3.3 Data Acquisition and Analysis

1. Set up experiment template on ASSIST calibrated ImageStream^X (ISx) (*see* **Note 11**).
2. Set cell classifiers to eliminate debris and large aggregates: area lower limit = 50; area upper limit = 250 (Fig. 1a, red box, and *see* **Note 27**).
3. Set cell classifiers to eliminate non-nucleated cells: Channel 7 Intensity Minimum = 1000 (Fig. 1a, green box, and *see* **Note 28**).
4. Enable only those channels used in the experiment to minimize the electronic file size. In this protocol channels 1 (bright field, camera 1), 2 (CEP-SpectrumGreen), 6 (scatter), 7 (DAPI-nucleus), and 9 (bright field, camera 2) are enabled (*see* Fig. 1a, yellow box).
5. Turn on 405 nm laser at 10 mW and 488 nm laser at 50 mW and 785 nm laser for scatter at the ASSIST-defined optimal intensity (usually between 1 and 5 mW) (Fig. 1b, green oval, and *see* **Note 29**).

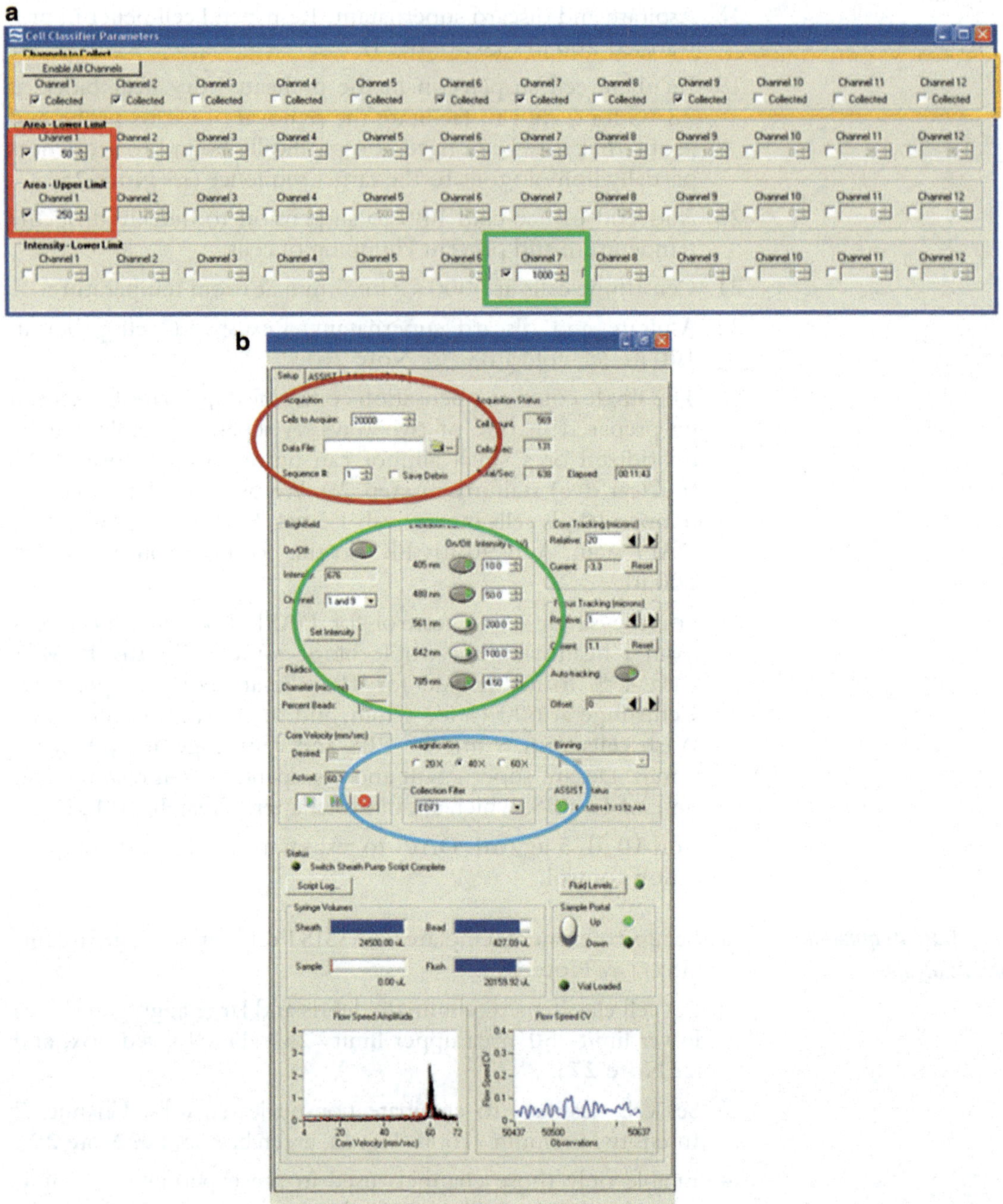

Fig. 1 Setting up the Inspire software. (**a**) Cell classifiers panel with the area lower limit and area upper limit for channel 1 bright field selected (*red box*). Intensity lower limit of channel 7 is selected to collect events with a positive nuclear signal (*green box*), and the desired channels selected (*yellow box*). (**b**) Experimental set-up panel on the left of the user interface is highlighted. Acquisition parameters (*red oval*), illumination (*green oval*), and magnification with EDF filter (*blue*) are shown

6. Ensure that the objective is 40× and the EDF function box is checked (Fig. 1b, blue oval, and *see* **Note 11**).
7. Collect 20,000 events that are "single cells" and nucleated (Fig. 1b, red oval, and *see* **Note 30**). Events are collected as a raw image file (*.rif) file.
8. To acquire compensation controls enable all channels and turn off the bright-field illumination and scatter laser. The 405 and 488 nm lasers should remain on at the power used to acquire the test sample (Fig. 1b, green oval). A minimum intensity cell classifier is set at 1000 for the channel of interest (for SpectrumGreen it would be channel 2, and for DAPI, channel 7) (Fig. 2). Collect 500 positive events (*see* **Notes 31** and **32**).
9. Perform compensation post-acquisition in IDEAS® software according to standard procedure by making a compensation matrix (*.ctm) file based on the single-color controls.
10. When opening the rif file for the test sample you will be prompted to apply the compensation (*.ctm) file to the analysis. Apply the *.ctm file prepared in **step 9**.

a

b

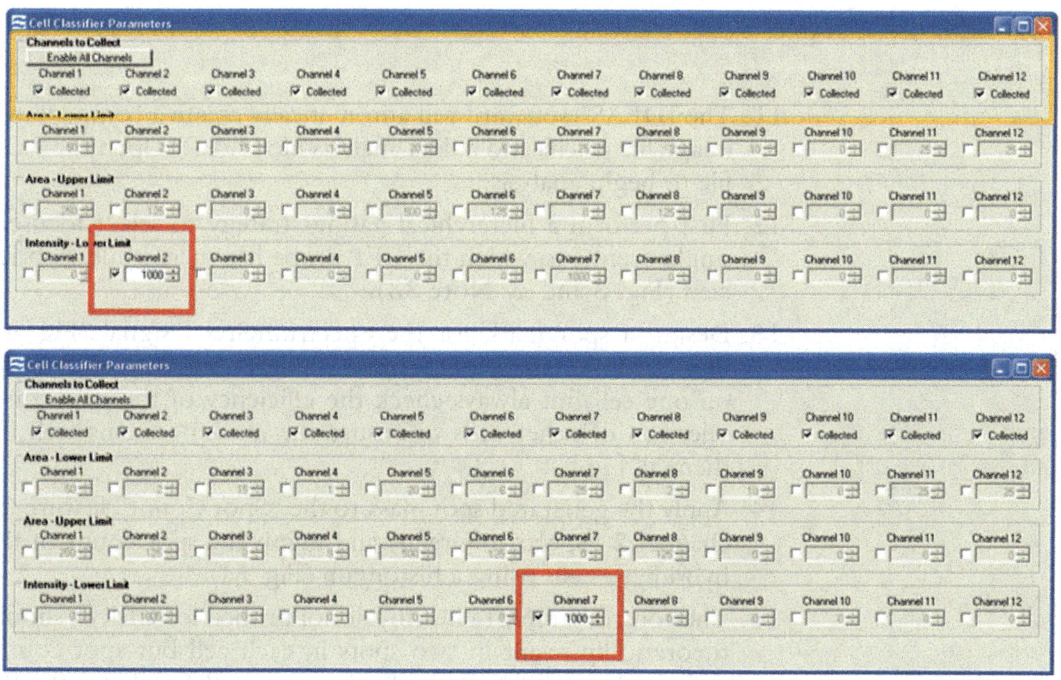

Fig. 2 Setting classifiers for compensation controls. (**a**) Compensation classifiers for the SpectrumGreen probe that is detected in channel 2. All channels are enabled (*yellow box*) and an intensity lower limit of channel 2 is selected to collect events with a positive signal (*red box*). (**b**) Compensation classifiers for the DAPI nuclear dye that is detected in channel 7. All channels are again enabled and an intensity lower limit of channel 7 is selected to collect events with a positive signal (*red box*)

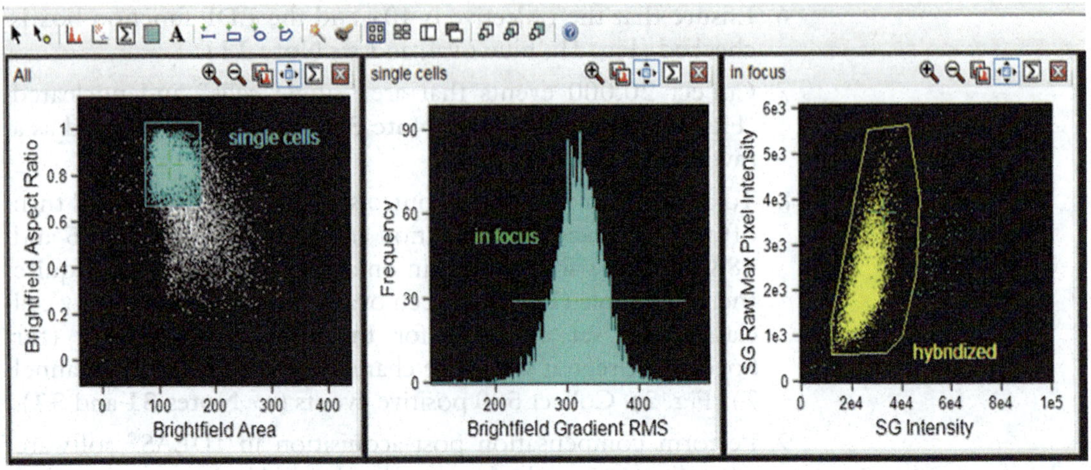

Fig. 3 Hierarchical gating strategy to select cells of interest. Single events are gated using area of the bright-field image (either in channel 1 or channel 9) versus aspect ratio of the bright-field image (aspect ratio close to 1 represents a single cell). Cells in focus are gated using gradient root mean squared of the bright-field image. Positively hybridized cells are gated using intensity of the SpectrumGreen signal versus raw max pixel of the SpectrumGreen signal (raw max pixel is a measure of the brightest pixel in the event). Very bright event to the right of the scatterplot are not included since, upon visual inspection, they represent doublet events that were insufficiently eliminated by the initial single cell gate

11. The IDEAS® software will automatically create a compensated image file (*.cif) and a data analysis file (*.daf). Open the *.daf file to begin analysis.
12. First perform a hierarchical gating strategy to select focused, single events, to which the CEP probe has successfully hybridized (Fig. 3 and *see* **Note 33**).
13. Design a spot mask for the SpectrumGreen signal (Fig. 4). Optimize the radius and spot-to-background ratio of the mask on one cell but always check the efficiency of the mask on a selection of other cells to ensure the mask performs well for them all [11].
14. Apply the generated spot mask to the "Spot Count" feature for channel 2 (SpectrumGreen) and graph the spot count of the hybridized cells using a histogram (Fig. 5a).
15. The control sample (a healthy donor probed for CEP 8) should theoretically result in two spots in each cell but spot counts and ploidy are not necessarily directly correlated. Since the ISx captures a 2D image of a 3D cell in flow one spot may be the result of two spots in such close proximity that the software cannot distinguish them (*see* **Note 34**). Three or even four

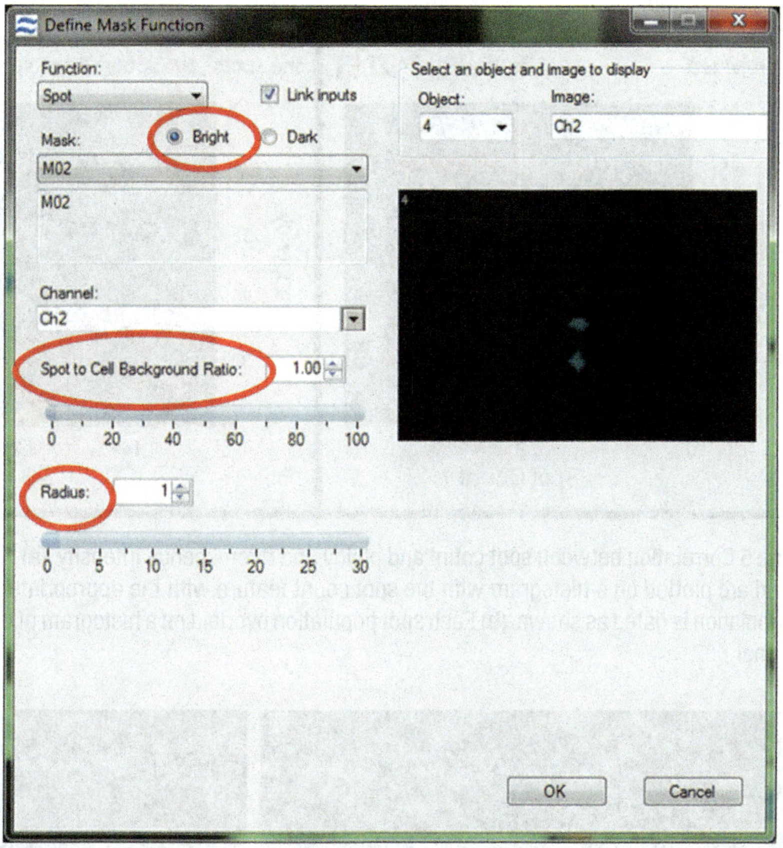

Fig. 4 Setting the spot mask. To accurately design a spot mask first the "bright" option must be checked. This ensures that the mask obtains the bright regions from an image regardless of any intensity differences from one spot to another. Next, alter the spot-to-background ratio and the radius of the spot mask to ensure that the mask covers the spots but does not include any background

spots may be the result of a segregated hybridization signal and is a common occurrence also observed in conventional FISH (Fig. 6). The next steps will apply a fluorescence intensity parameter to correct spot counting deviations from ploidy.

16. Draw a gate on each spot population (Fig. 5a).
17. Draw an Intensity histogram of the SpectrumGreen signal for each population. True disomies whether counted as one, two, or three spots should have a similar fluorescence distribution (Fig. 5b and *see* **Note 35**).

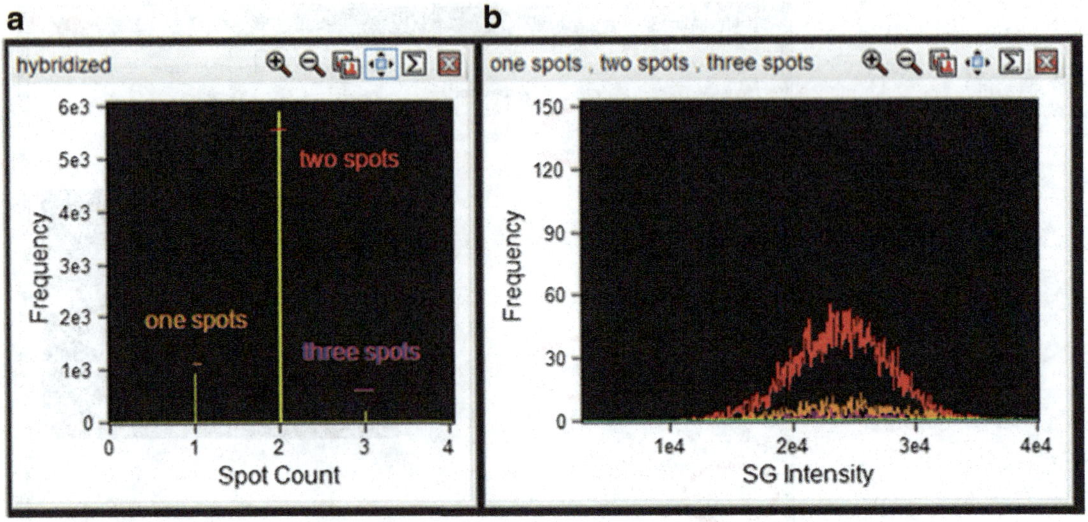

Fig. 5 Correlation between spot count and ploidy and fluorescence intensity. (**a**) Cells that are positively hybridized are plotted on a histogram with the spot count feature with the appropriate spot mask applied. Each spot population is gated as shown. (**b**) Each spot population overlaid on a histogram of intensity of the SpectrumGreen signal

Fig. 6 Segregated hybridization signal. The images show CEP8 SpectrumGreen probe from a healthy donor where all events should contain two spots. Three spots (*left image*) or four spots (*right image*) can be observed but the intensity and proximity of the spots indicate that segregated hybridization has occurred. These examples would be classified as a disomy by conventional FISH

4 Notes

1. This step should always be performed in a fume hood due to the vapors released and since methanol is flammable and acetic acid is corrosive to skin. Wear appropriate safety equipment including gloves, lab coat, and safety goggles.

2. Solutions containing BSA should be kept and stored at 4 °C due to the potential for bacterial growth in the protein-containing solution. This is also why the wash buffer should be used within 1 week. To aid prevention of bacterial growth solutions can be sterile filtered to remove contaminants.
3. Concentrated NaOH (10 N) can be used first to narrow the gap from the starting pH to the required pH. From then on a lower ionic strength should be used (5 N) to avoid a sharp increase in pH above 7.0.
4. We have found that using SSC buffer more than 1 week old has negatively affected fluorescent staining of the CEP probes. Buffers can be prepared fresh for each individual experiment if desired.
5. Sodium heparin tubes are used without bias. It is not expected that the protocol would be affected by the use of other anti-coagulants such as EDTA.
6. Ficoll-Hypaque can be purchased from local suppliers. A protocol for density-gradient centrifugation protocol should be performed according to manufacturer's instructions. For reference, we use Histopaque®-1077 (Sigma-Aldrich, St. Louis, MO).
7. We have also used FISH probes supplied by Cytocell, an Oxford Gene Technology company (Tarrytown, NY). These probes are available in FITC or TexasRed.
8. In this protocol we use the commercially available CEP hybridization buffer (Abbott Molecular, Des Plaines, IL). There is no reason to believe efficient hybridization could not be achieved using a home-made buffer. However, it would be essential to optimize the buffer formulation with the hybridization temperature. Changes in formamide concentration will affect hybridization efficiency and melt temperature will need to be altered.
9. Cell counts in this study were performed using a manual hemocytometer. It is not expected that cell concentrations would differ beyond the normal range using an automated method.
10. A thermocycler can also be used for incubation steps. However, rigorous testing that the desired hybridization temperature is achieved quickly (within 1 min) should be performed. Suboptimal function will negatively affect hybridization efficiency.
11. ImageStream-100 and ImageStream^X MkII are suitable alternatives. The EDF function is required to accurately distinguish spots (Fig. 7). The 40× magnification is standard on the IS^X with optional 60× and 20× magnification. Samples can be run

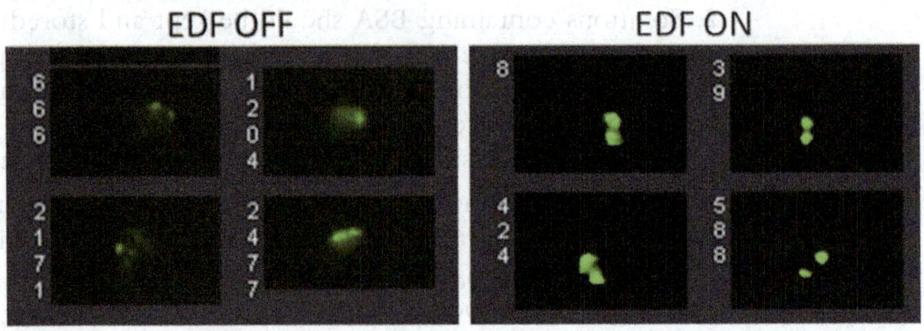

Fig. 7 Extended depth of field resolves spot imagery. Examples of images acquired with the EDF OFF are shown on the *left*, and with the EDF ON on the *right*

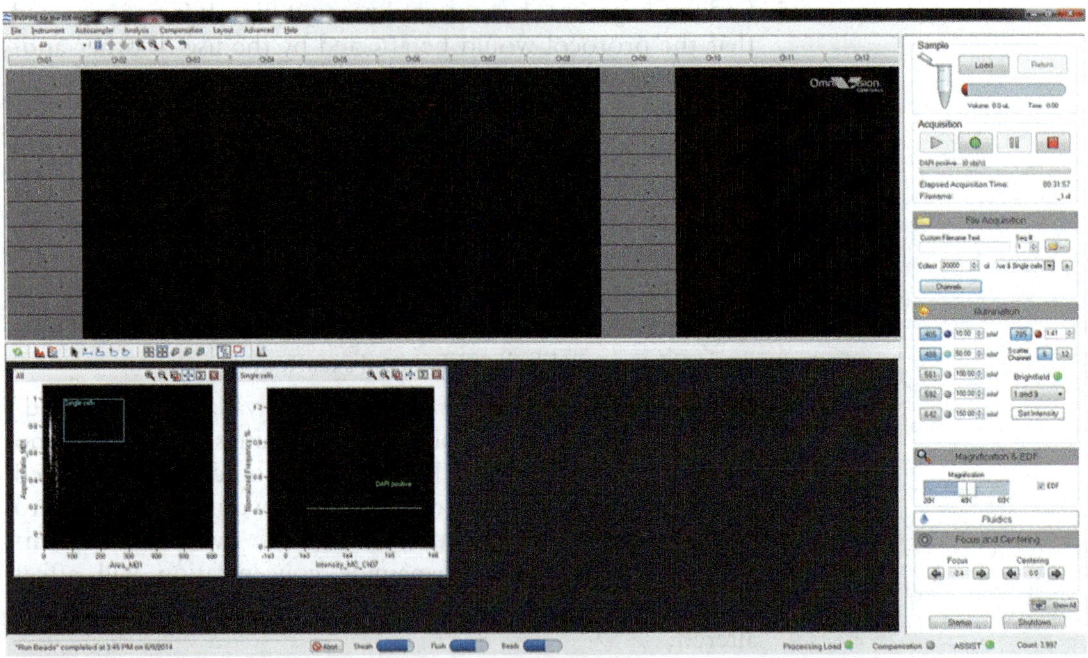

Fig. 8 The user interface in the Inspire software of an ImageStream^{X} Mark-II

at 60× but we have not observed an added advantage of this. We would not recommend using the 20× magnification. The MkII has a slightly different user interface with live gating (Fig. 8).

12. We have found that a fast addition of fixative solution has had a negative effect on hybridization and acquisition efficiency.

Care should be taken at this early stage to ensure a single cell suspension is achieved.

13. Fixed cells can be stored up to 3 months without effecting cellular integrity. Although we have not tested hybridization after 3 months, a loss of hybridization is expected due to DNA degradation over time. Storage at −80 °C may extend the shelf life of samples further.
14. One tube PBL contains approximately $2–4 \times 10^7$ PBMCs. This protocol is designed to perform 1 test sample. 3 mL fixed cells contains approximately 5×10^6 cells. Since each hybridization should be performed on 1.5×10^6 cells, and cells will be lost during processing, the volume washed at this step should be altered as appropriate for your experiment. It is recommended to over-estimate the volume of fixed cells needed if it is possible.
15. Cells cannot be accurately counted in fixative since it will evaporate, so it must be washed out and cells suspended in SSC buffer I before placing on a hemocytometer.
16. 1.5×10^6 cells are needed for each individual hybridization. The total number of cells placed in a fresh 15 mL tube at this step should be altered as appropriate for your experiment.
17. A small Eppendorf tube is required to ensure small surface area over which hybridization takes place. Any thin-walled 0.5–0.65 mL tube is sufficient, and if a thermocycler is used, cells should be placed in tubes that will fit exactly into each position. Thin-walled tubes are necessary to ensure correct heat distribution during hybridization. 1.5 mL tubes are not recommended for use. 0.2 mL tubes have not been tested.
18. Removal of supernatant in small tubes should always be performed by aspiration to ensure minimal loss of yield.
19. Prepare enough hybridization mix for all samples being used in the experiment.
20. 2 μL is the volume of CEP probe recommended by Abbott Molecular, and the optimal volume for CEP 8 from our own titration experiments. Each probe will need to be titrated and the optimal volume used.
21. 40 μL has reproducibly given efficient hybridization. Altering the volume up or down has decreased hybridization efficiency in our hands.
22. Incubation at 42 °C for between 9 and 14 h yields similar results. Below 9 h and above 14 h begins to negatively affect hybridization, either by incomplete hybridization, or increase of nonspecific background signal, respectively.

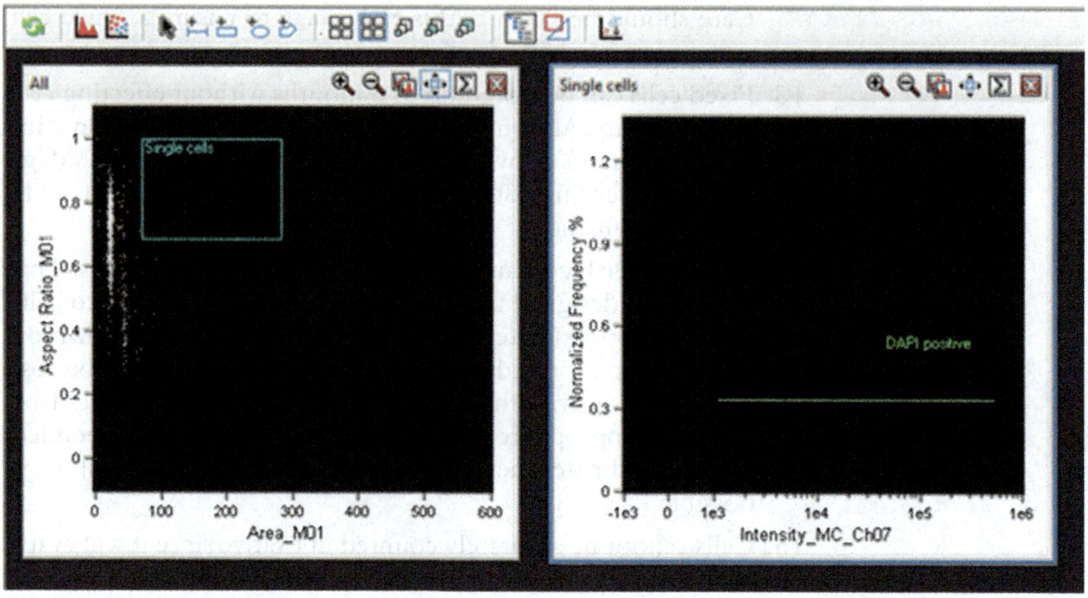

Fig. 9 Live gating on the ImageStream^X Mark-II. A gate is drawn on an area versus aspect ratio of the bright-field image (channel 1) to select single cells, eliminating aggregates, debris, and speed beads (*left graph*). A histogram of the nuclear dye fluorescence intensity is drawn (channel 7) and a line gate for positive events, i.e., those with a fluorescence intensity of 1000 or higher is selected

23. 200 μL SSC buffer II is needed for each test sample so a sufficient volume of buffer will need to be warmed to 73 °C for all samples in your experiment.
24. Take note of the change in centrifugation speed at this step. This is to ensure maximum possible yield.
25. If background signal from non-specific binding of the probe remains after this incubation, alternatively a formamide wash solution (50 % formamide, 2× SSC buffer) recommended by Abbott Molecular can be used but we have not come across the need to test this.
26. Resuspension of cells in FBS greatly reduces cell aggregates and results in easier acquisition on the ImageStream. It is not recommended to resuspend cells in 1× PBS.
27. When using the ISX MkII, live gating is performed. A scatter-plot of area versus aspect ratio, with a gate set from 50 to 250 for area and 0.7 to 1 in aspect ratio, will yield a majority of single cells (Fig. 9).
28. With the ISX MkII a histogram of Channel 7 Intensity is made and a gate set to collect all events above 1000 (Fig. 9).
29. Laser intensities should be optimized for your own machine.

30. 20,000 events are usually sufficient but the number of events collected should be increased according to the expected incidence of the target cells when analyzing rare populations.
31. 500 events are probably the minimum required according to our own experience. More events can be collected if desired but it is not necessary.
32. When using the ISX MkII live gating for compensation is performed.
33. Single events are required since two cells will lead to overestimation of spot count. Focused cells are required to accurately detect spots but since cells are acquired with the EDF function most cells will be focused.
34. As mentioned in the introduction, FISH is used not only for numerical aberrations but also structural aberrations such as translocation abnormalities. FISH determines translocations by dual-color co-localization; that is, when BCR is labeled green and ABL labeled red, the BCR/ABL fusion would appear yellow. Since FISH-IS cannot currently distinguish between true fusion events and events captured that are simply in close proximity, it cannot be used for detection of structural aberrations at this time.
35. It is important to have a control healthy donor sample for each experiment when assessing chromosomal abnormalities to determine the "normal" fluorescence distribution of the signal. Spiking a disomy sample with a monosomy should result in a measurable different distribution with a twofold difference in mean fluorescence, e.g., taking male and female healthy donor samples and preparing each sample with CEP X will result in controlled monosomies and disomies of X [11].

References

1. Garcia-Sagredo JM (2008) Fifty years of cytogenetics: a parallel view of the evolution of cytogenetics and genotoxicology. Biochim Biophys Acta 1779:363–375. doi:10.1016/j.bbagrm.2008.05.003
2. Mrozek K, Heerema NA, Bloomfield CD (2004) Cytogenetics in acute leukemia. Blood Rev 18:115–136. doi:10.1016/S0268-960X(03)00040-7
3. Grimwade D (2001) The clinical significance of cytogenetic abnormalities in acute myeloid leukaemia. Best Pract Res Clin Haematol 14:497–529
4. Marcucci G, Mrozek K, Ruppert AS, Archer KJ, Pettenati MJ, Heerema NA, Carroll AJ, Koduru PR, Kolitz JE, Sterling LJ, Edwards CG, Anastasi J, Larson RA, Bloomfield CD (2004) Abnormal cytogenetics at date of morphologic complete remission predicts short overall and disease-free survival, and higher relapse rate in adult acute myeloid leukemia: results from cancer and leukemia group B study 8461. J Clin Oncol 22:2410–2418
5. Döhner H, Estey EH, Amadori S, Appelbaum FR, Büchner T, Burnett AK, Dombret H, Fenaux P, Grimwade D, Larson RA, Lo-Coco F, Naoe T, Niederwieser D, Ossenkoppele GJ, Sanz MA, Sierra J, Tallman MS, Löwenberg B, Bloomfield CD, LeukemiaNet E (2010) Diagnosis and management of acute myeloid leukemia in adults: recommendations from an international expert panel, on behalf of the

European LeukemiaNet. Blood 115:453–474. doi:10.1182/blood-2009-07-235358

6. Cheson BD, Bennett JM, Kopecky KJ, Büchner T, Willman CL, Estey EH, Schiffer CA, Doehner H, Tallman MS, Lister TA, Lo-Coco F, Willemze R, Biondi A, Hiddemann W, Larson RA, Löwenberg B, Sanz MA, Head DR, Ohno R, Bloomfield CD (2003) Revised recommendations of the International Working Group for diagnosis, standardization of response criteria, treatment outcomes, and reporting standards for therapeutic trials in acute myeloid leukemia. J Clin Oncol 21:4642–4649
7. Balleisen S, Kuendgen A, Hildebrandt B, Haas R, Germing U (2009) Prognostic relevance of achieving cytogenetic remission in patients with acute myelogenous leukemia or high-risk myelodysplastic syndrome following induction chemotherapy. Leuk Res 33:1189–1193. doi:10.1016/j.leukres.2009.03.004
8. Sinclair PB, Green AR, Grace C, Nacheva EP (1997) Improved sensitivity of BCR-ABL detection: a triple-probe three-color fluorescence in situ hybridization system. Blood 90:1395–1402
9. Campbell LJ, Oei P, Brookwell R, Shortt J, Eaddy N, Ng A, Chew E, Browett P (2013) FISH detection of PML-RARA fusion in ins(15;17) acute promyelocytic leukaemia depends on probe size. Biomed Res Int 2013:164501
10. Buccisano F, Maurillo L, Del Principe MI, Del Poeta G, Sconocchia G, Lo-Coco F, Arcese W, Amadori S, Venditti A (2012) Prognostic and therapeutic implications of minimal residual disease detection in acute myeloid leukemia. Blood 119:332–341. doi:10.1155/2013/164501
11. Minderman H, Humphrey K, Arcadi JK, Wierbicki A, Maguire O, Wang ES, Block AW, Sait SN, George TC, Wallace PK (2012) Image cytometry-based detection of aneuploidy by fluorescence in situ hybridization in suspension. Cytometry A 81:776–784. doi:10.1002/cyto.a.22101

Chapter 8

Analysis of Nucleocytoplasmic Protein Shuttling by Imaging Flow Cytometry

Elizaveta Fasler-Kan, Yeldar Baiken, Ivan A. Vorobjev, and Natasha S. Barteneva

Abstract

Many intracellular signal transduction events involve the reversible shuttling of proteins between the cytoplasm and the nucleus. Study of these processes requires imaging information on the protein localization in a given cell and a large number of measurements to obtain sufficient statistics on the protein localization in the whole population. The protocol describes method for quantitative imaging flow cytometry analysis of intracellular distribution of NF-kappaB in ARPE-19 cells stained with specific fluorochrome-conjugated antibodies. The described technique alone or in combination with standard flow cytometry methods can be applied to study any protein undergoing translocation from cytoplasm into the nucleus in a variety of cell lines as well as in heterogeneous primary cell populations.

Key words Nucleocytoplasmic translocation, Nucleocytoplasmic shuttling, Imaging cytometry, Flow cytometry, NF-kappaB, ARPE-19 cell line, Retinal pigment cells

1 Introduction

The nucleocytoplasmic translocation of transcription factors, splicing factors and other proteins plays a fundamental role in controlling messenger RNA metabolism and gene expression both in the nucleus and in the cytoplasm. The regulatory output of nucleocytoplasmic trafficking is conditioned by an array of parameters such as the abundance and phosphorylation profile of proteins, their association with regulatory factors, binding sites with mRNA molecules, and other parameters. Conventional methods to assess nucleocytoplasmic translocation utilize a semiquantitative electrophoretic mobility shift assay (EMSA), which requires large amount of cells (10^6 and more), application of radioactive probes and separate preparation of cytoplasmic and nuclear extracts [1, 2]. Moreover, EMSA does not provide information on the subpopulations that might respond differently to stimulation of nucleocytoplasmic translocation.

Natasha S. Barteneva and Ivan A. Vorobjev (eds.), *Imaging Flow Cytometry: Methods and Protocols*, Methods in Molecular Biology, vol. 1389, DOI 10.1007/978-1-4939-3302-0_8, © Springer Science+Business Media New York 2016

To analyze the nucleocytoplasmic transport a traditional approach applies different variants of fluorescent microscopy, often in combination with western blotting and recently, with flow cytometry [3–6]. A number of publications on STATs [7, 8], proto-oncogene SET/I2PP2A [9], and other proteins focused on improving image analysis and microscopy techniques using FRAP [10], special software tools [11], and 4D analysis [8] Microscopic approaches have two major limitations: (1) statistical limitations—amount of cells analyzed by advanced microscopy normally do not exceed tenths-hundreds [10], (2) image processing and segmentation can be difficult using standard software (ImageJ, Metamorph, SlideBook, and others). To obtain larger cohorts of cells (up to 5000 per sample) laser scanning cytometry (LSC) was used to quantify NF-kappaB translocation [12]; however, application of this method was rather tedious.

This chapter provides a method for quantitative assessment of nucleocytoplasmic translocation of proteins by imaging flow cytometry. We used as an example nucleocytoplasmic translocation of NF-kappaB protein in retinal pigment epithelial cell line ARPE-19 upon cell stimulation with different cytokines. NF-kappaB belongs to a family of dimeric transcription factors such as RelA-p50 heterodimer that are shuttling between the cytoplasm and nucleus. In resting cells, NF-kappaB dimers are found within the cytoplasm complexed with IkB (the inhibitor kappa B kinase). In response to activating stimuli such as TNF-alpha or IL-1β, the NF-kappaB complex with IKK is activated, which leads to phosphorylation of NF-kappaB [13, 14] and IkB [15]. Phosphorylated IkB proteins are liberating NF-kappaB dimers that get translocated into the nucleus and participate in the regulation of transcription for genes targets. It is important to measure NF-kappaB dynamics in large cell populations, since cells are heterogeneous and answer to stimuli asynchronously and only part of cells respond to NF-kappaB stimuli at certain time period [16]. By adopting imaging flow cytometry the large populations including 10–20,000 events were analyzed and the statistical variations were decreased. We had used quantification of protein translocation events identified by intracellular staining with polyclonal antibodies to NF-kappaB.

2 Materials

2.1 Cell Culture

1. Human retinal pigment epithelial cell line ARPE-19 could be purchased from ATCC (Manassas, VA, USA). Complete DMEM/F12 medium supplemented with L-glutamine, HEPES, and penicillin/streptomycin (all from Gibco-Life Technologies, Grand Island, USA) and 10 % newborn or fetal inactivated bovine serum (NBS or FCS).

2.2 Cell Stimulation

1. Cell stimulation is achieved using different activators: lipopolysaccharide (LPS) from *E. coli* or *S. typhimurium*, TNF-alpha, phorbol esters, interleukin-1α (IL-1α), and IL-1β. Stock solutions should be 1000× and kept at −20°C (*see* **Note 1**).

2.3 Cell Detachment, Fixation, Permeabilization, and Staining

1. Accutase solution for cell detachment is kept frozen (< −20°C), or could be kept at +4 °C (not longer than 1 month).
2. The primary rabbit polyclonal antibodies against NF-kappaB (Santa Cruz Biotech, Santa Cruz) are stored at +4 °C (working dilution 1:50–1:100) (*see* **Note 2**).
3. Secondary FITC-conjugated $(Fab')_2$ fragment of goat anti-rabbit IgG antibody (Jackson ImmunoResearch Labs, West Grove) (*see* **Note 3**).
4. Centrifugation buffer. Calcium- and magnesium-free Dulbecco's phosphate-buffered saline with 1 % of FCS and 0.01 % sodium azide. Stored at room temperature.
5. 16–20 % ultrapure paraformaldehyde (PFA) stock solution (such as electron microscopy grade from Polysciences, Warrington, PA) should be used for preparation of fresh fixative just before the experiment.
6. Fixation buffer. Can be purchased from company or prepared in house. Add PFA from stock vial to phosphate-based solution (PBS) to a final concentration of 1–4 % (*see* **Note 4**). Prepare fresh just before the experiment.
7. Permeabilization buffer. Add Triton X-100 from stock solution (10–15 %) to the final concentration of 0.25 % (w/v) to PBS. Permeabilization buffer can be stored at 4 °C at least for 1 week. Alternatively ready-to use permeabilization buffer, can be purchased from company, and can be stored for up to several months (contains additional stabilizers) (*see* **Note 5**).
8. Alternatively, ice-cold MeOH–acetone (50:50) solution can be used substituting permeabilization or for both fixation and permeabilization. You can fix in formalin, then permeabilize with Triton X-100 or MeOH, second possibility: to use MEOH–acetone for fixation/permeabilization.
9. Blocking solution: 10 % goat serum in PBS. The choice of serum depends on the origin of secondary antibodies. We used goat anti-mouse or goat anti-rabbit secondary antibodies, and therefore goat serum was chosen.
10. NF-kappaB inhibitor BAY11-7085 (Bayer) and/or LPS inhibitor polymyxin (Sigma-Aldrich, St Louis, MO).
11. Centrifuge (Sorvall-Thermo Scientific, Waltham, MA) or similar, tabletop.
12. Microcentrifuge (Eppendorf) or similar.

13. 15-ml conical centrifuge tubes.
14. 1.5 ml Eppendorf centrifuge tubes.

2.4 Imaging Flow Cytometry Analysis

1. 0.6 ml siliconized microcentrifuge tubes.
2. Speed beads from Amnis-Millipore for ImageStream imaging cytometer.
3. Imaging cytometer ImageStream 100 or ImageStream X Mark II.

3 Methods

3.1 Cell Culture

1. Culture ARPE-19 cells in 75 cm^2 flasks using DMEM/F12 medium supplemented with 10 % fetal calf serum (FCS), HEPES, 100 U/ml penicillin, and 100 μg/ml streptomycin. Place flasks in CO_2-incubator with 5 % CO_2 at 37°C.
2. Typically splitting cells two to three times per week is necessary. The number of cell passages in the assay can be critical. Use as low passage of cells from ATCC as possible. Do not allow cells to overgrow in the tissue culture flasks (to become over confluent). For detachment of cells Accutase is preferable versus trypsin–EDTA. After harvesting cells filter suspension through a cell strainer (45–70 μm mesh) to remove large clamps.

3.2 Cell Stimulation

1. Splitting and seeding cells in 6-well plate or small flasks (12.5–25 cm^2 bottom area). ARPE-19 cells are seeded in 6-well plates at a density 1×10^6 cells per well. Include non-stimulated wells or flasks as a negative control. Sufficient amount of cells for single-stained antibody controls, as well as secondary antibody conjugates should be included. Single-stained controls will be needed for spectral compensation when using ImageStream X.
2. Before the experiment cells should be starved in the complete culture medium without FCS for at least 6 h.
3. Stimulate cells with LPS from E. coli or S. typhimurium (10–100 ng/ml) or TNF-alpha (1–10 ng/ml). Cells are stimulated with appropriate cytokine for 20–30 min at 37^{0}C in a humidified CO_2 incubator. After incubation cells are washed three times with PBS (*see* **Note 6**).

3.3 Cell Fixation, Permeabilization, and Staining

1. Detach cells using Accutase and transfer them into 15-ml conical centrifuge tube. Centrifuge cells for 10 min at $800 \times g$, 4 °C. All centrifugation procedures are better to perform in the siliconized tubes (*see* **Note 7**).

2. Discard the supernatant, add 5 ml of centrifugation buffer and centrifuge again 10 min at 800 × *g*, 4 °C.
3. Discard supernatant, resuspend cell pellet in 3 ml fixation buffer, vortex and incubate for 30 min at room temperature. Spin down 10 min at 800 × *g*, 4 °C.
4. Discard the supernatant, vortex and resuspend cell pellet in 5 ml of centrifugation buffer. Spin down 10 min at 800 × *g*, 4 °C. Repeat this wash step 2 times.
5. Permeabilization: to the cell pellet add 3 ml of permeabilization buffer. Incubate for 20 min at room temperature. Spin down 10 min at 800 × *g*, 4 °C.
6. Discard the supernatant, vortex and resuspend the cell pellet in 5 ml of centrifugation buffer. Spin down 10 min at 800 × *g*, 4 °C. Repeat this wash step 2 times.
7. Blocking: pelleted cells are incubated with 3 ml 10% goat serum in PBS for 1 hr at room temperature. Spin down (centrifuge 10 min at 800 × *g*, 4 °C). *See* **Note 8**.
8. Incubation with primary antibodies: discard supernatant and resuspend cell pellet in 100 μl blocking buffer mixed with primary anti-NF-kappaB antibodies (appropriate concentration of antibodies is 1:250 for imaging cytometry experiments; 1:500 for confocal microscopy experiments). Incubate for 45 min at room temperature or overnight at 4 °C. The duration of incubation should be tested and antibody has to be titrated before the experiment (*see* **Note 9**).
9. Add 5 ml of centrifugation buffer and centrifuge again 10 min at 800 × *g*, 4 °C.
10. Discard supernatant and resuspend cell pellet in 100 μl blocking buffer, mixed with appropriate concentration of secondary antibody (FITC- or Alexa 488-conjugated $F(ab')_2$ anti-rabbit IgG). Incubate not longer than 1 h at room temperature.
11. Add 5 ml of centrifugation buffer and centrifuge 10 min at 800 rpm, 4 °C. Discard supernatant and resuspend cell pellet in 100 μl PBS. Vortex and transfer cell suspension to the tubes for ImageStream acquisition (0.6 ml siliconized microcentrifuge tubes).
12. 5–8 min before the start of acquisition the nuclear dye (DRAQ5) has to be added to the cells (working concentration of DRAQ5 is 1 μM) (*see* **Note 10**).
13. The specificity of NF-kappaB staining should be validated by preparing appropriate negative control samples. Negative control #1. Pre-incubate cell cultures with LPS-inhibitor (polymyxin B) at 10 μg/ml or 20 μg/ml for 30 min at 37 °C, 5 %

CO_2 before adding LPS to cells [17]. Negative control #2. Pre-incubate cell cultures with NF-kappaB inhibitors such as BAY11-7085 (Bayer) before stimulation by TNFα (inhibits TNFα-stimulated IkBα phosphorylation) [18, 19]. Negative control #3—unstimulated cells.

3.4 Imaging Flow Cytometry (IFC) Data Acquisition and Analysis

1. Imaging cytometry analysis is performed on ImageStream 100 (Amnis-Millipore, Seattle) equipped with 488, 658 and 405 nm laser sources and bright-field light source using 40× objective as described elsewhere [20]. For subsequent analysis with IDEAS software (Amnis-Millipore, Seattle, USA) at least 5000 cell events has to be collected (*.rif* file format). Examples of nucleocytoplasmic translocation for NF-kappaB in stimulated ARPE-19 cells are provided in Fig. 1.

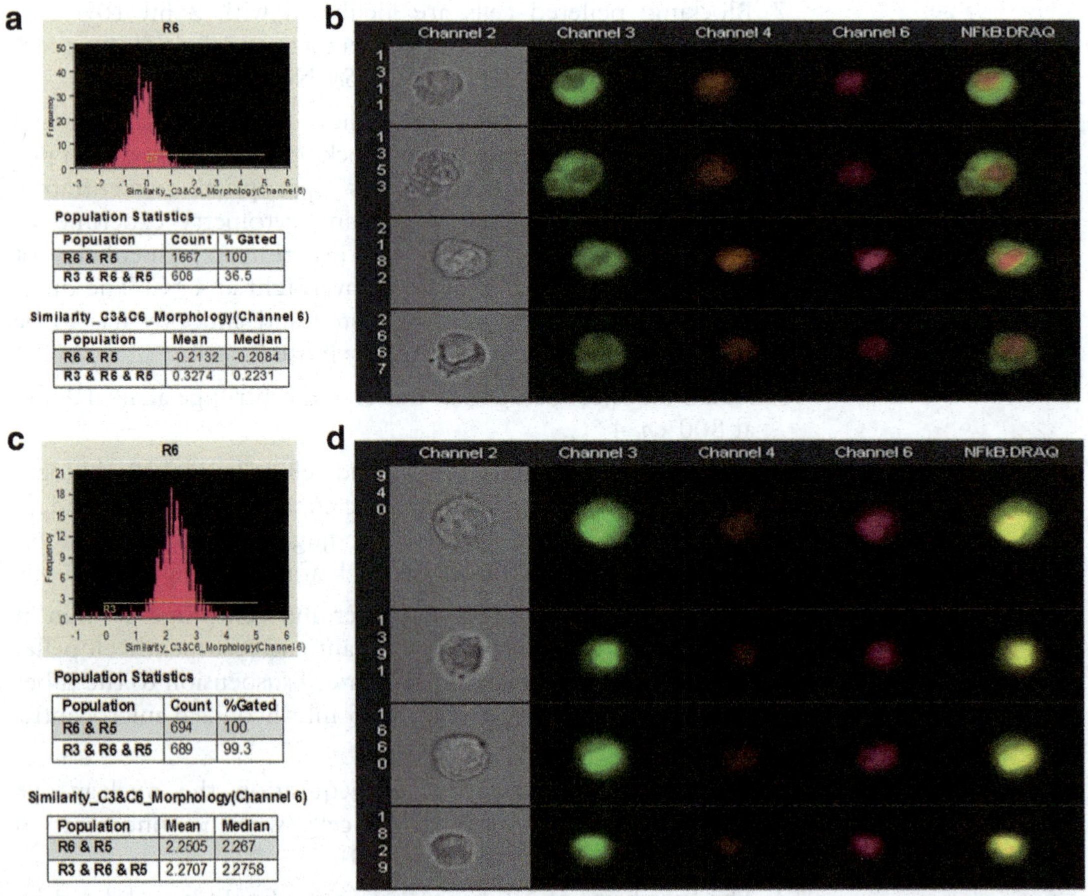

Fig. 1 The similarity histogram and a representative imagery panel of ARPE-19 cells non-stimulated (**a**, **b**) and stimulated with 10 ng/ml TNF-alpha (**c**, **d**). Ch2—Bright-field; Ch3—staining with NF-kappa Beta antibodies, conjugated with FITC; Ch6—DRAQ5 staining; Similarity score for unstimulated ARPE-19 cells equals 0.33 and for stimulated with TNF-alpha equals 2.27

2. Apply Compensation wizard under IDEAS software in order to create compensation matrix. For this purpose, use single controls (raw image files or *.rif* files) acquired in the absence of bright-field illumination. For example, we created single controls for cells stained only with nuclear dye and for cells, stained with antibody against translocated molecule (NF-kappaB). Once the compensation matrix process is finished, IDEAS software will perform corrections on the imagery and apply calculated compensation values to raw files creating new compensated image files (file extension *.cif*). The IDEAS application then calculate features values as specified by created template that can be saved as data analysis files (file extension *.daf*). The compensation template can be saved for future use as .ctm file. Details of the spectral compensation sequence are described in the manual for ImageStream (available for internet downloading as IDEAS version and from company website).
3. Next, use the Nuclear Translocation wizard or perform the analysis of experimental *.daf* files. Alternatively, perform next steps to analyze the experimental data: identify single cells based on nuclear imagery and staining with DRAQ5; exclude out-of-focus events; select positively stained subpopulation for the translocated molecule (NF-kappaB) and DRAQ5.
4. Using IDEAS analytical software feature ("Similarity" feature) calculate similarity scores for cells based on a logarithmic transformation of Pearson's correlation coefficient [21]. The user defines the gate on fluorescence intensity histogram of cell population stained with antibodies against translocated protein and analogous histograms of negative and positive controls. Cells with low similarity score correspond to predominantly cytoplasmic distribution of NF-kappaB, whereas the higher similarity score defines higher level of translocation. The difference (shift) between stimulated and not stimulated cell populations can be evaluated with nonparametric statistical criteria such as Mann–Whitney test.

4 Notes

1. Cells for NF-kappaB translocation can be activated with a number of stimuli, such as LPS, TNF-alpha, phorbol esters, and interleukin-1. Preliminary titration of LPS may require. TNF-alpha can be used in the physiological range (0.2–1.5 ng/ml) [22] as well as in the supra-physiological range (higher than 1.5 ng/ml) [23, 24].

We purchased lipopolysaccharide (LPS) from *E. coli* or *S. typhimurium* from Sigma (St. Louis, MO, USA). Other activators for NF-kappaB, namely TNF-alpha, phorbol esters, interleukin-1α (IL-1α), and IL-1β were purchased from R&D Systems (Minneapolis, MN).

2. The polyclonal and monoclonal antibodies from different companies should be checked by flow cytometry in the beginning of the protocol in order to identify a correct antibody and, perhaps, a correct clone which will show significant increase in a fluorescence after activation of cells with LPS or TNF-alpha. Choosing the right antibody clone is critical, since translocation may involve a conformational change of the epitopes of translocated molecule.

3. We use small molecular weight fluorochromes for antibody detecting the translocated molecule (FITC, Alexa 488, Alexa 647, Cy5, and other small molecules from Alexa and cyanine families). Large proteins, such as multi-subunit R-phycoerythrin (PE) (M.W. 240 KDa) and PE-based tandem dyes have difficulties to traverse inside nuclei. The fluorochromes sensitive to fixation and permeabilization should be also excluded (such as Pacific Blue fluorochrome, tandem dyes, etc.). It is recommended to determine the stability of fluorescent signal after fixation.

4. High concentrations of PFA (4 % and such) in the fixation buffer can diminish a fluorescence of some fluorochromes since it has an effect on conformational protein epitopes. Compatibility of fixation buffer with translocated molecule should be evaluated in the preliminary experiments.

5. Permeabilization buffer is critical for successful intracellular staining and proper visualization of nuclear translocation. With a variety of available reagents we recommend to check a few variants of commercial buffers and choose the best suitable for the antibody against NF-kappaB molecule. In house prepared buffers also can be used.

6. Significant increase in NF-kappaB translocation can be seen after 20 min after simultaneous stimulation with both LPS and TNF-alpha.

7. Importantly, to use for centrifugation siliconized plastic tubes. Siliconization involves placing a thin layer of siliconizing solution (dimethyldichlorosilane) or Sigmacote (Sigma-Aldrich, St Louis) onto glass or plastic surfaces to make them extremely hydrophobic. "Low-adhesion" plastic is not a good replacement for siliconized tubes. Siliconized tubes can be purchased or prepared by short-time incubation with Sigmacote. Shortly, fill

a centrifuge tube with Sigmacote all the way to the top. Collect the Sigmacote from the tube (with glass Pasteur pippete) and transfer back to the original bottle or pour in the next tube. Rinse each time with ddH_2O and let it completely dry. The adequate siliconization is retained by plastic and glass for a long time (years).

8. Blocking step is important. The sequence for staining events with anti-mouse antibodies would include next steps: incubating cells with anti-Fc antibodies for 30 min on ice; a quick spin to remove anti-Fc-antibodies supernatant; staining for surface antigens; fixation and permeabilization for intracellular staining. The selection of blocking serum for antibodies originated from other species (rat, rabbit, others) depends on the species of used antibodies. Generally, blocking with serum from the same animal species as a secondary antibody is best. The alternative is to use 1–3 % of bovine serum albumin (BSA). Optimization of blocking conditions is necessary in order to decrease background of staining on images.
9. Titration of antibodies (primary antibodies and secondary conjugates) and nuclear dyes is necessary in order to prevent saturation of images. Staining volume is very important and titration of antibodies should be performed in the same volume as sub sequential staining. Shortly, start at 1:50 dilution and dilute by twofold steps. Incubate with primary antibody at least 60 min at room temperature. Longer incubation times may be required for some antibodies, such as overnight incubation at 4 °C. Verify the staining with antibodies of different cellular compartments by fluorescent microscopy.
10. DNA dyes should be titrated to give balanced signal with fluorescent staining on other channels. Possible options include but are not limited to: DAPI, DRAQ5, 7-AAD, propidium iodide, SYTOX Blue. Our preferential dye is DRAQ5 (added 5–7 min before running experiment on ImageStream).

Acknowledgements

The work was supported in part by NIH S10 RR023459 grant, Harvard Pilot grant, the Russian Foundation for Basic Research grants and 13-04-40189-H, Swiss IBD grant and PI Nuris. We are also grateful to Aleksandra Gorelova (Harvard University) for help with the editing and preparation of the manuscript.

References

1. Lahiri DK, Ge YW (2000) Electrophoretic mobility shift assay for the detection of specific DNA-protein complex in nuclear extracts from the cultured cells and frozen autopsy human brain tissue. Brain Res Protoc 5:257–265
2. Ruscher K, Reuter M, Kupper D, Trendelenburg G, Dirnagl U, Maisel A (2000) A fluorescence based non-radioactive electrophoretic mobility assay. J Biotechnol 78:163–170
3. Blaecke A, Delneste Y, Herbault N, Jeannin J, Bonnefoy J-Y, Beck A, Aubry J-P (2002) Measurement of nuclear factor-kappa B translocation on lipopolysaccharide-activated human dendritic cells by confocal microscopy and flow cytometry. Cytometry 48:71–79
4. Cognasse F, Sabido O, Beniguel L, Genin C, Garraud O (2003) A flow cytometry technique to study nuclear factor-kappa B (NFkappaB) translocation during human B cell activation. Immunol Lett 90:49–52
5. Nicolas FJ, De Bossher K, Schmerer B, Hill CS (2004) Analysis of Smad nucleocytoplasmic shuttling in living cells. J Cell Sci 117:4113–4125
6. Brandt S, Kwok T, Hartig R, Koenig W, Backert S (2005) NF-kappaB activation and potentiation of proinflammatory responses by the *Helicobacter pylori* CagA protein. Proc Natl Acad Sci U S A 102:9300–9305. doi:10.1073/pnas.0409873102
7. Jerke U, Tkachuk S, Kiyan J, Stepanova V, Kusch A, Hinz M, Dietz R, Haller H, Fuhrman B, Dumler I (2009) Stat1 nuclear translocation by nucleolin upon monocyte differentiation. PLoS One 4, e8302. doi:10.1371/journal.pone.0008302
8. Han F, Liang P, Wang F, Zeng L, Zhang B (2011) Automated analysis of time-lapse imaging of nuclear translocation by retrospective strategy and its application to STAT1 in HeLa cells. PLoS One 6, e27454. doi:10.1371/journal.pone.0027454
9. Lam BD, Anthony EC, Hordijk PL (2011) Analysis of nucleo-cytoplasmic shuttling of the proto-oncogene SET/I2PP2A. Cytometry A 81:81–89. doi:10.1002/cyto.a.21153
10. Costa M, Marchi M, Cardarelli F, Roy A, Beltram F, Maffei L, Ratto GM (2006) Dynamic regulation of ERK2 nuclear translocation and mobility in living cells. J Cell Sci 119:4952–4963
11. Shen H, Nelson G, Nelson DE, Kennedy S, Spiller DG, Griffiths T, Paton N, Oliver SG, White MRH, Kell DB (2006) Automated tracking of gene expression in individual cells and cell compartments. J R Soc Interface 3: 787–794
12. Deptala A, Bedner E, Gorczyca W, Darzynkiewicz Z (1998) Activation of nuclear factor kappa B (NF-kappaB) assayed by laser scanning cytometry (LSC). Cytometry 33:376–382
13. Sakurai H, Chiba H, Myoshi H, Sugita T, Toriumi W (1999) IkappaB kinases phosphorylate NF-kappaB p65 subunit on serine 536 in the transactivation domain. J Biol Chem 274:30353–30356
14. Jiang X, Takahashi N, Matsui N, Tetsuka T, Okamoto T (2003) The NFkappaB activation in lymphotoxin beta receptor signaling depends on the phosphorylation of p65 at serine 536. J Biol Chem 278:919–926
15. DiDonato JA, Hayakawa M, Rothwarf DM, Zandi E, Karin M (1997) A cytokine-responsive IkappaB kinase that activates the transcription factor NFkappaB. Nature 388:548–554
16. Covert MW, Leung TH, Gaston JE, Baltimore D (2005) Achieving stability of lipopolysaccharide-induced NF-kappaB activation. Science 309:1854–1857
17. Cardoso LS, Araujo MI, Goes AM, Pacifico LG, Oliveira RR, Oliveira SC (2007) Polymyxin B as inhibitor of LPS contamination of *Schistosoma mansoni* recombinant proteins in human cytokine analysis. Microb Cell Fact 6:1. doi:10.1186/1475-2859-6-1
18. Ni H, Ergin M, Huang Q, Qin JZ, Amin HM, Martinez RL, Saeed S, Barton K, Alkan S (2001) Analysis of expression of nuclear factor kappa B in multiple myeloma: downregulation of NF-kappa B induces apoptosis. Br J Haematol 115:279–286
19. Pierce JW, Schoenleber R, Jesmok G, Best J, Moore SA, Collins T, Gerritsen ME (1997) Novel inhibitors of cytokine-induced IkappaBeta phosphorylation and endothelial cell adhesion molecule expression show anti-inflammatory effects in vivo. J Biol Chem 272:21096–21103
20. Ponomarev ED, Veremeyko T, Barteneva NS (2011) Visualization and quantitation of the expression of microRNAs and their target genes in neuroblastoma single cells using imaging cytometry. BMC Res Notes 4:517. doi:10.1186/1756-0500-4-517
21. George TC, Fanning SL, Fitzgerald-Bocarsly P, Medeiros RB, Highfill S, Shimizu Y, Hall BE, Frost K, Basiji D, Ortyn WE, Morrissey PJ, Lynch DH (2006) Quantitative measurement of

nuclear translocation events using similarity analysis of multispectral cellular images obtained in flow. J Imm Methods 311:117–129

22. Aderka D, Sorkine P, Abu-Abid S, Lev D, Setton A, Cope AP, Wallach D, Klausner J (1998) Shedding kinetics of soluble tumor necrosis factor (TNF) receptors after systemic TNF leaking during isolated limb perfusion: relevance to the pathophysiology of septic shock. J Clin Invest 101:650–659
23. Staal FJT, Roederer M, Herzenberg LA, Herzenberg LA (1990) Intracellular thiols regulate activation of nuclear factor kB and transcription of human immunodeficiency virus. Proc Natl Acad Sci U S A 87:9943–9947
24. Ginis I, Hallenbek J, Liu J, Spatz M, Jaiswal R, Shohami E (2000) Tumor necrosis factor and reactive oxygen species cooperative cytotoxicity is mediated via inhibition of NF-kB. Mol Med 6:1028–1041

Chapter 9

Using Image-Based Flow Cytometry with a FISH-Based FlowRNA Assay to Simultaneously Detect Intracellular TNF-α Protein and mRNA in Monocytes Following LPS Stimulation

Adam S. Venable, Andrea L. Henning, Eric A. Prado, and Brian K. McFarlin

Abstract

Existing methods of assessing monocyte inflammatory cytokine (IL-1β, IL-6, IL-8, and TNF-α) response to in vitro lipopolysaccharide (LPS) stimulation lack the ability to simultaneously detect intracellular mRNA and protein. This procedure takes advantage of new methodologies and instrumentation to simultaneously measure intracellular TNF-α mRNA and protein in $CD14^+$ monocytes after 1, 3, and 6 h of LPS stimulation. By assessing multiple timepoints, we are able to discern how LPS stimulation affects the temporal relationship between TNF-α mRNA and protein. By using image-based flow cytometry it is possible to co-localize mRNA and protein signals to identify the length of incubation that is needed to initiate protein translation.

Key words Flow cytometry, Lipopolysaccharide, Peripheral blood mononuclear cells, TNF-α, Cytokines, mRNA, FlowSight, QuantiGene, FlowRNA

1 Introduction

Monocyte inflammatory cytokine (IL-1β, IL-6, IL-8, and TNF-α) response to LPS-stimulation in vitro is a commonly used model in the literature [1–8]. However, researchers must choose between measuring mRNA or protein and cannot measure both simultaneously due to the limitations of current methods. For example, TNF-α mRNA is commonly assessed by quantitative RT-PCR, where monocytes are isolated after stimulation so that total RNA can be extracted and then undergo Taqman or SYBR Green RT-PCR reactions [3, 5, 9]. To measure TNF-α protein in cell-culture supernatants, one can use ELISA, bead-based capture assays, or western blotting [1]. In both cases, the results are based

Natasha S. Barteneva and Ivan A. Vorobjev (eds.), *Imaging Flow Cytometry: Methods and Protocols*, Methods in Molecular Biology, vol. 1389, DOI 10.1007/978-1-4939-3302-0_9, © Springer Science+Business Media New York 2016

on bulk cell preparations. Alternatively, one may use antibodies and intracellularly stain for cytokines to identify the specific cells producing TNF-α [2, 4]. Here we describe a novel FISH-based method using the FlowRNA assay to detect mRNA by image-based flow cytometry. Further, we use this new method to study the poorly characterized relationship between intracellular TNF-α mRNA and protein in monocytes after 0, 1, 3, and 6 h of in vitro LPS-stimulation. The use of image-based flow cytometry allows for the co-localization of mRNA and protein and, thus, the ability to detect the timepoint at which protein translation begins. The utilization of the FlowRNA assay together with image-based flow cytometry surmounts the limitations of bulk assays such as RT-PCR and ELISA by allowing the simultaneous detection of both TNF-α mRNA and protein at the single-cell level. To our knowledge, no other current method is capable of providing robust, simultaneous detection of intracellular mRNA and protein.

2 Materials

2.1 Cell Culture Medium

All cell culture medium reagents are prepared in assay-specific volumes and made in bulk; store according to manufacturer's recommendations between uses. Discard any unused buffer after 3 months.

1. Dulbecco's Modified Eagle's Medium Base (Sigma-Aldrich, St. Louis, MO): Dissolve powder in ultrapure water (18 MΩ) and vacuum filter (see below) for sterilization.
2. L-glutamine (Sigma-Aldrich): Aliquot into individual 1 mL fractions.
3. Sodium pyruvate (Sigma-Aldrich): Aliquot into individual 1 mL fractions.
4. Fetal bovine serum (Sigma-Aldrich): Aliquot into individual 10 mL fractions.

2.2 Cell Surface and Intracellular Staining Materials

All staining materials should be diluted in PBS and stored at 2–8 °C in the dark or in a brown, light-protecting tube.

1. Brefeldin A: dilution factor = 50.
2. CD14-PE-eFluor 610 (Clone#61D3): dilution factor = 10.
3. TNF-α-FITC (Clone#MAb11): dilution factor = 15.

2.3 PBMC Fixation and Permeabilization

1. All materials provided by eBioscience and are stored at 2–8 °C unless otherwise noted.
2. Fixation Buffer 1A.
3. Fixation Buffer 1B.

4. Permeabilization Buffer (10×).
5. Fixation Buffer 2 (10×).
6. Wash Buffer.
7. RNasin®: Store at −20 °C.

2.4 mRNA Hybridization, Amplification, and Labeling

All materials provided by eBioscience and are stored at 2–8 °C unless otherwise noted.

1. Target Probe Diluent.
2. PreAmp Mix.
3. Amp Mix.
4. Label Probe Diluent.
5. Storage Buffer.
6. Label Probe Mix (100×): Store at −20 °C.
7. TNF-α and positive control gene Target Probes (20×): Store at −20 °C.

2.5 Additional Materials

1. Sterile vacuum filtration system, 0.22 μm pore size.
2. 50 mL Leucosep tube (Greiner Bio-One).
3. 50 mL conical tube.
4. Ficoll-Paque™ PLUS.
5. Viacount Reagent (EMD Millipore).
6. Phosphate buffered saline: Dissolve powder in 1 L ultrapure water. Store at 4 °C.
7. Phosphate buffered saline with 1 % bovine serum albumin: Dissolve powder in 1 L ultrapure water. Store at 4 °C.
8. LPS from *Salmonella minnesota* R595: Reconstitute 1 mg LPS in 1 mL sterile/endotoxin-free water. Dilute in 4 mL ultrapure water. Store at −20 °C.
9. Ethylenediaminetetraacetic acid: Dilute in PBS to 50 mM concentration. Store at 4 °C.
10. Flow cytometry staining buffer (eBioscience).
11. FlowSight Calibration Beads (EMD Millipore).
12. 1.5 mL microcentrifuge tubes.

2.6 Instrumentation

1. easyCyte 8HT flow cytometry system (EMD Millipore -Amnis).
2. FlowSight imaging flow cytometer (EMD Millipore-Amnis).
3. Bead Bath Incubator.

3 Methods

The following procedure has been broken into 2 days due to the time requirement for the complete assay.

3.1 Isolation and Concentration of Peripheral Blood Mononuclear Cells (Day 1)

1. 30 min prior to blood collection, warm Leucosep tubes, PBS, and Viacount Reagent to room temperature.
2. Collect venous blood into 9 mL evacuated tube containing lithium heparin and invert eight to ten times.
3. Pour whole blood into a Leucosep tube prepared with Ficoll-Paque (*see* **Note 1**). Dilute whole blood with 12 mL pre-warmed PBS. Centrifuge for 20 min at 300 × *g*.
4. Decant upper layer of Leucosep tube into a 50 mL conical tube and add 24 mL PBS/BSA. Centrifuge for 10 min at 400 × *g*.
5. Decant into a waste receptacle and wipe the inside of the 50 mL conical tube with a lab wipe.
6. Resuspend PBMC pellet in 1 mL flow cytometry staining buffer.
7. In a 96-well, round bottom plate add 145 μL Viacount Reagent.
8. Add 5 μL PBMC suspension into the same well containing Viacount Reagent. Let sit at room temperature for 2–3 min.
9. Acquire samples on easyCyteTM 8HT flow cytometer.
10. Dilute 1 mL PBMC suspension in flow cytometry staining buffer to a final concentration between 1×10^6 and 3×10^6 cells/mL.
11. Store diluted PBMC on ice between incubations.

3.2 Cell Culture Medium Preparation

1. Complete these step during the abovementioned centrifugation steps.
2. Under sterile conditions in a laminar flow biological safety cabinet, combine 100 mL Dulbecco's Modified Eagle's Medium Base, 1 mL L-glutamine, 1 mL sodium pyruvate, and 10 mL fetal bovine serum.
3. Filter using vacuum-driven filtration system. Store at 4 °C.

3.3 LPS Time-Course Stimulation of Monocytes and Retention of Intracellular TNF-α Protein

1. Four 1.5 mL microcentrifuge tubes will be prepared at the same time for LPS stimulations of 6, 3, and 1 h and a 0 h control. Combine 322.5 μL culture medium, 20 μL diluted Brefeldin A (DF 1:50), and 7.5 μL LPS (final concentration 15 μg/mL). Mix by vortexing and store at 2–8 °C until use.

2. Transfer 150 μL diluted PBMC into premade (*see* **step 1**) 1.5 mL microcentrifuge tube and mix by vortexing.
3. Culture PBMC for 6, 3, or 1 h at 37 °C in a 5 % CO_2/95 % air, humidified incubator (*see* **Note 2**).
4. Remove stimulated PBMC cultures from CO_2 incubator and add 20 μL of diluted EDTA (*see* **Note 3**), mix by vortexing, and incubate in the dark at room temperature for 15 min.
5. Add 150 μL PBMCs to 0 h control tube.
6. Wash EDTA treated PBMC in 500 μL PBS and centrifuge for 10 min at 400 ×*g*. Aspirate to a final volume of approximately 100 μL (*see* **Note 4**) and resuspend cells in 800 μL PBS and repeat centrifugation with the same conditions.

3.4 Cell Surface Antibody Staining

1. Aspirate and vortex to resuspend cells in residual PBS (*see* **Note 5**). Add 10 μL of diluted CD14-PE-eFluor® 610 mix by vortexing. Incubate cells for 1 h on ice and in the dark (*see* **Note 6**).
2. Add 1 mL flow cytometry staining buffer and centrifuge for 5 min at 600 ×*g*.
3. Resuspend PBMC in residual buffer.

3.5 PBMC Fixation and Permeabilization

1. Add 1 mL of Fixation Buffer 1 and close microcentrifuge tube. Invert the tubes two to three times to adequately mix the sample. Incubate for 30 min in the dark and in the refrigerator at 2–8 °C. Centrifuge for 5 min at 800 ×*g* (*see* **Note 7**). Aspirate the supernatant and resuspend in the residual buffer.
2. Add 1 mL 1× Permeabilization Buffer with RNasin (DF 1:1000) and centrifuge for 5 min at 800 ×*g*. Aspirate the supernatant and resuspend the cells in residual buffer. Repeat the wash with 1 mL 1× Permeabilization Buffer with RNasin.

3.6 Intracellular TNF-α Staining

1. After resuspending the cells in the residual buffer, add 10 μL diluted TNF-α-FITC and mix by inversion. Incubate for 45 min on ice and in the dark (*see* **Note 8**).
2. After intracellular staining, add 1 mL 1× Permeabilization Buffer with RNasin and centrifuge at 800 ×*g* for 5 min. Aspirate the supernatant and resuspend the cells in the residual buffer.
3. Add 1 mL of 1× Fixation Buffer 2 and mix by inversion. Incubate for 60 min in the dark and at room temperature.
4. Centrifuge for 5 min at 800 ×*g*. Aspirate the supernatant.

5. Add 1 mL Wash Buffer. Mix by inversion and centrifuge for 5 min at 800 × *g* (*see* **Note 9**). Aspirate the supernatant and resuspend the cells in the residual buffer.
6. Add 1 mL Wash Buffer with RNasin (DF 1:1000). Mix by inversion and centrifuge for 5 min at 800 × *g*. Aspirate the supernatant and resuspend the cells in the residual buffer and store in the dark and overnight at 2–8 °C (*see* **Note 10**).

3.7 Target Probe Hybridization (Day 2)

1. Warm samples, TNF-α Target Probes, and Wash Buffer to room temperature.
2. Pre-warm Target Probe Diluent to 40 °C.
3. Dilute TNF-α Target Probe (DF 1:20) in Target Probe Diluent (100 μL per sample; *see* **Note 11**).
4. Add 100 μL diluted Target Probe (1:20) directly into the cell suspension (*see* **Note 12**). Mix by vortexing and incubate for 2 h at 40 °C (*see* **Note 13**). Invert samples after 1 h (*see* **Note 14**).
5. Add 1 mL Wash Buffer and invert to mix. Centrifuge for 5 min at 800 × *g*. Aspirate supernatant to 100 μL (all aspirations for Day 2 should be to final volume of 100 μL) and resuspend in residual buffer. Repeat wash step under the same conditions (*see* **Note 15**).

3.8 Signal Amplification

1. Add 100 μL PreAmp mix directly to the cell suspension (*see* **Note 12**). Vortex to mix and incubate for 90 min at 40 °C (*see* **Note 16**).
2. Add 1 mL Wash Buffer and invert to mix. Centrifuge for 5 min at 800 × *g*. Aspirate and repeat wash.
3. Add 100 μL Amp mix directly to the cell suspension (*see* **Note 12**). Vortex to mix and incubate for 90 min at 40 °C in the dark (*see* **Note 17**).
4. Add 1 mL Wash Buffer and invert to mix. Centrifuge for 5 min at 800 × *g*. Aspirate and repeat wash (*see* **Note 18**).
5. Add 100 μL diluted Label Probes (DF 1:100) to each sample and vortex to mix (*see* **Note 12**). Incubate for 1 h at 40 °C in the dark.
6. Add 1 mL Wash Buffer and invert to mix. Centrifuge for 5 min at 800 × *g*. Aspirate and repeat wash.
7. If acquiring samples immediately, add 100 μL PBS and acquire on the FlowSight imaging flow cytometer. If samples are being stored, repeat an additional wash step using Storage Buffer.
8. Prior to acquisition, transfer all samples to a 96-well, round-bottom plate, add 25 μL FlowSight Calibration Beads (*see* **Note 19**) and cover with a pierceable seal.

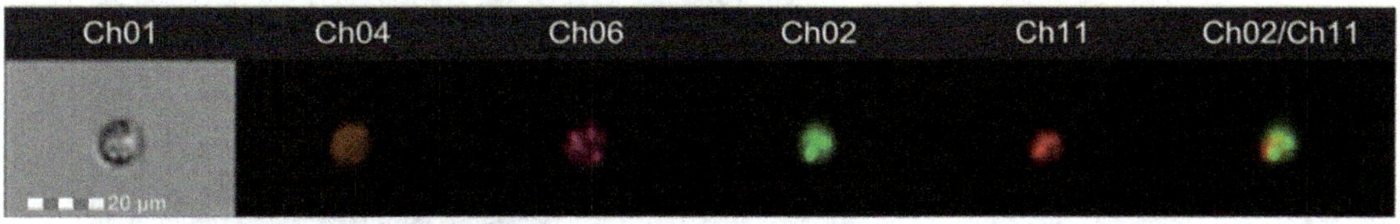

Fig. 1 Representative fluorescent cell image. Image gallery for one monocyte (CD14+) depicting all channels of fluorescence used for the present method. Ch01: Bright-field; Ch04: CD14-PE-eFluor610; Ch06: SSC; Ch02: TNF-α-FITC; Ch11: QuantiGene Probe TNF-α mRNA—AF647; and merge mRNA/Protein image

3.9 Sample Acquisition

All samples are acquired on FlowSight Imaging flow cytometer.

Collect 20,000 single cell events with the following laser settings: 488 nm = 60 mW, 642 nm = 75 nW, and 785 nm = 8.06 mW. Collect bright-field images in Channels 1 and 9, TNF-α protein staining in Channel 2, CD14 in Channel 4, Side Scatter in Channel 6, TNF-α mRNA in Channel 11, and positive control mRNA in Channel 12 (Fig. 1).

3.10 Intracellular TNF-α Protein and mRNA Analysis

1. Open the 0 h control raw image file (.rif) and apply a predefined compensation matrix (.ctm). Save as compensated image file (.cif).
2. Separate single cell events from doublets and debris by creating a dot plot of bright-field aspect ratio (cell width/cell height) vs. bright-field area (Fig. 2a).
3. Generate a daughter dot plot of single cells based on CD14 vs. SSC to identify monocytes (CD14+, Fig. 2b) (*see* **Note 20** and Fig. 2).
4. Generate a daughter dot plot of monocytes based on TNF-α mRNA vs. TNF-α protein. Use the 0 h (control) incubation as a reference (Fig. 2c) to set the double positive gate for 1, 3, and 6 h incubations (Fig. 2d).
5. Run the co-localization wizard in the IDEAS software to identify cells that are co-localizing mRNA and protein in the same location. A high co-localization index is indicative of active protein translation (*see* **Note 21**).
6. Save the template (.ast) and use batch processing to apply to remaining time points (1, 3, and 6 h) for the same patient sample.

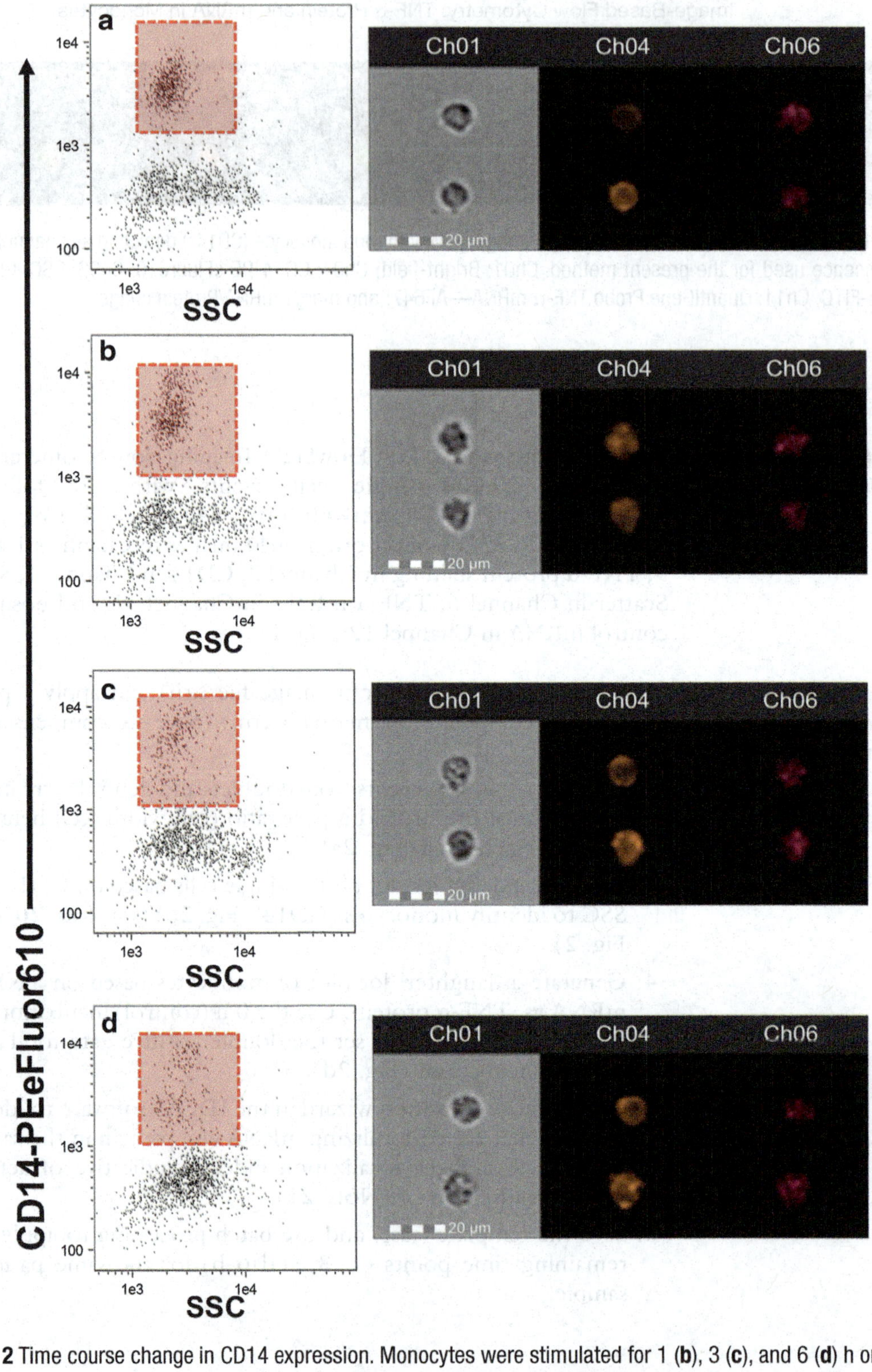

Fig. 2 Time course change in CD14 expression. Monocytes were stimulated for 1 (**b**), 3 (**c**), and 6 (**d**) h or held on ice as a control (**a**). Images to the *right of each plot* are representative of the cells from the corresponding time point. While we observed no change in the relative abundance of CD14 expression over time, the distribution of monocytes and the SSC appeared to change. Care should be taken when identifying CD14+ monocytes to ensure that the gate is placed properly

4 Notes

1. PBMC were prepared according to a previously used method in our laboratory (1–3). Prior to PBMC isolation, prepare Leucosep tubes by adding 15 mL Ficoll-Paque and centrifuge for 5 min at 200 × *g*. Decant additional Ficoll-Paque into original container.
2. PBMC need to be added starting with the longest incubation period so that all incubations finish at the same time. In other words, begin the incubation with the 6 h stimulation. Then add the 3 h stimulation, followed by the 1 h stimulation. When the incubation period finishes, remove all sample preparations at the same time, then add PBMCs to the 0 h control tube.
3. EDTA should be added to sample preparations to release adherent monocytes from the walls of the tube. It has been our experience that failure to use EDTA after LPS stimulation results in fewer monocytes per mL, and ultimately fewer monocytes collected.
4. With any mention of "resuspend cells in residual buffer," this simply means to vortex the samples after aspirating.
5. Although not required for cell surface staining, a residual volume as close to 100 μL is critical to the success of this assay during probe hybridization and signal amplification. When aspirating, use the 100 μL mark on the microcentrifuge tube as a guide.
6. During cell surface staining, prepare Fixation Buffer 1 (1 mL per sample) by combining equal parts of Fixation Buffer 1A and Fixation Buffer 1B. Mix by inversion. Do not vortex. Prepare Fixation Buffer 1 fresh for each day and discard any unused buffer.
7. During Fixation Buffer 1 incubation, prepare 1× Permeabilization Buffer (3 mL per sample) with RNasin by diluting 10× Permeabilization Buffer with ultrapure water. RNasin should be added to a final concentration of 1:1000. Mix by inversion. Do not vortex. Keep at 2–8 °C. Prepare 1X Permeabilization Buffer fresh for each day and discard any unused buffer.
8. During intracellular antibody staining, prepare 1× Fixation Buffer 2 (1 mL per sample) by diluting 10× Fixation Buffer 2 in Wash Buffer. Prepare 1× Fixation Buffer 2 fresh for each day and discard any unused buffer.
9. Prepare Wash Buffer with RNasin (1 mL per sample; DF 1:1000). Mix by inversion. Do not vortex. Prepare Wash Buffer with RNasin fresh for each day and discard any unused buffer.

10. Alternatively, you may proceed through the Target Probe Hybridization, and to do so, skip this step and repeat the previous wash step with Wash Buffer only.
11. Diluted Target Probes (DF 1:20) should be prepared fresh for each sample. If more than one target probe is being measured, adjust Target Probe Diluent so that each is diluted to a final concentration of 1:20.
12. It is critical to the success of the assay that this volume be pipetted directly into the cell suspension. Avoid pipetting onto the walls of the tube.
13. Prior to beginning this assay, the incubation equipment and setup for Target Probe hybridization must be validated so that the temperature of the volume being incubated reaches and maintains 40 ± 1 °C within 5 min of being placed into the incubator.

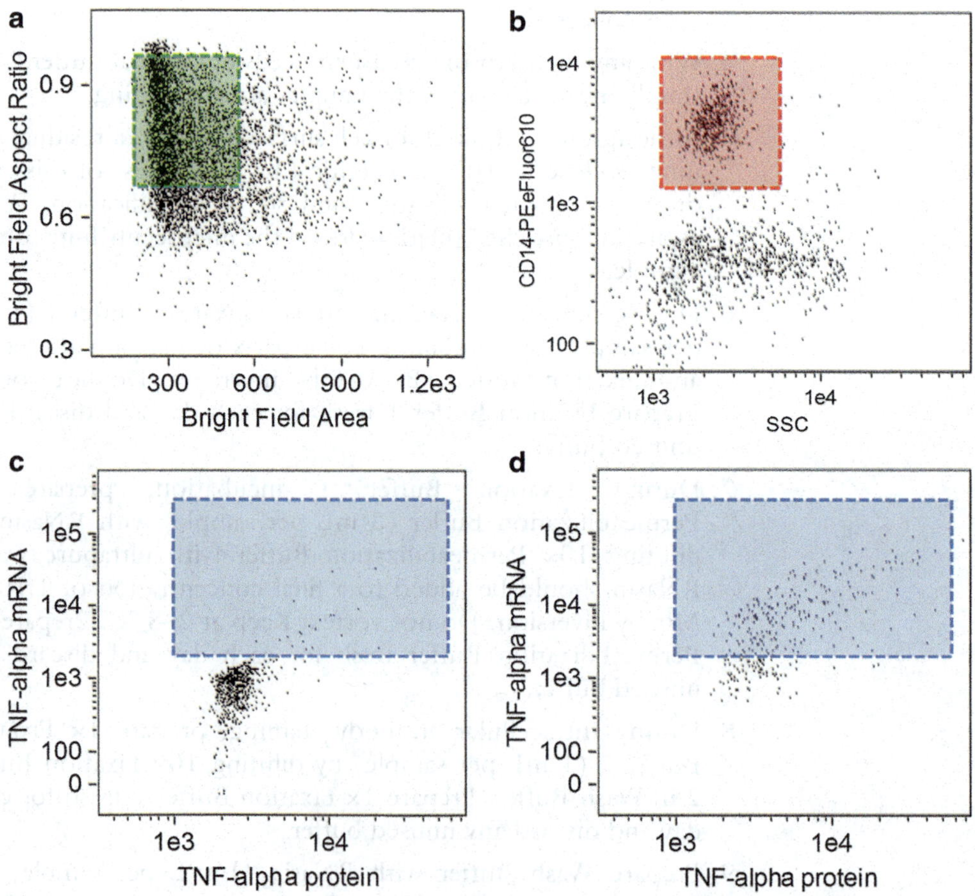

Fig. 3 Gating strategy for monocytes. Single cell events were identified by plotting bright-field area vs. bright-field aspect ratio on all collected events (**a**). Monocytes were identified by gating CD14 vs. SSC (**b**). Plots of TNF-α protein vs. mRNA were used to separate inactive (**c**) from active (**d**) monocytes

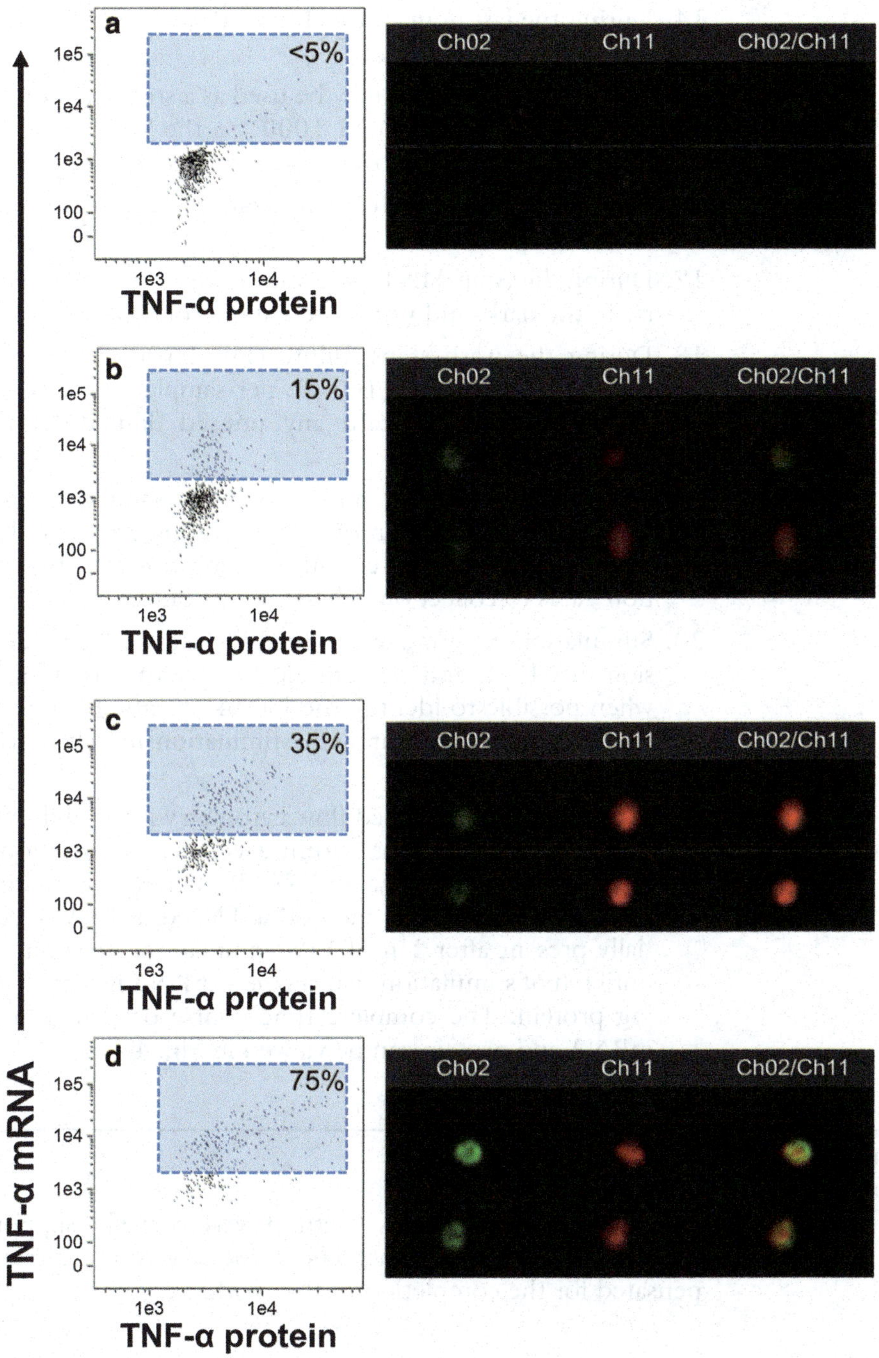

Fig. 4 Identification TNF-α protein and mRNA expression. Monocytes (CD14+) were further classified for dual expression of TNF-α protein and mRNA. In the control sample (**a**) there was less than 5 % cells positive for both markers. The positive percentage progressively increased from 1 (**b**) to 3 (**c**) to 6 (**d**) h of LPS-stimulation. Representative images demonstrated the expression of TNF-α protein (Ch02; *green*), TNF-α mRNA (Ch11; *red*), and merged co-localization (Ch02/Ch11; *yellow*). It should be noted that even the control sample stains for TNF-α protein, but no mRNA is detected until 1 h of stimulation. Also, there is no co-localization of protein and mRNA until 6 h of stimulation, indicating that at the current LPS dose, the stimulation must last at least 6 h to produce new TNF-α protein

14. During the last hour of the Target Probe hybridization, warm PreAmp Mix to 40 °C.
15. This point in the assay may be used as a stopping point, and to do so, add RNasin (DF 1:1000) to the Wash Buffer for the final wash step, then cover and store in at 4 °C overnight.
16. During the PreAmp Mix hybridization, warm the Amp Mix to 40 °C.
17. During the Amp Mix hybridization, thaw the Label Probes on ice in the dark, and warm the Label Probe Diluent to 40 °C.
18. During the wash steps, dilute (DF 1:100) Label Probes in Label Probe Diluent (100 μL per sample). Prepare fresh for each sample and discard any unused diluted Label Probe solution.
19. Calibration beads are added to each sample to assist the FlowSight in validation of sample during automated acquisition. The beads are gated out of acquisition by setting acquisition gates to collect on $CD14^+$ events only.
20. Stimulation of monocytes with LPS alters cell-surface expression of CD14, and care should be taken to use visual gating when possible to identify the $CD14^+$ events. Figure 3 depicts the changes that occur with stimulation in cell morphology and CD14 expression.
21. A benefit of image-based flow cytometry is the ability to identify co-localized events within a cell. Co-localization in this method indicates that mRNA is actively translating new protein. Based on our method a TNF-α mRNA signal is initially present after 1 h of LPS stimulation; however, it is not until 6 h of stimulation that the new mRNA is actively translating protein. The complete time course of change in TNF-α mRNA and protein can be viewed in Fig. 4.

Acknowledgements

The development of this method was partially supported by Affymetrix-eBioscience, Inc. The authors were not directly compensated for the completion of this work.

References

1. Gualdoni GA, Kovarik JJ, Hofer J, Dose F, Pignitter M, Doberer D, Steinberger P, Somoza V, Wolzt M, Zlabinger GJ (2014) Resveratrol enhances TNF-alpha production in human monocytes upon bacterial stimulation. Biochim Biophys Acta 1840:95–105. doi:10.1016/j.bbagen.2013.09.009
2. Raspé C, Czeslick E, Weimann A, Schinke C, Leimert A, Kellner P, Simm A, Bucher M, Sablotzki A (2013) Glutamine and alanine-induced differential expression of intracellular IL-6, IL-8, and TNF-alpha in LPS-stimulated monocytes in human whole-blood. Cytokine 62:52–57. doi:10.1016/j.cyto.2013.02.020

3. Bastami S, Norling C, Trinks C, Holmlund B, Walz TM, Ahlner J, Uppugunduri S (2013) Inhibitory effect of opiates on LPS mediated release of TNF and IL-8. Acta Oncol 52:1022–1033. doi:10.3109/0284186X.2012.737932
4. Dimitrov S, Shaikh F, Pruitt C, Green M, Wilson K, Beg N, Hong S (2013) Differential TNF production by monocyte subsets under physical stress: blunted mobilization of proinflammatory monocytes in prehypertensive individuals. Brain Behav Immun 27:101–108. doi:10.1016/j.bbi.2012.10.003
5. Zgair AK (2012) The effect of high temperature on the kinetics of lipopolysaccharide (LPS)-induced human monocytes activity in vitro. Cell Immunol 275:55–60. doi:10.1016/j.cellimm.2012.03.001
6. Carpenter KC, Breslin WL, Davidson T, Adams A, McFarlin BK (2013) Baker's yeast beta-glucan supplementation increases monocytes and cytokines post-exercise: implications for infection risk? Br J Nutr 109:478–486. doi:10.1017/S0007114512001407
7. Kueht ML, McFarlin BK, Lee RE (2009) Severely obese have greater LPS-stimulated TNF-alpha production than normal weight African-American women. Obesity (Silver Spring) 17:447–451. doi:10.1038/oby.2008.552
8. McFarlin BK, Flynn MG, Campbell WW, Craig BA, Robinson JP, Stewart LK, Timmerman KL, Coen PM (2006) Physical activity status, but not age, influences inflammatory biomarkers and toll-like receptor 4. J Gerontol A Biol Sci Med Sci 61:388–393
9. Flynn MG, McFarlin BK (2006) Toll-like receptor 4: link to the anti-inflammatory effects of exercise? Exerc Sport Sci Rev 34:176–181

Part III

IFC in Immune Response and Drug Screening

Chapter 10

Multiparametric Characterization of Human T-Cell Immune Synapses by InFlow Microscopy

Guido H. Wabnitz and Yvonne Samstag

Abstract

Immune cells need to communicate with each other via direct cell contact formation. The contact zone has similar functions as a neuronal synapse and is therefore named immune synapse. Supramolecular activation clusters consisting of a variety of surface receptors and cytoplasmic proteins are formed within the immune synapse, which are pivotal for T-cell activation. Thus, a malfunction of immune synapse formation has detrimental effects on the healthiness of the individual.

Classical confocal microscopy to analyze the supramolecular cluster formation and maturation of the immune synapse between primary human T-cells and antigen-presenting cells is time consuming and the number of cells that can be analyzed is limited. Therefore, we have established an InFlow microscopy approach for the analysis of immune synapses. InFlow microscopy is a hybrid method combining fluorescence microscopy and flow cytometry. Our InFlow microscopy method allows quantifying protein distribution in immune synapses of several hundred or even thousand cell couples in one sample. Importantly, comparisons of different samples with a strong statistical power are possible with InFlow microcopy.

Key words InFlow microscopy, Immune synapse, Imaging cytometry, Primary human T-cells, Supramolecular activation cluster

1 Introduction

T-cells are activated through binding of their T-cell receptor (TCR) to the cognate antigen presented on major histocompatibility complexes on antigen-presenting cells (APCs). In addition, T-cells receive progression, differentiation and modulatory signals to induce and control T-cell mediated immunity (four signal model of T-cell activation, reviewed in ref. [1]). In order to integrate these signals, T-cells have to form close contacts with their APCs. The contact zone is of special importance because it represents the area in which the cells communicate with each other. Since this contact zone resembles the neuronal synapse it was called immune synapse [2, 3]. Several signaling and scaffolding proteins are transported and enriched in the immune synapse in order to receive signals and

Natasha S. Barteneva and Ivan A. Vorobjev (eds.), *Imaging Flow Cytometry: Methods and Protocols*, Methods in Molecular Biology, vol. 1389, DOI 10.1007/978-1-4939-3302-0_10, © Springer Science+Business Media New York 2016

to stabilize the T-cell/APC contact. Two types of clusters form on the T-cell surface: (1) Microclusters containing the TCR arise at the periphery of the immune synapse and are constantly transported to the central area. TCR-dependent signaling takes places mainly in these microclusters [4, 5]. (2) Macroclusters are formed in concentric rings and are called "supramolecular activation clusters (SMACs)". The inner cluster—central SMAC or cSMAC—is enriched with TCRs [2]. It is surrounded by the peripheral SMAC (pSMAC) which contains LFA-1 and F-actin as well as the actin regulatory proteins cofilin and L-plastin [6, 7]. The pSMAC with LFA-1 enriched is important for the stability of the T-cell/APC contacts [8]. The outermost ring (distal SMAC or dSMAC) is composed of receptors with a large extracellular part (e.g., CD45). It is believed that immune synapse formation is crucial for cellular adhesion, signal transduction, T-cell activation, cytotoxicity, and memory development as well as asymmetric cell division [9, 10]. These essential functions of the immune synapse demand reliable imaging techniques to explore immune synapse formation, stability, and function.

Immune synapses can be analyzed using confocal laserscan (CLSM) [6, 11], total internal reflection (TIRF) [12], and InFlow microscopy [6, 13]. Direct T-cell/APC couples that are formed by different types of T-cells and APCs (e.g., dendritic cells (DCs) or B-cells) can be analyzed using CLSM. The image perspective gives a lateral orientation of the cells. The cellular interface can be reconstructed from optical sections, allowing the analysis of SMACs. A detailed image is limited by lower resolution of CLSM in the z-axis. Moreover, due to the elaborative and time-consuming acquisition, analysis and 3D-reconstruction the number of individual experiments and cell couples that can be analyzed per sample is limited. TIRF microscopy is performed with supported lipid bilayers containing embedded fluorescently labeled stimulatory ligands. The immune synapse is perpendicular to the imaging axis which gives an en face view of the T-cells interface. It is suitable for detailed imaging of dynamic microcluster or F-actin movements in the immune synapse. The drawback is the usage of artificial lipid bilayers as surrogate APCs. As described for CLSM, the sample number as well as the amounts of cells per sample that can be analyzed is likewise restricted using TIRF.

InFlow microscopy is a hybrid technology that combines fluorescence microscopy with flow cytometry. Thus, multicolor images of several thousand cells can be acquired within a brief period of time. The subsequent analysis of images is carried out using the IDEAS software. This tool allows application of masks that define regions of interest according to the particular fluorescences on the single cell level. A multitude of features can be calculated including mask morphology, fluorescence intensities, and colocalizations. Results are visualized by histograms or dot plots allowing hierarchical gating strategies. This enables unbiased analysis with high statistical power as well as investigation of rare events as occurring in heterogeneous (primary) cell populations. We successfully established a workflow to analyze immune synapses of primary human T-cells using InFlow microscopy [6, 8, 13–19]. The human T-cells are activated by Raji cells carrying *staphylococcus aureus* enterotoxin B (SEB) to trigger the TCR on T-cells. In addition, ligands for costimulatory receptors (e.g., CD80 and CD86) are highly expressed on Raji cells [11]. The degree of the enrichment of TCR/CD3 and LFA-1 in the T-cell/APC contact zone serves as measure for immune synapse maturation. Notably, a staining for CD3, CD18 (β2-subunit of LFA-1), and nuclei is sufficient to quantify and characterize T-cell immune synapses. Thus, the remaining fluorescence channels of the InFlow microscope (2 channels for IS100™ and up to 7 channels for IsX™) can be used to include additional readouts as for example (co-)localization of surface receptors, F-actin dynamics, protein phosphorylation, or nuclear translocation of transcription factors. Moreover, human T-cells that transiently express proteins with a fluorescence-tag can be analyzed. The high number of images that can be acquired with InFlow microscopy allows to characterize the immune synapse of primary human T-cells, even if their frequency is significantly lower as compared to experimental systems in which TCR transgenic mouse T-cells are analyzed. Together, this renders InFlow microscopy to a multiparametric tool to characterize human and other T-cell immune synapses.

2 Materials

2.1 Media and Buffers

1. Staining buffer (FW-buffer and FWS-buffer): For preparation of FW-buffer add 1 % BSA and 0.07 % NaN_3 to 1× PBS. Mix thoroughly and let stand until the foam is dissolved. To produce FWS add 0.1 % Saponin. Keep buffers at 4 °C until use.
2. 4 % Paraformaldehyde (PFA): Mix 4 g PFA in 80 ml PBS (with Ca^{2+} Mg^{2+}) and heat to 60 °C. Add one droplet of NaOH (1 N) and stir until the dissolving of PFA. Adjust pH to 7.4 using hydrochloric acid (1 N). Filter to remove residual unsolved PFA.
3. Culture medium: the complete medium is composed of RPMI1640 supplemented with 10 % FCS. To produce T-cell blasts PHA-L (Sigma-Aldrich, Taufkirchen, Germany) and recombinant human IL-2 (hIL-2; R&D Systems, Wiesbaden, Germany) is added to the medium as described below.

2.2 Cells, Antibodies and Reagents

1. Cells: Human T-cells are obtained from the peripheral blood of healthy volunteers as described below. Raji B-cells (ATCC accession No. CCL-86) serve as APCs.
2. Antibodies and staining reagents: anti-human CD3-PeTxR (Caltag Medsystems Limited, Buckingham, UK), anti-human CD18-FITC (BDBioscience, Heidelberg. Germany), Hoechst33342 (Life Technologies, Darmstadt, Germany).
3. Stimulating reagents: *Staphylococcus aureus* enterotoxin B (SEB, Sigma-Aldrich, Taufkirchen, Germany).
4. Miscellaneous Reagents: FicoLite H (Linaris, Wertheim, Germany); PBS with and without Mg^{2+}/Ca^{2+} (Cell Concepts, Umkirchen, Germany); MACS Puffer: PBS with 2 mM EDTA and 0.25 % human albumin; Pan T Cell Isolation Kit II (Miltenyi Biotec GmbH, Bergisch Gladbach, Germany).

2.3 Instruments

1. ImageStream IS100™ (or IsX) (Amnis corp., Seattle, USA).
2. The images are acquired with Inspire software and are analyzed with IDEAS version 6.0. Both software packages are from Amnis corporation (Seattle, USA).
3. QuattroMACS LS MACS Separation Columns (Miltenyi Biotec GmbH, Bergisch Gladbach, Germany).
4. Heraeus (Hanau, Germany) Multifuge 3 (Rotor 75006445); IKA (Stauffen, Germany, MS1 Mini Shaker vortex).

3 Methods

All procedures are carried out at room temperature unless otherwise specified.

3.1 Isolation of Primary Human T-Cells

1. Pipet 15 ml PBS (+2 mM EDTA) into a 50 ml reaction tube. Add 20 ml heparinized whole blood and pipet carefully 15 ml FicoLite H underneath the PBS/whole blood mixture to form two layers. Centrifuge with slow and constant acceleration (2500 rpm/24 °C/15 min.). Decelerate without break to stop centrifugation.
2. Transfer leukocytes that formed a visible ring on the FicoLite H into a new 50 ml reaction tube containing 15 ml PBS (+2 mM EDTA). Pellet cells by centrifugation with 660×*g* at 10 °C for 15 min. Acceleration and deceleration can be performed at maximum speed. Decant supernatant and suspend pellet in 50 ml PBS (+2 mM EDTA) and spin cells down again by centrifugation at 1200 rpm (4 °C, 10 min.). Repeat this step twice and count cells (diluted in 3 % acidic acid) before the last washing step.

3. Isolate T-cells using the pan T-cell isolation kit (negative isolation) exactly according to the manufactures description (*see* **Note 1**). Suspend purified T-cells in complete medium at a density of 3×10^6/ml until further use.

3.2 Generation of T-Cell Blasts

Incubate T-cells at a density of 3×10^6/ml in complete medium with PHA (2 μg/ml) at 37 °C for 24 h. Wash cells by pelleting (290 × *g*/10 min) and suspending in equal volume of complete medium. Add 80 U/ml of recombinant hIL-2. Let the cells grow in hIL-2 for 6 days. Add hIL-2-containing complete medium if necessary. One day before the experiment, the hIL-2 is removed by centrifugation at 290 × *g* for 10 min and suspension of the pellet in complete medium (density: 3×10^6/ml) (*see* **Note 2**).

3.3 Cell Mixture and Induction of Immune Synapses

1. Spin down 9×10^6 T-cells at 290 × *g* for 10 min. Remove supernatant and suspend pellet in 1350 μl culture medium.
2. Spin down 6×10^6 Raji B-cells at 290 × *g* for 10 min. and aspirate the supernatant. Suspend the cell pellet in 1.5 ml culture medium. Transfer 750 μl in a new tube and incubate these cells with 3.75 μl of a 1 mg/ml SEB solution at 37 °C for 30 min. Add 1 ml culture medium in each tube and centrifuge at 290 × *g* for 10 min (repeat this washing procedure). Suspend pellets in 600 μl culture medium and prepare three aliquots á 200 μl to obtain triplicates. Spin down Raji B-cells of each aliquot and aspirate the supernatant.
3. Suspend Raji B-cells with 150 μl of prepared T-cells and incubate at 37 °C for 45 min. Plan four additional aliquots to prepare compensation controls (Cells stained with only one dye-tagged antibody—single stain). To disrupt unspecific cell couples, the samples are vortexed for 10 s (800–1000 rpm). Fix cells by adding 1.5 ml paraformaldehyde (1.5 %) for 10 min at room temperature. Pellet cells by centrifugation (290 × *g* for 5 min) and wash with FW-buffer.
4. Permeabilize cells by incubating with FWS for 15 min at room temperature. Stain cells with 50 μl antibody mix (CD18-FITC (15 μl), CD3-PE-TxR (10 μl), Phalloidin-AF647 (3 μl), Hoechst33342 (3 μl of a 1/100 dilution), FWS (269 μl)) for 30 min in a 96-well plate on a shaking device. Include single stains of T-cells accordingly. Wash cells twice by repeated centrifugation (290 × *g*/5 min) and suspension in 200 μl FWS. Following centrifugation remove supernatant completely, suspend cells in 50 μl PBS and transfer cell suspension in 0.5 ml reaction tube.

3.4 Image Acquisition

Images of cells will be acquired with an ImageStream™. This system produces uncompensated images that are stored in a raw image file (.rif) (Fig. 1a). Compensation controls will be used to

a

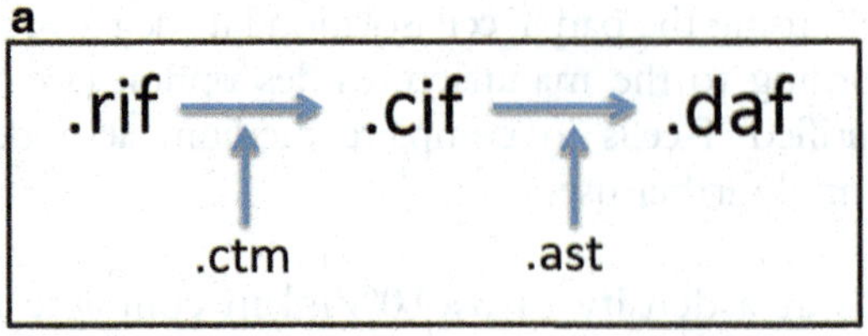

b

	Ch01 (Side Scatter)	Ch02 (Hoechst33342)	Ch03 (CD18-FITC)	Ch04 (empty)	Ch05 (CD3-PeTxR)	Ch06 (Brightfield)
Ch01	0	0	0	0	0	0
Ch02	0	1	0.044	0	0.042	0
Ch03	0	0.875	1	0	0.159	0
Ch04	0	0.291	0.269	1	0.559	0
Ch05	0	0.097	0.064	0	1	0
Ch06	0	0.057	0.035	0	0.303	1

Fig. 1 (**a**) Types of files. The ImageStream™ produces *.rif files, in which the uncompensated images are stored. A compensation matrix can be produced and stored in a *.ctm file. Applying the compensation matrix to the *.rif file produces a compensated image file (*.cif) and a default data analysis file (*.daf). The analysis can be modified according to the experiment specific requirements. Modifications are stored in the *.ast file. The *.ctm and the *.ast file are the basis for batch analysis to get unbiased data from different samples. (**b**) Compensation matrix. The figure shows a representative compensation matrix of triple stained cells. The channels are as following: Ch1: Side scatter or dark field; Ch2: Hoechst 33342: Ch3: CD18-FITC; Ch4: not used; Ch5: CD3-PeTxR; Ch6: bright field

produce a compensation matrix (stored as compensation matrix file [.ctm] and, subsequently, compensated image files are generated [.cif]). The latter are the basis of data analysis which itself is saved as data analysis file (.daf). The usage of the ImageStream™ is described in the systems manual and will not be discussed here unless it is specific for the present method.

1. Set the cell classifiers for compensation controls as following: Area upper limit Ch1: 2500; area lower limit Ch1: 100; intensity lower limit of the respective channel: 50; intensity upper limit of all channels: 1022. Channel loading is described in Fig. 1b.
2. Acquire 500 images for each compensation control. Adjust laser power to achieve a bright signal without saturated pixels for each channel. Switch off bright field.
3. Set the cell classifiers for samples as following: Area upper limit Ch1: 2500; area lower limit Ch1: 100; intensity lower limit

of Ch2 (Hoechst33342): 50; intensity upper limit of all channels: 1022 (*see* **Note 3**). Acquire 25,000 images per sample. Note that images are acquired uncompensated (raw image file: .rif).

3.5 Compensation, Gating Strategy, and Statistics

3.5.1 Compensation

1. Produce a compensation matrix (.ctm) using the compensation controls containing single stained cells according to the manufacturer's description or by using the compensation wizard of the IDEAS software. A typical compensation matrix is shown in Fig. 1b.
2. Apply the compensation matrix (.ctm) to the .rif file of one sample to create a .cif and the corresponding .daf file. The resulting .daf file will be used to generate the template data analysis file (.ast), in which the data analysis tools and gating strategy will be saved. The latter will be used in a batch analysis to create data analysis files (.daf) for each sample.

3.5.2 Gating Strategy

1. Use afore created compensated image file (.cif) and the data analysis file (.daf) to define the following gating strategy. Eventually this strategy is saved in an .ast file.
2. Cells that are not in focus need to be excluded from further evaluations. The gradient RMS feature represents the sharpness quality of the images. Thus, plot the gradient RMS feature within the default mask of the Hoechst33342 staining (M2) and fluorescence channel of the Hoechst33342 staining (Ch2) (encoded as "Gradient RMS_M2_Ch2") on the x-axis of a histogram (Fig. 2a). Note that Raji B-cells (Fig. 2a, image #1)—that serve as APCs—are bigger than primary human T-cells (Fig. 2a, image #2) and show smaller Gradient "RMS_M2_Ch2" values. Cells out of focus (Fig. 2a, image #3) have "Gradient RMS_M2_Ch2" values smaller than 15. Thus, set a gate that includes all events with "Gradient RMS_M2_Ch2" values bigger than 15.
3. Create a dot plot which includes cells in focus with the feature "Aspect ratio_M2" (linear scale) on the y-axis. This feature is calculated according to the nuclei stain (Hoechst33342) and a measure of the shape of cells and cell clusters. The feature "Intensity_MC_Ch1", which corresponds to the side scatter (SSC) is plotted on the x-axis (log scale). This dot plot allows distinguishing T-cell/APC couples (Fig. 2b, image #3) from single T-cells (Fig. 2b, image #1), single APCs (Raji B-cells) (Fig. 2b, image #2), and cell clusters containing 3 or more cells (Fig. 2b, image #4). Define a gate that contains cell couples and control gate borders by clicking on different spots (*see* **Note 4**).

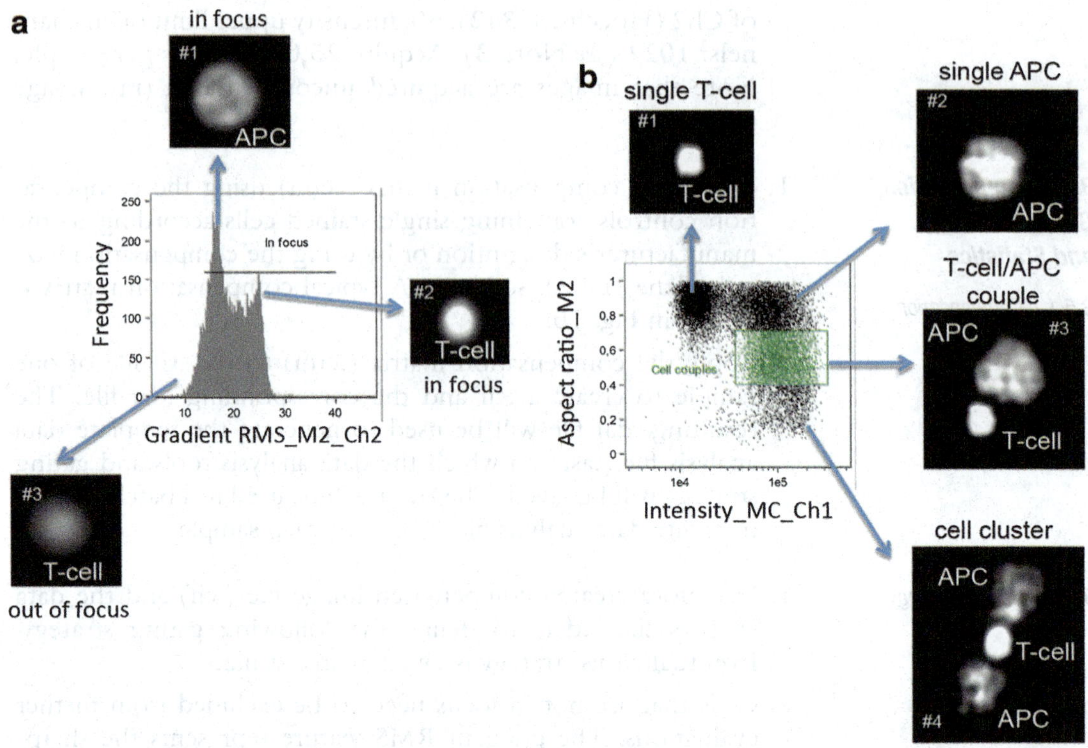

Fig. 2 Gating strategy to find T-cell/APC couples. (**a**) To gate on cells in focus the "Gradient RMS_M2_Ch2" feature is plotted as a histogram. Values lower than 15 indicate that cells are out of focus. The border has to be defined by manual observation of the respective bins. The images show representative images of the nuclear stain of the bins indicated by the *arrow*. (**b**) The *plot* shows cell in focus. Cell couples display a higher side scatter signal ("Intensity_MC_Ch1") and a reduced aspect ratio ("Aspect ratio_M2"). Cell couples are gated in *green*. The images show typical nuclei of cells or cell couples of the respective region of the plot

4. Define a T-cell mask according to the upper 60 % of the fluorescence signal of the CD3 staining ("Threshold(M05, Ch5, 60)", called T-cell-mask) and another mask that determines the immune synapse (called synapse mask, Fig. 3, turquoise areas in images). The latter is a combination of the minimum intensity between the nuclei stains of the T-cell and the APC (Valley-mask) and the aforedefined T-cell mask ("Valley(M02, Ch02, 3) And Threshold(M05, Ch05, 60)"). Calculate the new feature "Area" of both new masks and create a dot plot containing these features (linear scales). This allows gating on T-cell/APC couples with correctly calculated synapse mask (Fig. 3, image #1) and excluding cell couples in which no synapse mask (Fig. 3, image #2) or too big synapse masks. Miscalculated synapse masks occur if the T-cells optically overlap with their APCs (Fig. 3, image #3), if the calculation of the valley-mask was wrong (Fig. 3, image #4) or if two T-cells bind

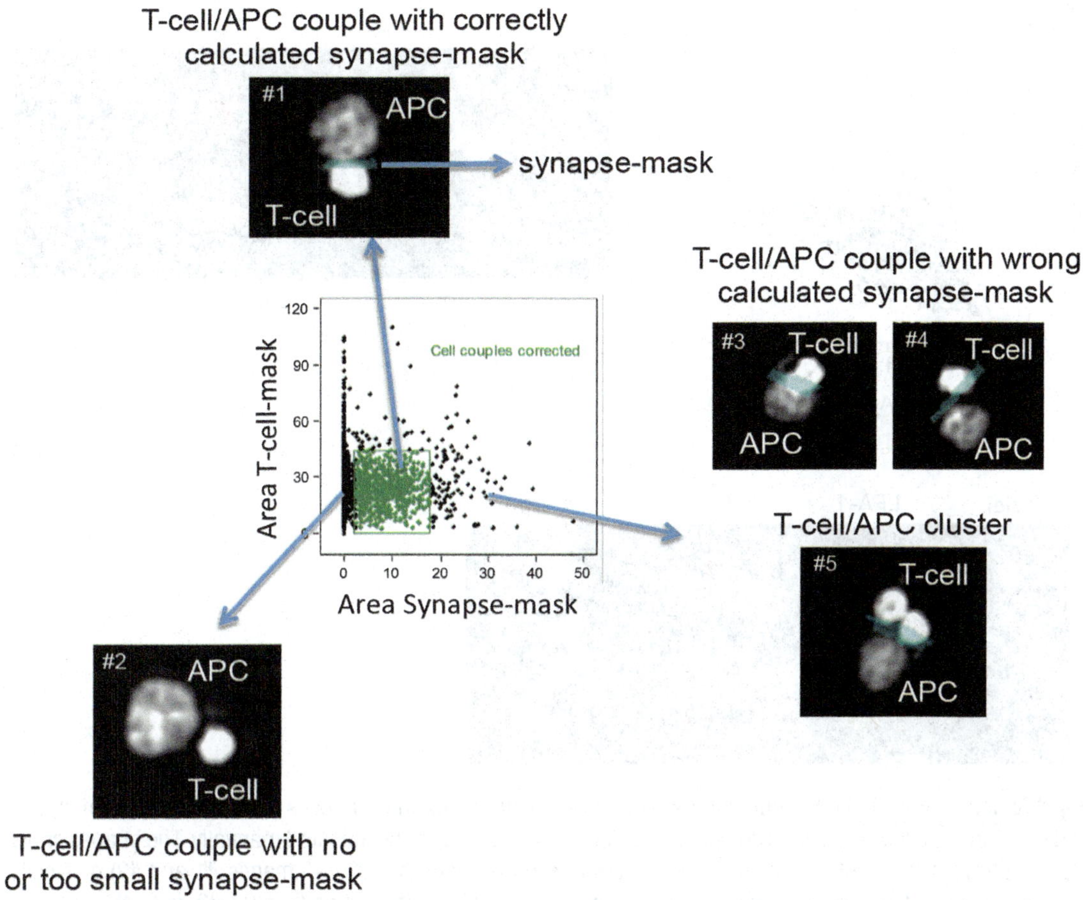

Fig. 3 Fail save gating strategy. The cell couples are corrected to reduce non analyzable cell couples. Thus, the areas of the T-cell and synapse mask are plotted in a *dot plot*. The images show representative nuclei stains of cell from the respective area of the plot. The *turquoise* are shows the synapse mask calculated for the depicted cell couple. A synapse mask, which is too big (Images #3 to #5), results from a miscalculated valley mask due to nuclei morphology. No synapse mask is present (Area Synapse mask = 0; Image #2) if the valley and the T-cell mask do not overlap. A huge increase in the T-cell-mask area is mostly due to multicellular clumps (Image #5)

to the same site of the APC (Fig. 3, image #5). The latter will not be excluded by the initial gating on cell couples (compare Fig. 2). The calculations of the area of the T-cell mask enable to exclude T-cell/T-cell couples (Fig. 3, no image shown). Set a gate named "Cell couples corrected" and verify the gating borders by visual evaluation of 10 % of the correct T-cell/APC couples. The error rate should be below 3 %.

5. Calculate the mean pixel intensity of the fluorescence of the CD3 (Mean Pixel_Synapse_Ch5) and the LFA-1 (Mean Pixel_Synapse_Ch3) staining in the synapse mask (Fig. 4). Calculate

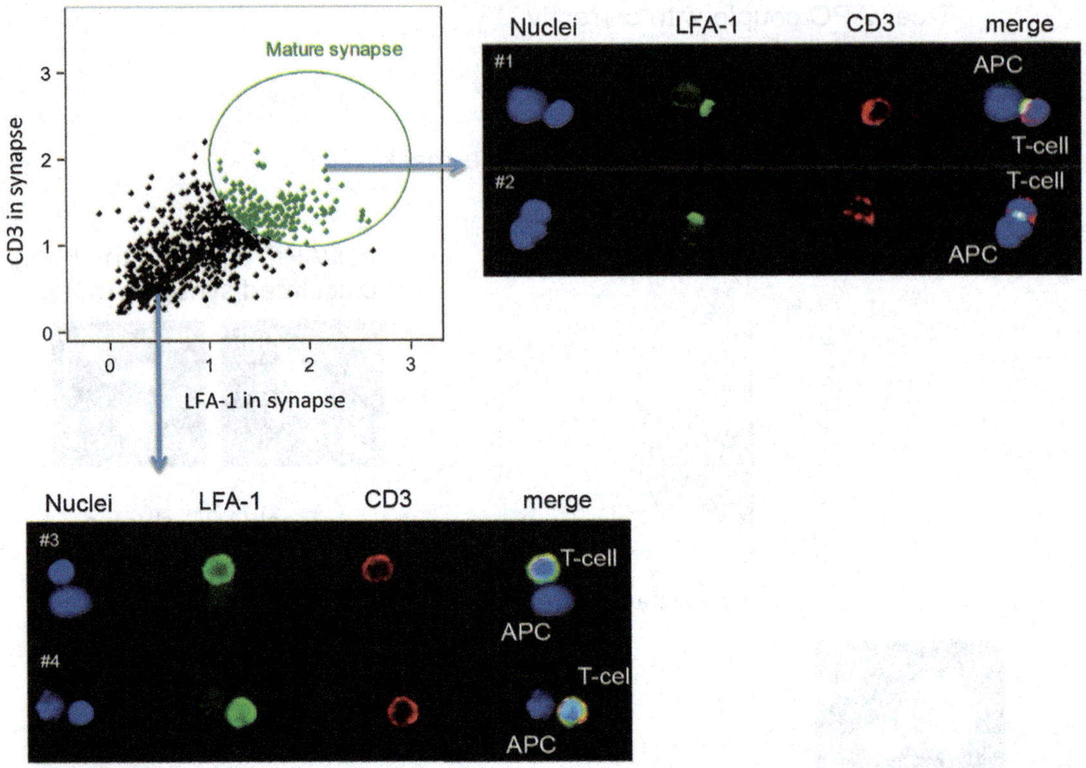

Fig. 4 Gating strategy to find mature immune synapses. Mature immune synapses show an enrichment of CD3 and LFA-1 in the synapse mask. The precise border for the gate has to be adjusted manually. The images show representative pictures of T-cells that form a mature synapse with their APCs (Images #1 and #2) or T-cells that form a contact with APCs without receptor enrichment in the contact zone (Images #3 and #4)

a ratio of the mean intensities of the CD3 fluorescence in the synapse and the T-cell mask (Mean Pixel_Synapse_Ch5/Mean Pixel_M5_Ch5) as well as the mean intensities of the fluorescences of LFA-1 in the synapse and the total mean of LFA-1 in T-cells (Mean Pixel_Synapse_Ch3/Mean Pixel_M5_Ch3). These new features are called "CD3 in synapse" and "LFA-1 in synapse" and are plotted in a dot plot (linear scale). Mature immune synapses have values bigger than 1 for both of these features (*see* **Note 5**). An oval gate should be defined (border 1–3 for each axis) and then controlled for 10 % of the events. The error rate highly depends on the quality of the staining and should not exceed 5 % enabling to find differences between samples of more that 15 %.

3.5.3 Statistical Approach

The number of cell couples analyzed should not be lower than 1000 and the coefficient of variance of triplicates should be lower than 5 %. The resulting numbers of the quantification of mature immune synapse are referring to the number of cell couples. Note that a reduced overall number of cell couples will not be recognized. Thus, a subsequent calculation of the number of mature immune synapses per 100 T-cell is advisable.

4 Notes

1. We recommend manual cell separation using MidiMACS™ or QuatroMACS™ devices.
2. The immune synapse protocol is valid for T-cell blasts as well as for freshly isolated T-cells. However, the frequency of mature immune synapses is considerable lower using freshly isolated T-cells. In addition, the protocol is also suitable to analyze immune synapses between TCR transgenic mouse T-cell lines and murine APCs, as for example 1934.4 T cells and L.Au.ICAM-1 [17].
3. An intensity lower limit of Ch5 (CD3-PE-TxR) can additionally be set to 50. This will decrease the amount of images containing sole APCs.
4. The gate may contain few single cells or cluster events, which will be removed in later gating steps. Note that morphological parameters change in a linear fashion. Therefore, it is hardly possible to find an objective way to draw the borders of a particular gate. Each gate is set by manual control of the cells inside and outside of the gate. However, although this gating strategy depends on the analyzing person, it will be the same gate for each sample. This ensures an unbiased comparison of differently treated samples.
5. Depending on the staining quality and the expression of the respective proteins, a sole calculation of the fluorescence strength within the synapse mask can be sufficient or even better to calculate the protein enrichment in the immune synapse.

Acknowledgements

The authors thank Henning Kirchgessner for his continuing support during the establishment of InFlow microscopy protocols and for critical reading of this method. This work was supported by the German research foundation (SFB938).

References

1. Samstag Y, John I, Wabnitz GH (2013) Cofilin: a redox sensitive mediator of actin dynamics during T-cell activation and migration. Immunol Rev 256:30–47. doi:10.1111/imr.12115
2. Grakoui A, Bromley SK, Sumen C, Davis MM, Shaw AS, Allen PM, Dustin ML (1999) The immunological synapse: a molecular machine controlling T cell activation. Science 285:221–227
3. Delon J, Germain RN (2000) Information transfer at the immunological synapse. Curr Biol 10:R923–R933
4. Yokosuka T, Sakata-Sogawa K, Kobayashi W, Hiroshima M, Hashimoto-Tane A, Tokunaga M, Dustin ML, Saito T (2005) Newly generated T cell receptor microclusters initiate and sustain T cell activation by recruitment of Zap70 and SLP-76. Nat Immunol 6:1253–1262. doi:10.1038/ni1272
5. Seminario MC, Bunnell SC (2008) Signal initiation in T-cell receptor microclusters. Immunol Rev 221:90–106. doi:10.1111/j.1600-065X.2008.00593.x
6. Wabnitz GH, Lohneis P, Kirchgessner H, Jahraus B, Gottwald S, Konstandin M, Klemke M, Samstag Y (2010) Sustained LFA-1 cluster formation in the immune synapse requires the combined activities of L-plastin and calmodulin. Eur J Immunol 40:2437–2449. doi:10.1002/eji.201040345
7. Eibert SM, Lee KH, Pipkorn R, Sester U, Wabnitz GH, Giese T, Meuer SC, Samstag Y (2004) Cofilin peptide homologs interfere with immunological synapse formation and T cell activation. Proc Natl Acad Sci U S A 101:1957–1962. doi:10.1073/pnas.0308282100
8. Hosseini BH, Louban I, Djandji D, Wabnitz GH, Deeg J, Bulbuc N, Samstag Y, Gunzer M, Spatz JP, Hammerling GJ (2009) Immune synapse formation determines interaction forces between T cells and antigen-presenting cells measured by atomic force microscopy. Proc Natl Acad Sci U S A 106:17852–17857. doi:10.1073/pnas.0905384106
9. Martin-Cofreces NB, Baixauli F, Sanchez-Madrid F (2014) Immune synapse: conductor of orchestrated organelle movement. Trends Cell Biol 24:61–72. doi:10.1016/j.tcb.2013.09.005
10. Chang JT, Palanivel VR, Kinjyo I, Schambach F, Intlekofer AM, Banerjee A, Longworth SA, Vinup KE, Mrass P, Oliaro J, Killeen N, Orange JS, Russell SM, Weninger W, Reiner SL (2007) Asymmetric T lymphocyte division in the initiation of adaptive immune responses. Science 315:1687–1691
11. Wabnitz GH, Kocher T, Lohneis P, Stober C, Konstandin MH, Funk B, Sester U, Wilm M, Klemke M, Samstag Y (2007) Costimulation induced phosphorylation of L-plastin facilitates surface transport of the T cell activation molecules CD69 and CD25. Eur J Immunol 37:649–662. doi:10.1002/eji.200636320
12. Dustin ML, Depoil D (2011) New insights into the T cell synapse from single molecule techniques. Nat Rev Immunol 11:672–684. doi:10.1038/nri3066
13. Wabnitz GH, Michalke F, Stober C, Kirchgessner H, Jahraus B, van den Boomen DJ, Samstag Y (2011) L-plastin phosphorylation: a novel target for the immunosuppressive drug dexamethasone in primary human T cells. Eur J Immunol 41:3157–3169. doi:10.1002/eji.201041366
14. Schulte B, John I, Simon B, Brockmann C, Oelmeier SA, Jahraus B, Kirchgessner H, Riplinger S, Carlomagno T, Wabnitz GH, Samstag Y (2013) A reducing milieu renders cofilin insensitive to phosphatidylinositol 4,5-bisphosphate (PIP2) inhibition. J Biol Chem 288:29430–29439. doi:10.1074/jbc.M113.479766
15. Feldmeyer N, Wabnitz G, Leicht S, Luckner-Minden C, Schiller M, Franz T, Conradi R, Kropf P, Müller I, Ho AD, Samstag Y, Munder M (2012) Arginine deficiency leads to impaired cofilin dephosphorylation in activated human T lymphocytes. Int Immunol 24:303–313. doi:10.1093/intimm/dxs004
16. Hoffmann SC, Wabnitz GH, Samstag Y, Moldenhauer G, Ludwig T (2011) Functional analysis of bispecific antibody (EpCAMxCD3)-mediated T-lymphocyte and cancer cell interaction by single-cell force spectroscopy. Int J Cancer 128:2096–2104. doi:10.1002/ijc.25556
17. Hoffmann S, Hosseini BH, Hecker M, Louban I, Bulbuc N, Garbi N, Wabnitz GH, Samstag Y, Spatz JP, Hammerling GJ (2011) Single cell force spectroscopy of T cells recognizing a myelin-derived peptide on antigen presenting cells. Immunol Lett 136:13–20. doi:10.1016/j.imlet.2010.11.005
18. Hochweller K, Wabnitz GH, Samstag Y, Suffner J, Hammerling GJ, Garbi N (2010) Dendritic cells control T cell tonic signaling required for responsiveness to foreign antigen. Proc Natl Acad Sci U S A 107:5931–5936. doi:10.1073/pnas.0911877107
19. Klemke M, Wabnitz GH, Funke F, Funk B, Kirchgessner H, Samstag Y (2008) Oxidation of cofilin mediates T cell hyporesponsiveness under oxidative stress conditions. Immunity 29:404–413

Chapter 11

Studying T Cells *N*-Glycosylation by Imaging Flow Cytometry

Ana M. Dias, Catarina R. Almeida, Celso A. Reis, and Salomé S. Pinho

Abstract

Imaging flow cytometry is an emerging imaging technology that combines features of both conventional flow cytometry and fluorescence microscopy allowing quantification of the imaging parameters. The analysis of protein posttranslational modifications by glycosylation using imaging flow cytometry constitutes an important bioimaging tool in the glycobiology field. This technique allows quantification of the glycan fluorescence intensity, co-localization with proteins, and evaluation of the membrane/cytoplasmic expression. In this chapter we provide the guidelines to analyze glycan expression, particularly the β1,6 GlcNAc branched N-glycans, on the membrane of intestinal T cells from inflammatory bowel disease patients.

Key words *N*-Glycosylation, T-Cell Receptor, T-cells, Ulcerative colitis, Imaging flow cytometry

1 Introduction

Glycosylation is an important protein posttranslational modification being fundamental for the regulation of proteins biological functions both in homeostasis and in pathological conditions. The development of novel accurate assays for analysis of this key biological process is of utmost importance in the glycobiology research field. Flow cytometry is a widely used technique that allows measuring fluorescence intensity in many events, but that lacks the possibility of analyzing imaging parameters. On the other hand, fluorescence microscopy is not ideal to analyze and quantify parameters from a large number of cells. The invention of the ImageStream system, which combines features of fluorescence microscopy and flow cytometry, overcomes these limitations opening new opportunities to explore a range of applications with utility in the glycoscience research field. This technique presents an extraordinary throughput, thus allowing fast acquisition of imaging data [1]. Furthermore the IDEAS data analysis software has the capacity to

Natasha S. Barteneva and Ivan A. Vorobjev (eds.), *Imaging Flow Cytometry: Methods and Protocols*, Methods in Molecular Biology, vol. 1389, DOI 10.1007/978-1-4939-3302-0_11, © Springer Science+Business Media New York 2016

calculate several quantitative features that allow the graphical identification of populations based in several variables (fluorescence intensity, cell size, shape, texture, co-localization of multiple probes, among others) [1]. Therefore, imaging flow cytometry technique becomes a remarkable tool for studies related with cellular morphology, internalization, cell signaling, co-localization, and trafficking (see specific applications in: https://www.amnis.com/applications.html). Additionally, we can even create/define masks to easily identify our targets, for instance, to identify co-localization of probes at the cell membrane.

Taking advantages of the combined parameters of imaging flow cytometry, we have analyzed the levels of membrane TCR branched N-glycans comparing different ulcerative colitis (UC) patients and controls. Furthermore, we provide here an example for analysis of co-localization between a lectin-recognizing glycan and a protein receptor (TCR). Using imaging flow cytometry we evaluated the co-localization of the L-PHA lectin (*Phaseolus Vulgaris Leucoagglutinin* lectin that specifically recognizes the β1,6 GlcNAc branched N-glycan structures) with surface TCR α/β on purified intestinal lamina propria T lymphocytes. Interestingly, this bioimaging technique has validated results obtained from the canonical molecular techniques used to assess proteins glycosylation (such as Western blot and immunoprecipitation approaches) [2]. In this chapter we provide a comprehensive approach on how to use imaging flow cytometry in glycoscience, particularly in studying *N*-glycosylation on the membrane of T cells.

2 Materials

2.1 Reagents for Cell Staining

1. Phosphate buffered saline (PBS) 1×.
2. FACS Buffer: PBS 1× with 0.1 % sodium azide and BSA 2 %.
3. Antibodies: anti-TCR α/β mAb (clone BW242/412, mouse IgG2b) conjugated with R-phycoerythrin (PE) (Miltenyi-Biotech); Fluorescein *Phaseolus vulgaris* Leucoagglutinin (L-PHA/FITC) (Vector Laboratories); mouse IgG2b anti-human conjugated with PE (Immunotools); Streptavidin-conjugated fluorescein isothiocyanate (FITC) (Caltag Laboratories).

2.2 Equipment

Imaging flow cytometer equipped with one laser and one camera (ImageStream^X, from Amnis, now part of Millipore). This equipment can have more lasers and cameras for more complex analysis.

3 Methods

This Imaging flow cytometry protocol involves three main parts: (1) Staining of cells with the probes of interest; (2) acquisition of data on an imaging flow cytometer; (3) data analysis and statistics.

3.1 Cell Staining

Cells can be stained with a protocol typically used for traditional flow cytometry, but a careful titration of the labeling agents (dyes or antibodies) is even more crucial. As with any imaging experiment, one should aim for good signal-to-noise ratio while avoiding saturation. This is particularly important when dealing with more than one fluorophore excited by the same laser: the settings that can be changed during acquisition (laser power and selected filters) will affect visualization of all stainings, and thus it is crucial to tune labeling for optimal imaging. The following protocol is described for already optimized dilutions of probes (*see* **Note 1**).

Each experiment must include different samples: the staining of interest, a negative control (isotype control if using antibodies) and single stained cells for compensation, when analyzing more than one color.

1. After isolating lamina propria T lymphocytes (LPLs), following an adapted protocol [3], wash cells with RPMI by centrifuging at 800 ×*g*, 5 min, 4 °C.
2. Remove supernatant.
3. Resuspend pellet with 1 ml FACS buffer and keep on ice.
4. Divide resuspended pellet in 4 Eppendorf tubes (*see* **Note 2**) for: (*a*) negative control, (*b*) double staining (anti-TCR and L-PHA), (*c*) single staining TCR, and (*d*) single staining L-PHA, for compensation.
5. Centrifuge at 300 ×*g*, 10 min, 4 °C.

Blocking

6. Discard supernatant.
7. Resuspend samples with 500 μl of FACS buffer.
8. Incubate for 30 min, on ice.
9. Centrifuge at 300 ×*g*, for 10 min, 4 °C.
10. Discard supernatant.

Staining

11. Resuspend samples in 100 μl FACS buffer.
12. Add 10 μl PE-conjugated anti-TCR antibody to tubes *b* and *c*, or the equivalent amount of isotype control to the negative control sample (*a*).
13. Incubate for 30 min, on ice and in the dark.

14. Wash with 500 μl FACS buffer by centrifuging at 300×*g*, 10 min, 4 °C.
15. Repeat washing step (*see* **Note 3**).
16. Prepare the lectin 4 μg/ml: 1 μl L-PHA-FITC in 500 μl FACS buffer.
17. Incubate samples *b* and *d* with 50 μl of prepared lectin and control *a* with 50 μl of FACS buffer.
18. Incubate on ice for 30 min, in the dark.
19. Prepare a streptavidin-FITC 1 μg/ml solution (from stock 1000 μg/ml): 1 μl Streptavidin-FITC in 1000 μl FACS buffer.
20. Incubate samples *a*, *b*, and *d* with 50 μl of streptavidin solution.
21. Incubate on ice for 30 min, in the dark.
22. Wash with 500 μl FACS solution by centrifuging at 300×*g*, 10 min, 4 °C.
23. Repeat washing step.

Fixation (after staining, see **Note 4***).*

24. After the previous step, discard supernatant.
25. Fix cells with 100 μl formaldehyde 4 %, 20 min, room temperature, in the dark.
26. Centrifuge at 300×*g*, 10 min, 4 °C.
27. Wash with 500 μl PBS 1×.
28. Centrifuge at 300×*g*, 10 min, 4 °C.
29. Repeat the washing step.
30. Discard supernatant.
31. Resuspend in 60 μl of PBS 1× (*see* **Note 5**) and maintain at 4 °C, in the dark until acquisition in the ImageStream cytometer (*see* **Note 6**).

3.2 Acquisition of Data by Imaging Flow Cytometry

Images of isolated cells can be acquired on a 6-Channel ImageStream^X imaging flow cytometer equipped with one laser (Amnis, EMD Millipore), using the INSPIRE software.

1. Initialization of equipment.
   Initialize the ImageStream by following your facility's rules: typically, you need to ensure all containers are filled with appropriate solutions (beads, sterilizer—you can use 10 % bleach, cleanser, debubbler—you can use 70 % isopropanol, rinse—ultrapure water, sheath—filtered PBS 1×, also add some bleach to the waste container) and you may need or not to calibrate.

Preparation of samples immediately before acquisition:

2. To filter samples, start by adding a 2 μl drop of PBS 1× to the inside edge of a 1.5 ml tube.
3. Place a 70 μm cell strainer on top of the tube.
4. While exerting some pressure, pipette the cell suspension on the cell strainer, aiming for the place where you added the PBS 1× drop (*see* **Note 7**).

3.3 Sample Acquisition

1. Start by acquiring the sample with all stainings, to allow you to adjust the equipment settings.
2. Create a classifier based on the bright-field image, so that only events with an area above a certain size will be included in the analysis. In this case, we choose an area higher than 20, measured in the bright-field image.
3. Adjust the laser power and choose the best filters for your image. These settings should be defined in such a way that images are not saturated (*see* **Note 8**).
4. Once all settings are defined, you can start acquiring your data. Ensure that you set up the number of events to acquire to at least 10,000 events.
5. Analyze firstly your stained samples and negative controls, and then proceed for the single stained controls for compensation.
6. Before acquiring samples for compensation enable the option "Comp settings," and change the classifier for the channel you will be visualizing (as you will not be able to see any bright-field image).
7. After finishing acquisition, proceed with sterilization and cleaning protocols for your equipment.

3.4 Data Analysis and Statistics

Analysis was performed with IDEAS 5.0 (Amnis, EMD Millipore) as follows (Fig. 1):

1. A compensation matrix can be created by choosing "create a new matrix" under "Compensation," and following the wizard. Select the files corresponding to single stained controls. At the end, verify whether images are properly compensated by using the "preview images" tool in the "create compensation matrix" dialog box.
2. For analysis, several wizards are available. Here, we advise starting with the co-localization wizard. Briefly, this wizard guides the user through the most common steps: uploading of raw data files (.rif) and compensation matrix to create compensated data files (.cif) and data analysis files (.daf); gating focused

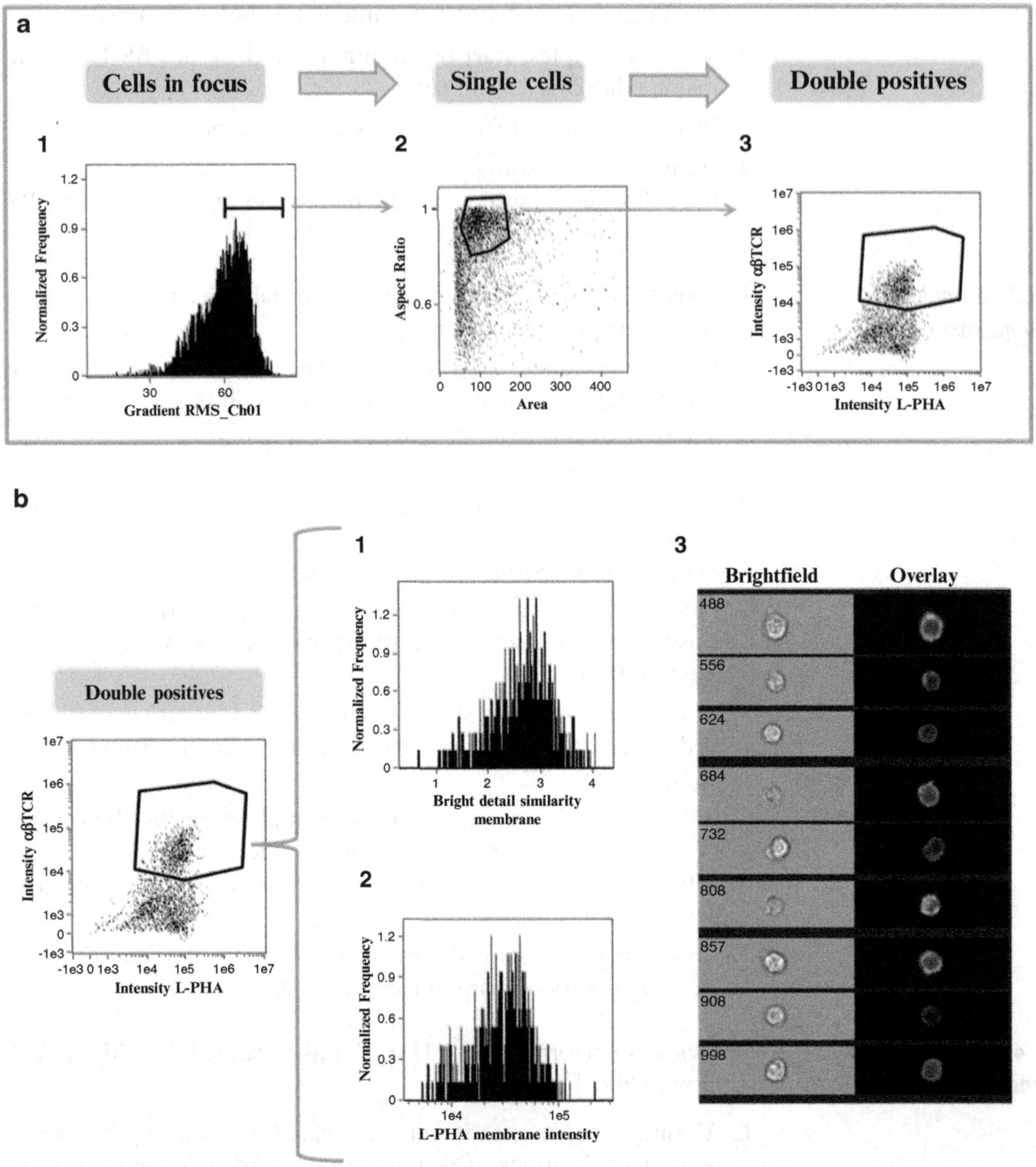

Fig. 1 Schematic representation of the analysis using an example of data acquired after staining for L-PHA and TCR. (**a**) Sequence of analysis steps. (1) A histogram for gradient RMS in bright-field images (Channel 01) can be used to gate for cells in focus. (2) A scatter plot of aspect ratio/area in Channel 1 (bright-field) is used to gate for single cells, within focused events. (3) The intensity of α/β TCR and L-PHA staining is used to gate on double positive cells within focused single cells. (**b**) Graphic representation of parameters analyzed from the double positive cells selected. (1) The level of co-localization at the membrane was quantified as the bright detail similarity. (2) Histogram represents the intensity of L-PHA staining on the membrane of α/β TCR[+] L-PHA[+] cells. (3) Panel showing examples of single cell bright-field (Ch01) and merged (TCR in *red* and L-PHA in *green*) images, showing co-localization between TCR α/β and L-PHA

events; gating single cells (*see* **Note 9**); gating in populations of interest (in this case, TCR-FITC+ cells); calculation of co-localization, or the Bright Detail Similarity between the two probes (L-PHA and anti-TCR) (*see* **Note 10**).

3. Besides calculating co-localization, one can proceed to other analysis in the same file. In this case, the intensity of L-PHA staining specifically located at the membrane of TCR α/β positive events can also be determined. For that, it is necessary to create a new mask by going to Analysis and then Masks (*see* **Note 11**). Here, a new mask to depict the cell membrane was created. This mask uses the functions dilate and erode ("dilate and not erode") and can be based in the bright-field or in a fluorescence image of the membrane.
4. Then create a new feature by going to Features under the Analysis menu. Here, we were interested in evaluating the intensity of L-PHA staining specifically in the cell membrane. A histogram plot for this new feature can then be generated.
5. Finally, create a statistics report to include all relevant information: percentage of TCR+ cells, mean bright detail similarity, mean fluorescence intensity of L-PHA at the membrane, etc..
6. An interesting feature of the IDEAS software is the possibility to easily analyze many files with one template. If you have many files to analyze, save the analysis file as a template and perform a batch analysis, by going to Tools, choosing batch data file and defining the input files and template to use.

4 Notes

1. If you do not have any information about using your probe with the available imaging flow cytometer, you should start by titrating it. Perform the protocol described herein, but with different concentrations of probe. The ideal concentration of probe(s) allow you to visualize cells with a good, but not saturated signal (*see* **Note 8**), using the same settings for all probes.
2. Ideally, cells should be counted and divided equally for each condition. But when having a limited number of cells, cell suspension can be divided in the 4 Eppendorfs, as following: 500 μl—double staining TCR-PE/L-PHA; 166 μl—single staining TCR-PE; 166 μl—single staining L-PHA- FITC; 166 μl—negative control (isotype + Streptavidin-FITC).
3. Samples single staining TCR-PE (tube *c*) should be resuspended in 60 μl of PBS 1× and kept on ice until FIXATION.

4. Fixation is a critical step that should be optimized to determine if it is better to fix before or after staining with the specific probes. Furthermore, it must be tested what is the best for fixation (paraformaldehyde, formaldehyde, methanol), since it may influence cells morphology and protein distribution, affecting the quality of images and consequently evaluation of the parameters of interest.
5. Samples are resuspended in 60 μl of PBS 1× because the ImageStream^X will run samples with a minimum of 50 μl. The number of cells per sample advised by the manufacturer is approximately 1×10^6 cells (up to 5×10^6 cells) in a final volume of 50 μl, in a 1.5 ml microcentrifuge tube, but we found that samples with a lower number of cells can still be analyzed, although acquisition will take a long time.
6. Ideally acquisition in the ImageStream cytometer should be performed immediately after staining. As this is not always practical, fixation allows keeping the cells in cold PBS 1× for a couple of days. Imaging should however be performed as soon as possible.
7. Samples with such a small volume are troublesome to filter. Add a 2 μl drop of PBS 1×, that may stay at the edge of the tube or flow down. Either way, it will create a channel for fluid to flow through by capillary action. It is important to avoid having fluid all around the tube as the sample will then not flow through. The first time you do this start by training with water.
8. During acquisition, you can plot in real time the highest intensity in one pixel of your event vs the area of each event on the bottom graphs of the acquisition windows. This allows you to visualize whether your images are saturated.
9. When performing analysis with the IDEAS software an interesting feature is the fact that you can always observe the images to which a dot in a dot plot corresponds. This becomes extremely useful when deciding on where to draw a gate, as it is possible to visualize whether that dot is a cell, cell debris, a doublet, etc.

   The bright detail similarity is one of the many built in features that can be used with IDEAS. This index gives a measure of the co-localization of two probes in a defined region. It computes the log transformed Pearson's correlation coefficient of the bright spots with a radius of 3 pixels or less within a masked area in the two input images. More details on how this feature functions can be found in [4].
10. It must be pointed out that this bright detail similarity gives a measure of whether two molecules are within the same area of the cells, but due to resolution limitations, it does not indicate

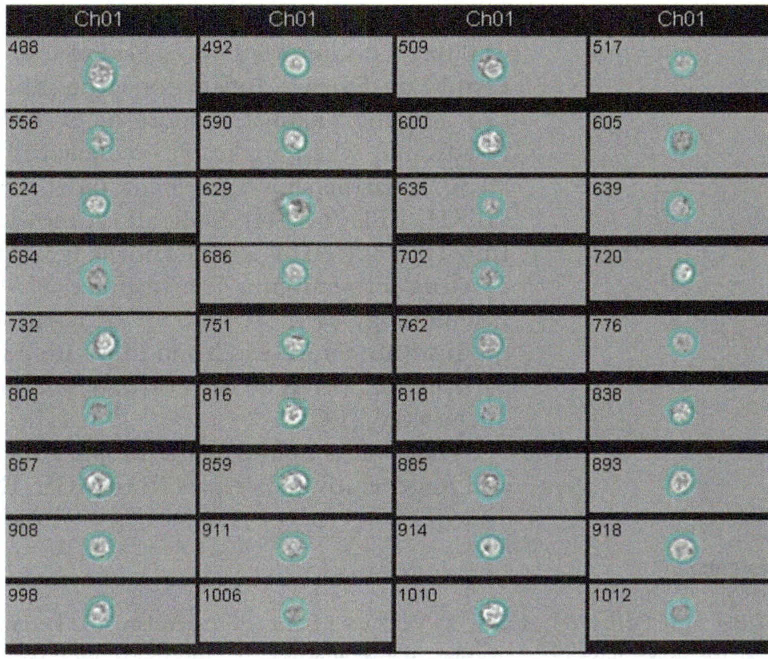

Fig. 2 Panel with examples of bright-field images (channel 1, Ch01) with indication of the membrane mask create (in *light blue*). By creating masks, with IDEAS 5.0 (Amnis, EMD Millipore), we can restrict the analysis to a specific location. The mask membrane created allows quantification of L-PHA staining on the cell membrane of TCR $\alpha/\beta+$ cells from ulcerative colitis patients and controls

whether these molecules interact with each other. For that, other techniques are necessary, such as immunoprecipitation [2] or FRET.

11. If you use a wizard that automatically creates a mask, it is advisable to confirm whether this mask is being correctly applied. For that, go to image properties (icon is the symbol for brightness and contrast manipulation), create a new view that includes your channel of interest with the mask created, go back to the global window and choose that view. Click then on the mask icon and scroll through your cells to see how the mask is being applied (Fig. 2).

Acknowledgments

This work was supported by grants from the Portuguese Foundation for Science and Technology (FCT), project grants (PTDC/DTP-PIC/0560/2014; PTDC/BBB-EBI/0786/2012; EXPL/BIM-MEC/0149/2012), "financiados no âmbito do Programa Operacional Temático Factores de Competitividade (COMPETE) e comparticipado pelo fundo Comunitário Europeu FEDER," e do Quadro de Referência Estratégia Nacional QREN. This work was further supported by a Portuguese grant from "Grupo de Estudo da Doença Inflamatória Intestinal" (GEDII). This work had also the finantial support of FCT/MEC through National Funds and, when applicable, co-financed by the FEDER via the PT2020 Partnership Agreement under the 4293 Unit I&D. S.S.P. (SFRH/BPD/63094/2009) also acknowledges FCT. A.M.D. PD/BD/105982/2014 also acknowledges FCT and BiotechHealth Doctoral Programme. The Institute of Molecular Pathology and Immunology of the University of Porto (IPATIMUP) integrates the Institute for Research and Innovation in Health (I3S), which is partially supported by the Portuguese Foundation for Science and Technology (FCT).

Data was acquired at the Bioimaging Center for Biomaterials and Regenerative Therapies (b.IMAGE, INEB, Porto, Portugal).

References

1. Basiji DA, Ortyn WE, Liang L, Venkatachalam V, Morrissey P (2007) Cellular image analysis and imaging by flow cytometry. Clin Lab Med 27:653–670. doi:10.1016/j.cll.2007.05.008
2. Dias AM, Dourado J, Lago P, Cabral J, Marcos-Pinto R, Salgueiro P, Almeida CR, Carvalho S, Fonseca S, Lima M, Vilanova M, Dinis-Ribeiro M, Reis CA, Pinho SS (2014) Dysregulation of T cell receptor N-glycosylation: a molecular mechanism involved in ulcerative colitis. Hum Mol Genet 23:2416–2427. doi:10.1093/hmg/ddt632
3. Sheridan BS, Lefrancois L (2012) Isolation of mouse lymphocytes from small intestine tissues. Curr Protoc Immunol, Chapter 3, Unit 3.19. Wiley, New York
4. Beum PV, Lindorfer MA, Hall BE, George TC, Frost K, Morrissey PJ, Taylor RP (2006) Quantitative analysis of protein co-localization on B cells opsonized with rituximab and complement using the ImageStream multispectral imaging flow cytometer. J Immunol Methods 317:90–99. doi:10.1016/j.jim.2006.09.012

Chapter 12

Assessment of Granulocyte Subset Activation: New Information from Image-Based Flow Cytometry

Brian K. McFarlin, Adam S. Venable, Andrea L. Henning, Randall R. Williams, and Eric A. Prado

Abstract

Analysis of granulocyte function can provide important information about the state of the body's innate immune system. Existing flow cytometry methods that lack image-based analysis capabilities fail to fully evaluate granulocyte function. In the present method, we combine simultaneous detection of phagocytosis and oxidative burst in granulocytes to identify unique subsets of activated granulocytes. This analysis method provides novel information about granulocytes that allows our lab and others to evaluate the effectiveness of nutritional and lifestyle countermeasures, designed to improve immunity.

Key words Imaging flow cytometry, Phagocytosis, Oxidative burst, Granulocyte activity, Innate immunity, FlowSight, Amnis, Exercise, Immunosuppression

1 Introduction

The body's innate immune system is considered the first line of defense against contagious agents [1]. Our laboratory and others have demonstrated that strenuous physical exercise can suppress the innate immune system response for up to 24-h after exercise [1–3]. This period of immunosuppression increases the likelihood that an athlete or physical laborer may become sick resulting in lost practice and workdays. Also, exercise-induced immunosuppression, may compromise the ability to recovery from injury as the innate immune system plays a key role in the inflammation/recovery process [1, 2].

The most common cells used to assess innate immune function are granulocytes, natural killer cells, and monocytes. The focus of this report is on the description of a novel technique that allows granulocytes to be separated into three functionally active phenotypes [4]. This phenotypic separation was accom-

Natasha S. Barteneva and Ivan A. Vorobjev (eds.), *Imaging Flow Cytometry: Methods and Protocols*, Methods in Molecular Biology, vol. 1389, DOI 10.1007/978-1-4939-3302-0_12, © Springer Science+Business Media New York 2016

plished by combining image-based flow cytometry [5] with simultaneous measurements of phagocytosis capacity and oxidative burst. The method described in this report represents revision of existing methods that lack an image-based analysis approach [6–9].

2 Materials

All stock solutions are prepared in a biological safety cabinet to ensure that sterility is maintained. Sterility is important to reducing unwanted sources of variation in assay performance. Once stock solutions are prepared, aliquots are prepared and frozen into assay-specific volumes for a specified number of blood samples (see below).

2.1 Bioparticles for Phagocytosis

1. *S. aureus* bioparticles labeled with pHrodo red dye are purchased (Life Technologies; Grand Island, NY).
2. Add 2 mL of sterile PBS to 2 mg of bioparticles to make a 1 mg/mL solution.
3. Aliquot 100–150 μL of diluted bioparticles into separate tubes (*see* **Note 1**).
4. Freeze at −20 °C until used in the assay (*see* **Note 2**).
5. Thaw at room temperature at least 20 min before used in assay.

2.2 Dihydroethidium (DHE) for Oxidative Burst

1. Dihydroethidium (DHE) is purchased from a commercial source and the stock powder is stored at −20 °C.
2. Dissolve DHE powder in DMSO to make a 32.5 μg/mL solution.
3. Aliquot 100–150 μL of diluted bioparticles into separate tubes.
4. 3Freeze at −20 °C until used in the assay (*see* **Note 3**).
5. Thaw at room temperature at least 20 min before used in assay.

2.3 N-ethylmaleimide for Blocking Additional Phagocytosis

1. N-ethylmaleimide was purchased from a commercial source and the stock powder was stored at −20 °C.
2. Add N-ethylmaleimide powder to methanol (*see* **Note 4**) to yield a 200 mg/mL solution (First dilution).
3. Further dilute the first dilution to a concentration of 17.5 mg/mL using sterile PBS.
4. Thaw in at 37 °C heat block at least 20 min before used in assay (*see* **Note 5**).

2.4 Antibodies for Cell-Surface Staining of Granulocyte Receptors

1. CD66b-APC (clone#G10F5; DF = 1:50) and CD45-APCeFluor780 (clone#2D1; DF = 1:50) were used for positive identification of granulocytes based on cells-surface staining rather than scatter (*see* **Note 6**).
2. Working dilutions are created by adding 50 μL of stock antibody to 2.450 mL of commercially purchased staining buffer. Dilutions are completed in a light-blocking 15 mL centrifuge tube and stored in the refrigerator between uses.

2.5 7-AAD for Nuclear DNA Detection

1. 7-AAD is stored in the refrigerator prior to assay.
2. Prior to use in the assay, add 100 μL of stock 7-AAD to 900 μL of sterile PBS to create a 1:10 dilution of 7-AAD (*see* **Note 7**). Dilutions are completed in light-blocking tubes.
3. Unused 7-AAD is stored in the refrigerator between uses.

3 Methods

3.1 Patient Blood Sample Collection

1. After signing an IRB-approved consent form, a patient is scheduled to report to the laboratory following an overnight fast (>8 h) and abstention from physical activity (>12 h) (*see* **Note 8**).
2. The patient is asked to sit quietly for 30 min and then 10 mL of venous blood was collected into an evacuated tube treated with sodium heparin (*see* **Note 9**).
3. A blood samples is held at room temperature until analysis and processed within 3 h of collection (*see* **Note 10**).

3.2 Phagocytosis Assay Technique

1. Blood samples are measured using assay incubations of 0, 10, 20, and 40 min. An additional tube will be held on ice and used as a negative control. Thus, a total of four assay tubes are required for each blood sample collected. Once the number of assay tubes is determined, remove an appropriate volume of bioparticles, DHE, and N-ethylmaleimide from the −20 °C freezer and thaw (*see* **Note 11**).
2. Label 1.2 mL assay tubes with the blood sample ID and incubation time. Add bioparticles (20 μL per tube) and DHE (40 μL) into each assay tube under sterile conditions (*see* **Note 12**).

   Pipet heparinized whole blood (100 μL) into the appropriately labeled assay tube containing the bioparticles and DHE (*see* **Note 13**). After blood is added, place assay tubes on ice and protect from light.
3. Complete assay incubations of 40, 20, and 10 min at 37 °C (*see* **Note 14**). After incubation, pipet N-ethylmaleimide solution (15 μL) in each assay tube and incubate for 30 min on ice (*see* **Note 15**).

3.3 Granulocyte Cell-Surface Staining

1. Pipet diluted CD66b-APC (10 μL) and CD45-APC-eFluor780 (10 μL) antibodies into all assay tubes and incubate for 60 min in the dark on ice.
2. Lyse red blood cells (RBC) and fix the white blood cells (WBC) by adding a commercially available WBC fix/RBC lyse solution (750 μL) to each assay tube and incubating for 60 min in the dark on ice (*see* **Note 16**).
3. After staining, pellet cells in a centrifuge (10-min at 400 × *g*). Remove supernatant, leaving approximately 100 μL of volume above the cell pellet (*see* **Note 17**).

3.4 DNA Staining and Final Preparation for Collection

1. Pipet diluted 7-AAD solution (10 μL) and sterile PBS (50 μL) to each assay tube.
2. Add FlowSight calibration beads (25 μL; EMD Millipore) to each assay tube (*see* **Note 18**).
3. Load the assay sample tubes into the autosampler of the FlowSight (or load tubes manually through the single tube loader) (*see* **Note 19**).
4. Collect 10,000 cellular events with the following laser settings: 488 nm = 60 mW, 640 nm = 100 nW, and 785 nm = 8.5 mW). Bright-field images are collected in Channel 1 and 9, bioparticles are collected in Channel 2, DHE is collected in Channel 4, 7AAD is collected in Channel 5, CD66b is collected in Channel 11, and CD45 is collected in Channel 12 (*see* **Note 20**).

3.5 Granulocyte Function Analysis

1. Open the negative control (0 min) raw image file (*.rif) and apply the predefined phagocytosis template and compensation matrix. This will yield a compensated image file (*.cif).

   Within the predefined phagocytosis template, a histogram for bright-field gradient RMS is used to identify cells that were in focus. From focused cells, single cells are separated from debris and doublets using a dot plot of bright-field aspect ratio (ratio of cell height vs. width) vs. bright-field area.
2. Granulocytes ($CD45^+/66b^+$) are separated from focused single cells by using a dot plot of CD45 vs. CD66b (Fig. 1).
3. Save template and using the batch-processing wizard, apply the negative control template to the 10, 20, and 40 min incubations for the same sample.
4. Open the 40 min *.cif file and create a dot plot of bright detail intensity for bioparticles (*x*-axis) vs. bright detail intensity for oxidative burst (DHE, *y*-axis) to identify subsets of activated granulocytes (*see* **Note 21**) (Fig. 2).

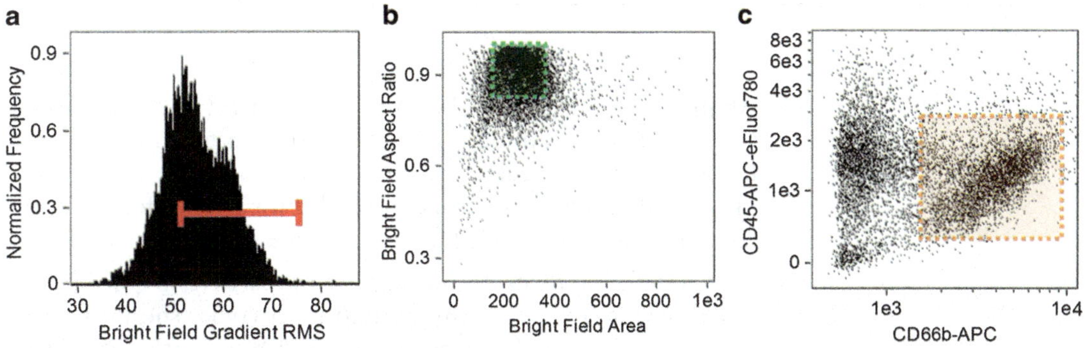

Fig. 1 Granulocyte Gating. Using the entire cell population, focused cells are identified using a histogram of bright-field gradient RMS (**a**), followed by single cells that are identified using a *dot plot* of bright-field aspect ratio vs. bright-field area (**b**), and finally granulocytes ($CD45^+/66b^+$) are identified using a *dot plot* of CD66b vs. CD45. This gating pattern results in the positive identification of at least 3000 granulocytes per patient sample

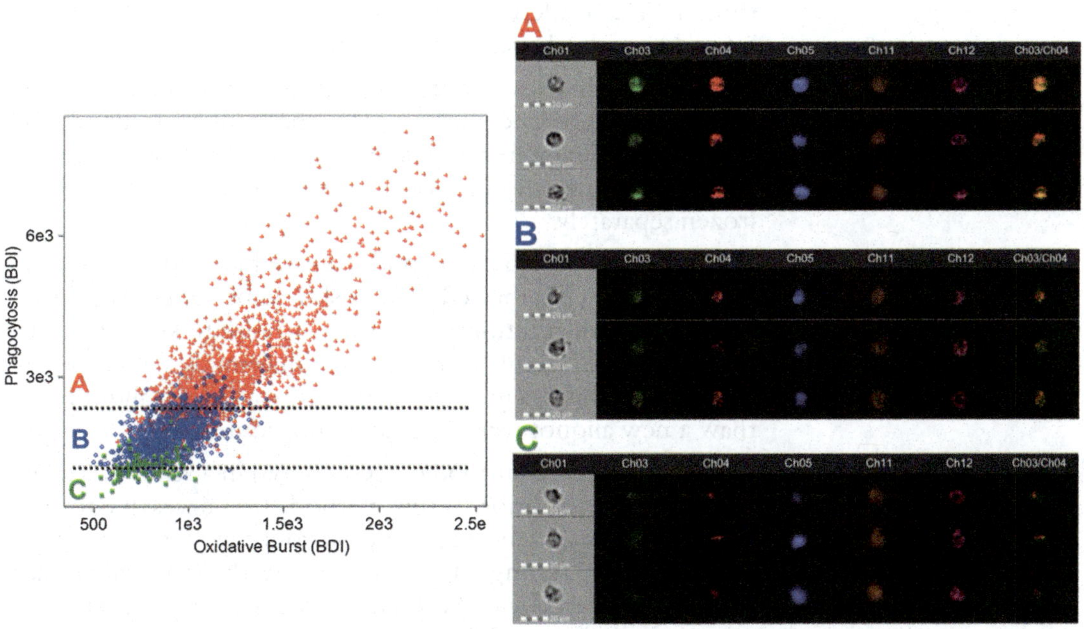

Fig. 2 Activated Granulocyte Subset Gating. After the identification of granulocytes using cell-surface markers ($CD45^+/66b^+$), a daughter plot is generated of bright detail intensity for phagocytosis vs. bright detail intensity for oxidative burst. From this plot it is possible to identify and quantify high-active (*red*, **a**), moderate-active (*blue*, **b**), and low-active (*green*, **c**) granulocytes. Additional analysis reveals that with each activation subset there are changes in the co-localization of phagocytosis and oxidative burst signals (the ch3/ch4 merge image demonstrates this effect). High-active granulocytes had the greatest co-localization, while low-active granulocytes had the least co-localization

4 Notes

1. Aliquot volume should be chosen based on the number of samples that are expect to be analyzed on a given day. In our laboratory, we typically analyze a minimum of six blood samples per day. Thus, we froze enough bioparticles for 24 assay tubes plus an extra 10 % (aliquot volume = 528 μL).
2. Care should be taken to not expose bioparticles to more than one freeze–thaw cycle. Any unused bioparticles should be discarded or held in the refrigerator. Any bioparticles stored in the refrigerator should be used within 7 days of thawing.
3. Similar to bioparticles, this should be aliquoted and frozen in increments that match the targeted experimental outcomes. Any unused DHE stock should be discarded each day. As with any fluorescent reagent, titration will be needed to confirm positive staining. The concentration suggested in this method should be considered a starting point that will need to be refined in each laboratory situation.
4. N-ethylmaleimide does not easily dissolve in solution. For this reason, it may be necessary to place the solution in the 37 °C bead bath. When freezing aliquots, identify a volume that matches the quantity of bioparticles and DHE that you have frozen separately.
5. When thawing N-ethylmaleimide, it should be done as quickly as possible and the thawed aliquot should be kept in the 37 °C bead bath until addition to the assay. Failure to keep the solution heated will result in the N-ethylmaleimide falling out of solution. If this occurs, you should discard that aliquot and thaw a new aliquot using the same procedure.
6. CD66b and CD45 antibodies were titered in a series of experiments completed prior to this method. It is necessary to titer antibodies in your laboratory and on your flow cytometer prior to completing testing of this nature. Additional cell-surface (i.e., CD197 etc.) can be included in the staining panel to assess other aspects of granulocyte function.
7. It is important that aliquots of 7-AAD be stored in a light sensitive (i.e., dark colored) tube. This can also be accomplished by wrapping a 1.5 mL centrifuge tube in aluminum foil.
8. All blood collection procedures described in this method were conducted in accordance with the Declaration of Helsinki and approved by the UNT institutional review board (IRB) for human subjects. All subjects gave written consent for blood collection, which was used in the present method to ensure that they were apparently healthy, of normal body weight, and disease free.

9. It is important to make sure that collected blood is mixed gently by inversion a minimum of ten times.
10. Blood samples were stored a room temperature on an automated blood mixer until assay.
11. Bioparticles and DHE were thawed at room temperature, while N-ethylmaleimide was thawed at 37 °C.
12. All assay steps completed prior to the assay incubation should be completed a Class II Biological Safety cabinet or similar equipment to maintain assay sterility.
13. Blood and reagent combinations should be made using an electronic pipet on a "pipet and mix" setting. This will allow through mixing of reagents that is less aggressive than a vortex mixer.
14. Any incubator that can hold a constant temperature of 37 °C is appropriate for this assay. In our laboratory we use a dry, bead bath incubator. As with any assay, it is important to calibrate your incubator so that a temperature of exactly 37 °C is realized.
15. N-ethylmaleimide is added to prevent any additional phagocytosis and stabilize the reaction. After this step, assay tubes can be processed on a standard laboratory bench as long as assay tubes are kept on ice.
16. When lysing is complete, the liquid in the tube should be a clear-red color. It should not be opaque-red. If the solution is opaque red then it has not lysed long enough. Continue lysing in 30 min intervals until a clear-red color is achieved. According to the manufacturer (eBioscience); this lysing solution could be left in contact with the cells for up to 24 h without harming the WBC.
17. There are several different techniques that can be used to remove the supernatant. In our laboratory we use a vacuum aspiration system; however, it is also possible to simply decant the tubes. The vacuum solution is likely preferred for high-throughput (HT) processing.
18. Given unique differences the cell WBC concentrations, when using the autosampler of the FlowSight it is important to add FlowSight calibration beads to each assay tube. FlowSight calibration beads are nonfluorescent beads that will not interfere with targeted measures, but will allow the FlowSight to complete auto-validation even if an assay tube has a low WBC count. Low WBC counts are common with this assay because it only uses 100 μL of whole blood per assay tube.
19. Given the quantity of sample tubes that we routinely process in the laboratory, we use an autosampler to acquire all assay tubes;

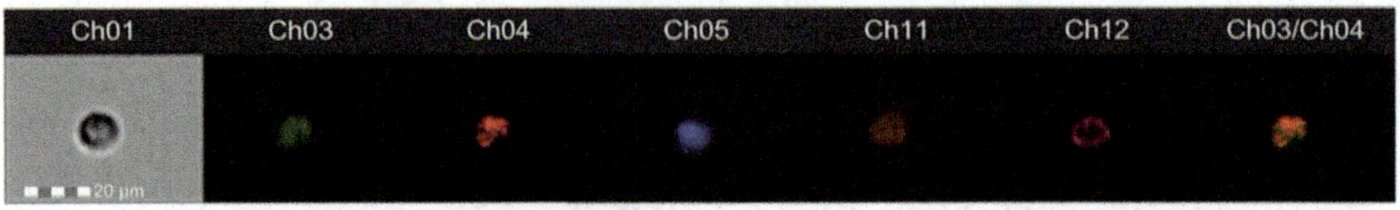

Fig. 3 Representative Images for Detection Channels. This figure denotes the position on the commons signals measured in this assay. Channel 1 (Ch01) is bright-field, channel 3 (Ch03) is *S. aureus*-DiI labeled bioparticles, channel 4 (Ch04) is oxidative burst as measured by DHE, channel 5 (Ch05) is the nucleus stained with 7-AAD, channel 11 (Ch11) is CD66b-APC, and channel 12 (Ch12) is CD45-APC-eFluor780. A final merge image of Ch03 and Ch04 was created to identify co-localized events that were considered indicative of granulocyte function

however, if your FlowSight was not equipped with an autosampler it is still possible to complete the assay, but sample collection would need to be completed manually.

20. When all the detection channels are being detected correctly the individual images should mirror these images (Fig. 3).
21. The FlowSight IDEAS software (EMD Millipore) includes an automated analysis wizard called the "Feature Finder." When we developed this assay we used the feature finder to compare granulocytes with the highest and lowest activation. The Feature Finder identified that the variable "Bright Detail Intensity" provided the best index to separate granulocytes of differing activation status. Bright detail Intensity was defined by the manufacture as the average intensity of all the 3 µm pixel spots that were within the cell membrane.

Acknowledgements

The present study was funded in part by a Research Initiation Grant (RIG) from the University of North Texas to Dr. McFarlin. The authors did not receive direct funding for the completion of this study and report no conflict of interest.

References

1. McFarlin BK, Carpenter KC, Davidson T, McFarlin MA (2013) Baker's yeast beta glucan supplementation increases salivary IgA and decreases cold/flu symptomatic days after intense exercise. J Diet Suppl 10:171–183. doi:10.3109/19390211.2013.820248
2. Carpenter KC, Breslin WL, Davidson T, Adams A, McFarlin BK (2013) Baker's yeast beta-glucan supplementation increases monocytes and cytokines post-exercise: implications for infection risk? Br J Nutr 109:478–486. doi:10.1017/S0007114512001407
3. McFarlin BK, Mitchell JB, McFarlin MA, Steinhoff GM (2003) Repeated endurance exercise affects leukocyte number but not NK cell activity. Med Sci Sports Exerc 35:1130–1138. doi:10.1249/01.MSS.0000074463.36752.87
4. McFarlin BK, Williams RR, Venable AS, Dwyer KC, Haviland DL (2013) Image-based cytometry reveals three distinct subsets of activated

granulocytes based on phagocytosis and oxidative burst. Cytometry A 83:745–751. doi:10.1002/cyto.a.22330

5. Ortyn WE, Hall BE, George TC, Frost K, Basiji DA, Perry DJ, Zimmerman CA, Coder D, Morrissey PJ (2006) Sensitivity measurement and compensation in spectral imaging. Cytometry A 69:852–862. doi:10.1002/cyto.a.20306
6. Kong M, Tian S, Shi H, Zhao J, Feng X, Zheng S, Duan Z, Chen Y (2012) The effect of alpha-fetoprotein on the activation and phagocytosis of granulocytes and monocytes. Hepatogastroenterology 59:2385–2388. doi:10.5754/hge12296
7. Prakash PS, Caldwell CC, Lentsch AB, Pritts TA, Robinson BR (2012) Human microparticles generated during sepsis in patients with critical illness are neutrophil-derived and modulate the immune response. J Trauma Acute Care Surg 73:401–406. doi:10.1097/TA.0b013e31825a776d, discussion 406–407
8. Salih HR, Husfeld L, Adam D (2000) Simultaneous cytofluorometric measurement of phagocytosis, burst production and killing of human phagocytes using Candida albicans and Staphylococcus aureus as target organisms. Clin Microbiol Infect 6:251–258
9. Tsuji S, Iharada A, Taniuchi S, Hasui M, Kaneko K (2012) Increased production of nitric oxide by phagocytic stimulated neutrophils in patients with chronic granulomatous disease. J Pediatr Hematol Oncol 34:500–502. doi:10.1097/MPH.0b013e3182668388

Chapter 13

Using Image-Based Flow Cytometry to Assess Monocyte Oxidized LDL Phagocytosis Capacity

Andrea L. Henning, Adam S. Venable, Eric A. Prado, and Brian K. McFarlin

Abstract

The examination of monocyte phagocytosis of modified lipoproteins is important to the understanding of plaque deposition and the development of atherosclerosis. Current methods lack the high-throughput image-based analysis capabilities, which may yield novel information concerning monocyte activity in disease processes. Specifically, this method identifies monocyte phagocytosis of oxidized LDL along with a change in adhesion molecules and scavenger receptors. Our laboratory is currently implementing this method as a means to study how acute dietary modifications alter risk of developing atherosclerosis.

Key words Imaging flow cytometry, Monocyte activity, Macrophage, Adhesion molecules, Phagocytosis, Oxidized LDL, Atherosclerosis, Foam cells, FlowSight

1 Introduction

Atherosclerosis is the result of a process, which begins as protective mechanism against arterial tissue damage caused by the accumulation of oxidized low density lipoproteins (oxLDL) within the subendothelial space [1]. Circulating monocytes and macrophages are recruited to the subendothelial space to remove the oxLDL; however, phagocytosis of these modified lipoproteins results in foam cell formation [2]. These foam cells contribute to the continuous cycle of monocyte recruitment, phagocytosis, and formation of additional foam cells, eventually leading to plaque arterial plaque formation. An understanding of how monocytes/macrophages respond in the presence of oxLDL is a key component to understanding the progression of plaque formation. In humans, the examination of subendothelial macrophages is impractical due to its tissue location. For this reason, many studies have examined phenotypic changes in circulating monocytes and consider their response to the presence of modified lipoproteins a crucial element in the etiology of atherogenesis [3].

Natasha S. Barteneva and Ivan A. Vorobjev (eds.), *Imaging Flow Cytometry: Methods and Protocols*, Methods in Molecular Biology, vol. 1389, DOI 10.1007/978-1-4939-3302-0_13, © Springer Science+Business Media New York 2016

In order to understand the dynamics of monocyte/macrophage phagocytosis of oxLDL, it is important to also measure adhesion molecules and scavenger receptors. CD36 is a commonly measured scavenger receptor associated with oxLDL uptake [4, 5], while CD11a, CD11b, and CD18 are adhesion molecules that mediate the transmigration of monocytes into the subendothelial space [6, 7]. Current methods for the measurement of monocyte/macrophage phagocytosis of oxLDL include western blots, ELISAs, confocal laser-scanning microscopy, and fluorescence microscopy [8]. One limitation of existing methods is that it is not possible to multiplex similar functional receptors together and localize them to the same cell. Thus, the method described below was designed to leverage the power of image-based flow cytometry to measure multiple aspects of monocyte function at the same time. Another benefit of this method over other common approaches is the ability to have a very high-throughput approach for processing large numbers of patient samples within a reasonable amount of time.

2 Materials

2.1 Buffers

1. Dulbecco's phosphate buffered saline (PBS) (pH = 7.0).
2. PBS containing 1 % bovine serum albumin (BSA).

2.2 Solutions

1. Ethylenediaminetetraacetic acid solution (EDTA): Dilute in PBS to 5 mg/mL concentration.
2. 1-Step Fix/lyse solution (1×) (eBioscience, Inc.; San Diego, CA, USA).
3. Rosiglitazone: Dilute in sterile PBS to 114 μg/mL concentration (*see* **Note 1**).

2.3 Fluorescent Antibodies (See Note 2)

1. oxLDL-Dil (Life Technologies): Dilute in sterile PBS to 120 μg/mL concentration (*see* **Note 1**).
2. CD11a-FITC: DF 1:20.
3. CD11b-BV605: DF 1:5.
4. CD14-APC eFlour780: DF 1:20.
5. CD16-PE/Atto594: DF 1:20 (*see* **Note 3**).
6. CD18-PE/Cy5: DF 1:20.
7. CD36-eFlour660: DF 1:80 (*see* **Note 4**).

2.4 Equipment and Supplies

1. Heparinized tubes: 9 mL.
2. Flow Sight (EMD Millipore-Amnis, Seattle, WA). Equipped with violet laser (405 nm, 100 mW), blue laser (488 nm,

60 mW), red laser (642 nm, 75 mW), and side scatter laser (785 nm, 8.25 mW).

3. AeraSeal breathable sealing films (Excel Scientific, St. Louis, MO).

3 Methods

3.1 Sample Preparation and Treatment

1. Obtain human subjects approval and written subject consent prior to blood collection.
2. Using a blood collection needle, collect venous blood (8 mL) into an evacuated tube containing lithium heparin and process within 30 min (*see* **Note 5**).
3. Transfer 150 μL of whole blood into the bottom of 1.2 mL polypropylene tubes (*see* **Note 6**).
4. Add 25 μL of PBS to the "non-inhibitor" tubes and 25 μL of Rosiglitazone (14.296 ng/mL final concentration) to the "inhibitor" tubes and vortex the samples (*see* **Note 7**).
5. Add 25 μL of oxLDL-Dil (15 μg/mL final concentration) to all tubes. Vortex the samples and seal tubes with an AeraSeal breathable sealing film.
6. Incubate the samples for 1 h in a 37 °C, 5 % CO_2 humidified incubator (*see* **Note 8**).
7. After incubation, remove the plate sealer and add 20.8 μL of PBS and 29.2 μL of EDTA (584 μg/mL final concentration) to all samples. Cap the tubes and vortex the samples. Incubate for 15 min at room temperature in the dark.
8. Remove the caps and add 750 μL of PBS/BSA to all tubes. Recap the tubes and slowly invert the samples two or three times to ensure they are thoroughly mixed.
9. Centrifuge tubes for 10 min at 400 × *g*. Remove the samples from the centrifuge without disturbing the pellet at the bottom of the tube. Aspirate the supernatant from each sample, leaving about 150 μL in the tube.

3.2 Cell-Surface Antibody Staining

1. Vortex the samples to break up the pellet and add 50 μL of PBS to all tubes.
2. Add 60 μL of the prepared, diluted antibody cocktail to each sample. Cap and vortex the samples before incubating for 1 h on ice and in the dark.

3.3 WBC Fix and RBC Lyse

1. Add 750 μL of 1-step fix/lyse (1×) to all tubes. Recap and slowly invert the samples to mix.
2. Incubate the samples for at least 1 h on ice and in the dark (*see* **Note 9**).

3. Centrifuge tubes for 10 min at 400 × *g*. Remove the samples from the centrifuge without disturbing the pellet at the bottom of the tube. Aspirate the supernatant from each sample, leaving about 150 μL in the tube.
4. Add 750 μL of PBS and vortex them to break up the pellet and thoroughly mix the samples. Centrifuge them again for 10 min at 400 × g. Remove the samples from the centrifuge without disturbing the pellet at the bottom of the tube. Aspirate the supernatant from each sample, leaving about 150 μL in the tube (*see* **Note 10**).
5. Add 100 μL of PBS and vortex them to break up the pellet and thoroughly mix the samples. Add one drop of FlowSight calibration beads to each sample and vortex (*see* **Note 11**).

3.4 Acquisition

1. Transfer 150 μL of each sample to a 96 well, round bottom plate and cover with a pierceable seal (*see* **Note 12**).
2. Load the sample plate into the autosampler of the FlowSight and collect 20,000 monocyte ($CD14^+$) events (*see* **Note 13**).

3.5 Monocyte oxLDL Phagocytosis Analysis (See Note 14)

1. Open the raw image file (.rif) from a sample that did not include the uptake inhibitor and apply the predefined compensation matrix to yield a compensated image file (*.cif).
2. Identify single cells using a dot plot of bright-field aspect ratio vs. bright-field area (Fig 1a). Confirm the presence of only single cells by visual inspection.
3. Create a daughter plot from single cells of CD14 vs. side scatter (SSC) to positively identify monocytes ($CD14^+$) (Fig 1b).
4. Create a daughter plot from $CD14^+$ cells of oxLDL vs. CD14. Identify the negative population in the inhibitor treatment sample (Fig 1c) and set a gate to the right of this population to identify monocytes ($CD14^+$) that have ingested oxLDL in the non-inhibitor sample (Fig 1d).
5. Save the file as a template and use the batch-processing wizard to apply the template and compensation matrix to the .rif files for all samples (*see* **Note 15**).

4 Notes

1. Stock solutions should be prepared in a biological safety cabinet to maintain sterility.
2. All fluorescent antibodies were titrated in a series of experiments completed prior to this method with the exception of the oxLDL. A final concentration of 15 μg/mL for oxLDL is recommended by the manufacturer.

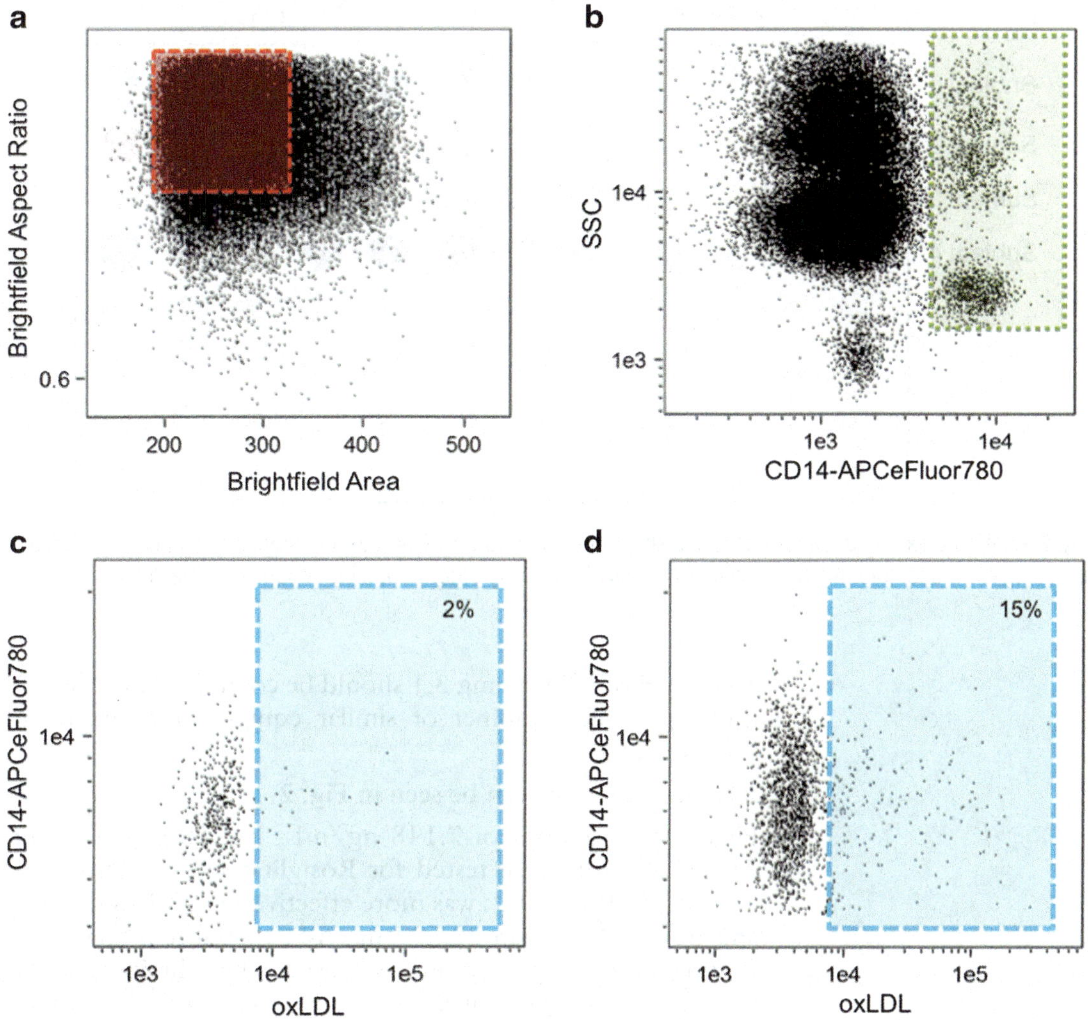

Fig. 1 Monocyte and phagocytosis gating. Using all collected events, single cells are identified using a dot plot of bright-field aspect ratio vs. bright-field area (**a**) followed by the identification of monocytes using a dot plot of CD14 vs. SSC (**b**) After the monocyte population is identified, a daughter plot is generated of oxLDL vs. CD14 for the "inhibitor" (**c**) and respective "non-inhibitor" (**d**) files. The "inhibitor" treated sample is used to set the position of the oxLDL-population. By applying this position to the "non-inhibitor" sample it is possible to identify $CD14^+/oxLDL^+$ events

3. Unconjugated CD16 was purchased from eBioscience and conjugated in our laboratory using a PE/Atto594 lighting-link kit from Innova Biosciences (Cambridge, UK). PE/Atto594 has a similar emission profile as PE/Texas Red.
4. Create an antibody cocktail by adding equal parts of each antibody to a light resistant tube. This will reduce variability by limiting the number of pipette transfers.

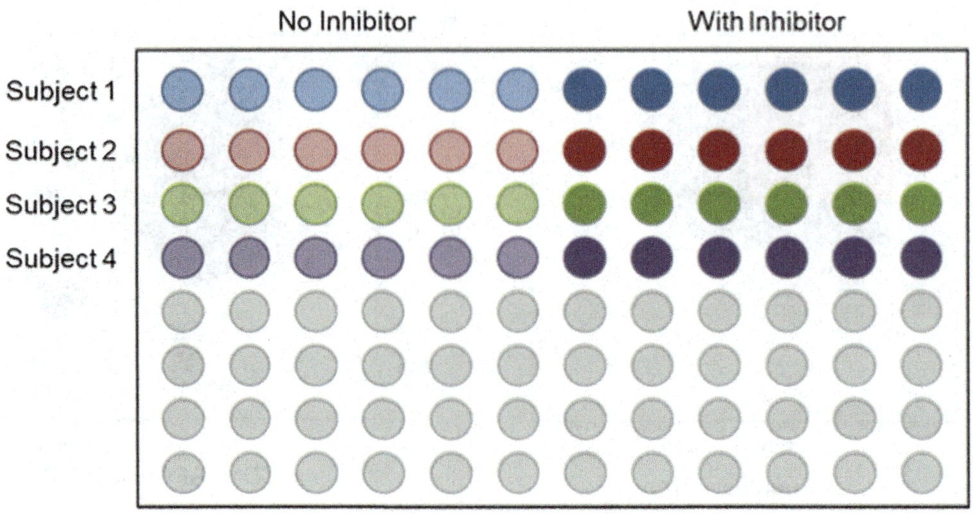

Fig. 2 Plate map example. Patient samples should be separated by treatment condition (inhibitor or no inhibitor) using rows and columns respectively. Organization of any other condition depends on the assay

5. **Steps 3–5** of Subheading 3.1 should be completed in a Class 2 Biological Safety cabinet or similar equipment to maintain assay sterility.
6. Plate map example can be seen in Fig. 2.
7. Final concentrations of 7.148 ng/mL, 14.296 ng/mL, and 28.592 ng/mL were tested for Rosiglitazone. A concentration of 14.296 ng/mL was more effective than 7.148 ng/mL and no different than 28.592 ng/mL, therefore a final concentration of 14.296 ng/mL was used. The inhibitor should be titered within your assay conditions to identify the correct dose.
8. Incubation times of 1, 2, and 4 h were compared and a similar oxLDL update was observed across all time points. Thus, for the purpose of this method, we used only a single 1 h assay incubation time. Additional testing may be needed to confirm the ideal incubation time for a given patient sample.
9. A minimum of 1 h is required for incubating the samples in the fix/lyse solution but an overnight incubation at 4 °C allows for optimal effectiveness of the fix/lyse solution resulting in cleaner samples while providing an opportunity for a break. If you are using a fix/lyse solution from another vendor, this step may have to be modified to reflect the requirements of that reagent.
10. A second wash will help remove excess debris from samples but it is not required. Acquisition is quicker and collected events look cleaner when the second wash is included.

11. FlowSight calibration beads assist the FlowSight in auto-validation when WBC counts are low without interfering with targeted measures.
12. Samples can be acquired manually using individual microcentrifuge tubes if your FlowSight is not equipped with an autosampler. We suggest using an autosampler if you are processing more than four patient samples at a time.
13. Bright-field images are collected in channels 1 and 9 respectively, CD11a-FITC is collected in channel 2, oxLDL-DiI is collected in channel 3, CD16-PE-Atto594 is collected in channel 4, CD18-PE-Cy5 is collected in channel 5, CD11b-BV605 is collected in channel 10, CD36-eFluor660 is collected in channel 11, and CD14-APC-eFluor780 is collected in channel 12.
14. Analysis is completed using IDEAS software v.6.1 (EMD Millipore).
15. The template may need to be adjusted relative to each patient sample to account for subtle differences between patient samples. If adjustments are made, save as a new template and batch the files for that subject using the modified template.

Acknowledgements

The development of this protocol was funded in part by a grant from the TACSM-SRDA to the University of North Texas on behalf of Ms. Henning. The authors were not directly compensated for their development of this method.

References

1. Ross R (1995) Cell biology of atherosclerosis. Annu Rev Physiol 57:791–804
2. Jialal I, Devaraj S (1996) The role of oxidized low density lipoprotein in atherogenesis. J Nutr 126:1053–1057
3. Zilversmit DB (1979) Atherogenesis: a postprandial phenomenon. Circulation 60:473–485
4. Ashraf MZ, Kar NS, Chen X, Choi J, Salomon RG, Febbraio M, Podrez EA (2008) Specific oxidized phospholipids inhibit scavenger receptor BI-mediated selective uptake of cholesteryl esters. J Biol Chem 283:10408–10414. doi:10.1074/jbc.M710474200
5. Podrez EA, Poliakov E, Shen Z, Zhang R, Deng Y, Sun M, Finton PJ, Shan L, Febbraio M, Hajjar DP, Silverstein RL, Hoff HF, Salomon RG, Hazen SL (2002) A novel family of atherogenic oxidized phospholipids promotes macrophage foam cell formation via the scavenger receptor CD36 and is enriched in atherosclerotic lesions. J Biol Chem 277:38517–38523. doi:10.1074/jbc.M205924200
6. Kostidou E, Koliakos G, Kaloyianni M (2009) Increased monocyte alphaL, alphaM, and beta2 integrin subunits in diabetes mellitus. Clin Biochem 42:634–640. doi:10.1016/j.clinbiochem.2008.12.007
7. Strohacker K, Breslin WL, Carpenter KC, Davidson TR, Agha NH, McFarlin BK (2012) Moderate-intensity, premeal cycling blunts postprandial increases in monocyte cell surface CD18 and CD11a and endothelial microparticles

following a high-fat meal in young adults. Appl Physiol Nutr Metab 37:530–539. doi:10.1139/h2012-034

8. Chiurchiu V, Lanuti M, Catanzaro G, Fezza F, Rapino C, Maccarrone M (2014) Detailed characterization of the endocannabinoid system in human macrophages and foam cells, and anti-inflammatory role of type-2 cannabinoid receptor. Atherosclerosis 233:55–63. doi:10.1016/j.atherosclerosis.2013.12.042

Chapter 14

Screening for Drugs Against the *Plasmodium falciparum* Digestive Vacuole by Imaging Flow Cytometry

Yan Quan Lee, Brian E. Hall, and Kevin S.W. Tan

Abstract

Phenotypic assays are increasingly employed to provide clues about drug mechanisms. In antimalarial drug screening, however, the majority of assays are designed to only measure parasite-killing activity. We describe here a high-content assay to detect drug-mediated perturbation of the digestive vacuole integrity in the trophozoite stage of *Plasmodium falciparum*, using the ImageStream imaging flow cytometer.

Key words Malaria, *Plasmodium falciparum*, Digestive vacuole, Drug discovery, Imaging flow cytometry, Programmed cell death, Apoptosis, Fluo-4

1 Introduction

The protozoan blood parasite *Plasmodium falciparum* is the major causative agent of malaria, a disease that threatens 3.4 billion people globally [1]. Traditionally, drug discovery efforts against the asexual blood stage of this parasite relied on growth inhibition assays, such as by assessing uptake of radiolabeled hypoxanthine, amount of *Plasmodium* lactate dehydrogenase, or reinvasion into fresh erythrocytes [2]. However, this approach does not provide information on the mode of drug action.

In the asexual blood stage of its life cycle, *P. falciparum* develops a large, unique organelle, the digestive vacuole [3]. This acidic compartment is essential to the survival of the parasite. In the digestive vacuole, 60–80 % of the host cell hemoglobin is digested, either to provide amino acids for parasite development and generate space for growth or to prevent premature lysis of the host erythrocyte [4–6]. Degradation of hemoglobin releases free heme, which is toxic to the parasite and sequestered as hemozoin in the digestive vacuole. Disrupting the integrity of the digestive vacuole results in the manifestation of apoptosis-like features in the parasite

Natasha S. Barteneva and Ivan A. Vorobjev (eds.), *Imaging Flow Cytometry: Methods and Protocols*, Methods in Molecular Biology, vol. 1389, DOI 10.1007/978-1-4939-3302-0_14, © Springer Science+Business Media New York 2016

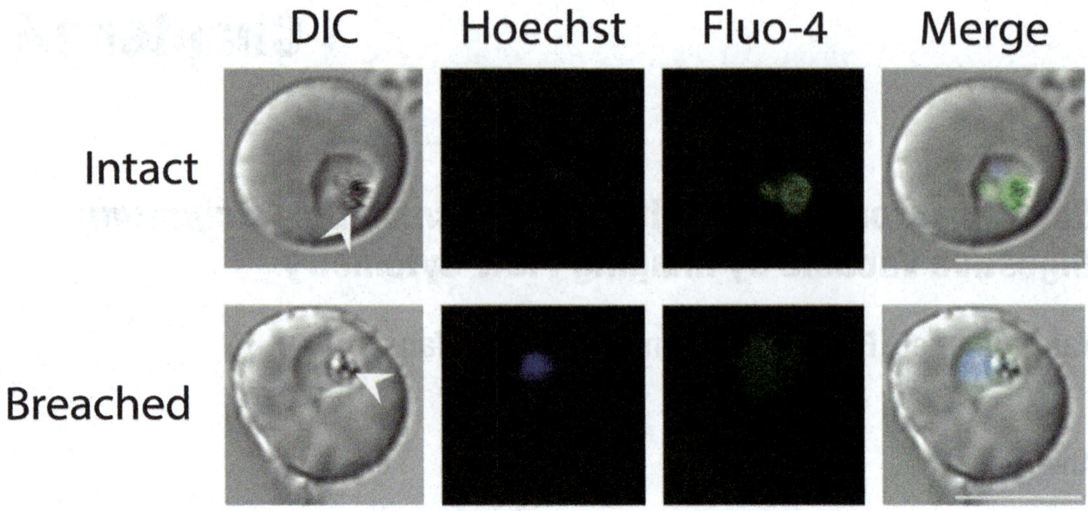

Fig. 1 Confocal micrographs of parasites stained with Fluo-4. Fluo-4 signal is redistributed to the parasite cytosol when the digestive vacuole is breached, resulting in an increased area of fluorescence. *Arrowheads* denote the hemozoin crystal, which is localized within the digestive vacuole. Scale bars represent 5 μm

[7–9]. Based on this, we have developed a high-content assay to screen for molecules that induce the permeabilization of the digestive vacuole [10]. The fluorescent calcium indicator dye Fluo-4 localizes to the digestive vacuole, a known store of Ca^{2+} in the parasite [11]. Upon lysis of the digestive vacuole, Fluo-4 is released into the parasite cytosol, resulting in an expansion of the fluorescent region (Fig. 1). By measuring the area of Fluo-4 signal with imaging flow cytometry, we were able to identify novel compounds that induce apoptosis-like death in the parasite.

2 Materials

2.1 Parasite Culture

1. Culture media: RPMI 1640 powder with phenol red and without HEPES and sodium bicarbonate (Life Technologies), 2.2 g/l sodium bicarbonate, 0.5 % (wt/vol) Albumax I (Life Technologies), 0.005 % (wt/vol) hypoxanthine, 0.03 % (wt/vol) L-glutamate, 25 μg/ml gentamicin. Dissolve hypoxanthine fully in a small volume of 1 M NaOH prior to addition; for 1 l of media, use 500 μl of NaOH to dissolve the required amount of hypoxanthine. Adjust the pH to 7.4 with NaOH or HCl. Sterilize media by filtration through a 0.22 μm membrane into a sterile container. Media may be stored at 4 °C for up to a month or until the pH changes (color of media becomes pink), whichever is sooner.

2. Red blood cells: These are stored at 4 °C for a period no longer than a month from the time of blood drawing.
3. Hypoxic gas: 3 % CO_2, 4 % O_2, 93 % N_2.

2.2 Giemsa Smear

1. Glass slides, grease-free.
2. 100 % methanol.
3. Giemsa stain: This is diluted to 15 % in tap water just prior to use; do not prepare large volumes of diluted stain for storage.

2.3 Treatment and Staining of Parasites

1. Sterile 96-well culture plates with lids, flat-bottom.
2. Hypoxic culture chamber.
3. Hoechst 33342: If purchased as a powder, dissolve in deionized water to obtain a 200 μg/ml stock solution. Sterilize with a 0.22 μm syringe filter and store at 4 °C in the dark.
4. Fluo-4, AM (Life Technologies). Dilute to 5 mM in DMSO prior to use.
5. Test compounds. If dissolved in DMSO, stock concentration should be at least 1000 times that of the screening concentration.

2.4 Imaging Flow Cytometry Data Acquisition and Analysis

1. Phosphate-buffered saline (PBS) for sheath fluid.
2. Amnis ImageStream system and INSPIRE software.
3. IDEAS analysis software.

3 Methods

3.1 Treatment of Parasites

Parasitemia of the culture used for the assay is relatively high to allow for the rapid acquisition of events during flow cytometry. Culture and synchronization techniques are described in detail elsewhere [12, 13] and will not be extensively discussed here. All steps are performed in an appropriate biological safety cabinet with proper aseptic techniques.

1. Culture synchronized parasites to approximately 10 % parasitemia (*see* **Note 1**).
2. Perform a thin Giemsa smear [13] to assess parasitemia and parasite stage. Parasites should be roughly 32 h post-invasion (hpi) at the start of the experiment (Fig. 2) (*see* **Note 2**). Transfer culture to a conical centrifuge tube and pellet cells by centrifugation at 800 × *g* for 5 min. Remove culture medium and resuspend in fresh medium to 2.5 % hematocrit. Excessive parasitemia can be adjusted to 10 % by the addition of fresh erythrocytes.

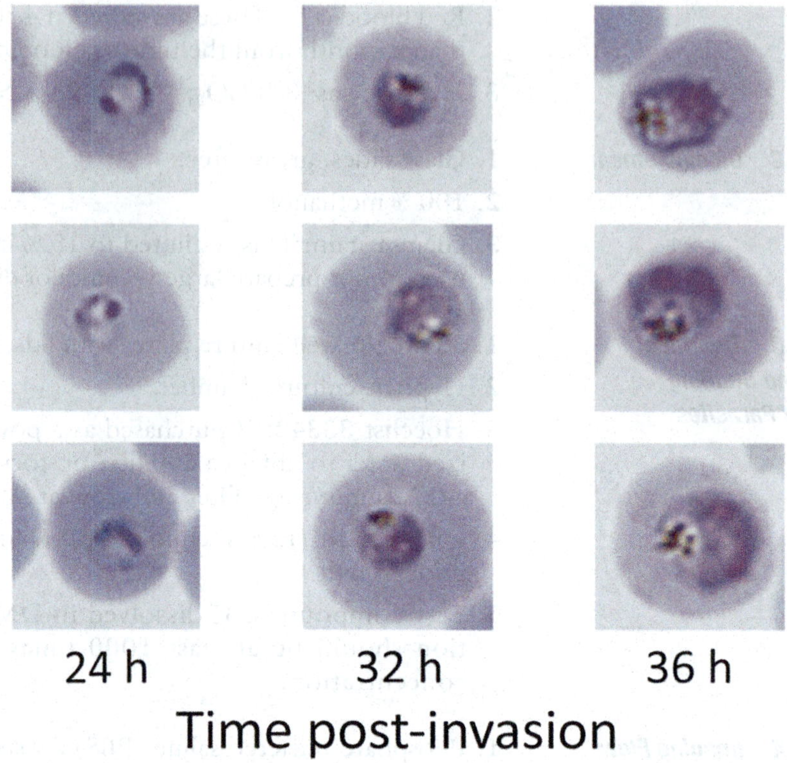

Fig. 2 Thin Giemsa smear of parasite culture. At the start of incubation with the test compounds, parasite stages should centre around 32 hpi, at the early to mid trophozoite stage. The parasite should occupy roughly a quarter of the area of the host erythrocyte, with prominent hemozoin pigment. Cultures with roughly 30 % of the parasites at the 24 or 36 hpi stages are considered well synchronized for this experiment; it is not necessary to obtain a very small window of synchrony

3. Prepare working concentrations of the test compounds by dilution in PBS. This concentration should be 10 times that of the test concentration, i.e., the concentration at which the screen is performed (typically 10 μM for the preliminary screen; a working concentration of 100 μM is prepared for this screen). If compounds were originally dissolved in dimethyl sulfoxide (DMSO), the final concentration of DMSO should not exceed 0.1 % v/v. For the positive control, prepare chloroquine at a working concentration of 30 μM by dissolving in PBS. Avoid exposing all compounds to light unless the compounds are known to be photostable.
4. In a sterile 96-well flat-bottomed microtiter plate, add 20 μl of the working concentration of each compound. Include a negative control (20 μl of PBS) and a positive control (20 μl of 30 μM chloroquine), as well as three additional wells for fluo-

rescence compensation controls (20 μl of PBS each). Add 180 μl of the culture prepared in **step 2** to each well and mix slowly to avoid generating bubbles (*see* **Note 3**)

5. Place the microtiter plate in a hypoxic chamber. Include a piece of tissue soaked in tap water in an open Petri dish to humidify the chamber. Flush the chamber with hypoxic gas for 1 min. Seal the chamber and incubate at 37 °C for 4 h (*see* **Notes 4** and **5**).

3.2 Staining of Parasites

1. Prepare a sufficient amount of staining solution by diluting both Fluo-4 AM to 1 μM and Hoechst 33342 to 1 μg/ml together in prewarmed culture medium. Also prepare 300 μl solutions of each stain individually in culture medium at the same concentrations; these are single-stain controls for fluorescence compensation.
2. Centrifuge microtiter plate at 800 × *g* for 1 min to pellet cells. Carefully aspirate supernatant so as to avoid disturbing the cell pellet. Wash cells by resuspending twice in 200 μl of prewarmed (37 °C) culture medium.
3. Pellet cells by centrifugation at 800 × *g* for 1 min. Resuspend cells in 200 μl of staining solution. For the three wells reserved for fluorescence compensation, resuspend the cell pellet in (a) culture media without stain, (b) culture media with Fluo-4 AM, and (c) culture media with Hoechst 33342.
4. Incubate the microtiter plate at 37 °C for 30 min.
5. Wash cells twice in 200 μl of PBS. Concentrate cells by resuspending in 100 μl of PBS. This is roughly 5 % hematocrit, barring cell loss from the washing steps. Proceed immediately to ImageStream data acquisition (*see* **Note 6**).

3.3 ImageStream Acquisition

1. Start the AMNIS ImageStream and the INSPIRE software as per the manufacturer's instructions, including the loading of SpeedBeads and automated calibration. The sheath fluid used should be PBS.
2. Acquire control samples for fluorescence compensation. Load the unstained samples. Use the Run/Setup function to make any minor adjustments such as the Core Tracking setting to centre cells in the field of view. Turn the bright field off by toggling the bright field on/off button in the Setup tab. Acquire at least 500 images of first the unstained cells, then the Fluo-4 single-stained cells, then the Hoechst 33342 single-stained cells for fluorescence compensation.
3. Turn the bright field back on. Choose a channel for the bright field apart from those used for the detection of Fluo-4 and Hoechst 33342 fluorescence (typically channels 3 and 2 respectively). Set the cell classifier to select for events that have

an area between 50 and 500 in the bright-field channel ("Object Area" lower limit and upper limit for the bright-field channel), and mean fluorescent intensity between 50 and 500 in the Fluo-4 channel ("Mean Intensity" lower limit and upper limit for channel 3). Acquire at least 10,000 cells per test sample (*see* **Note 7**).

3.4 Analysis with the IDEAS Software

1. Create a compensation matrix with the single-stained controls (file names are suffixed with "-noBF.rif"). Save the compensation matrix.
2. Open your negative control sample file (*.rif) and apply the compensation matrix.
3. Create a scatter plot with the *X*-axis as the area and the *Y*-axis as the aspect ratio of the bright-field object ("Area_brightfield" and "Aspect Ratio_brightfield" respectively). Apply a polygonal region, R1, on single round cells (*see* **Notes 8** and **9**). For erythrocytes, single cells have areas approximately between 40 and 90, while aspect ratio should be above 0.7 to gate for round cells (Fig. 3).
4. Create a histogram on region Rl with Gradient RMS of the bright field on the *X*-axis to select for images that are in focus. Identify images that are in focus by manually checking

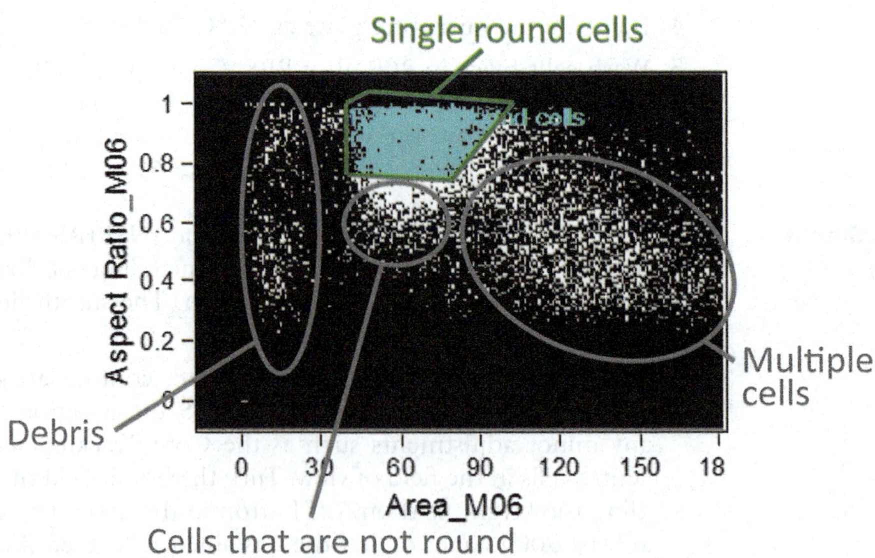

Fig. 3 Selecting for round single cells. The discoid shape of uninfected erythrocytes tends to produce images that have lower aspect ratios in the bright field, while infected erythrocytes at the trophozoite stage are rounder and thus have higher aspect ratios. Larger bright-field areas correspond with images of multiple cells, while smaller areas are typically debris

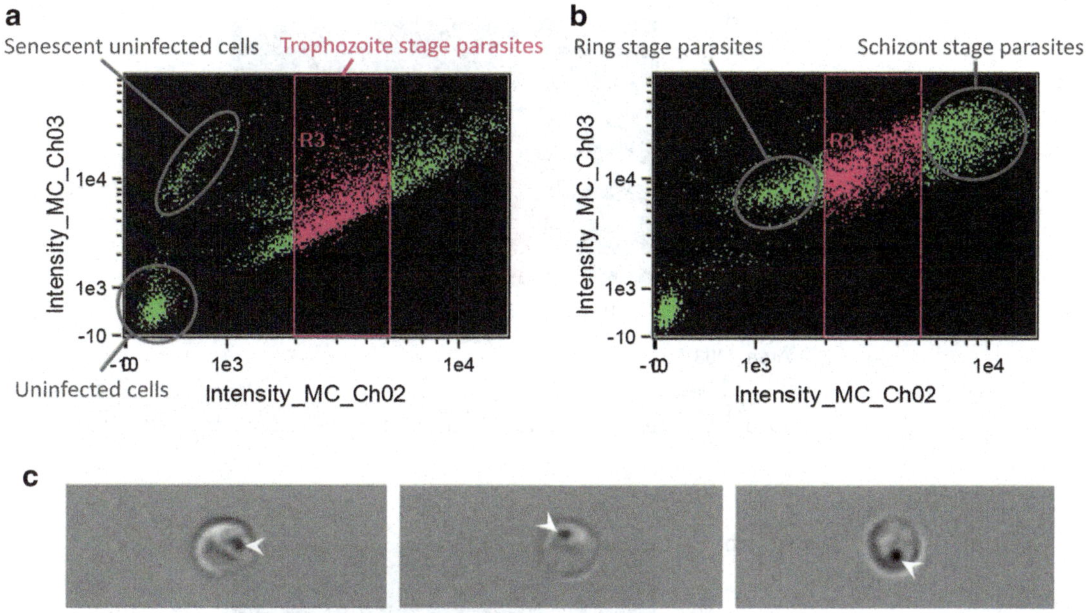

Fig. 4 Selecting for parasites in the mid-trophozoite stage. Hoechst 33342 is a cell-permeable dsDNA stain. Since mature erythrocytes do not contain DNA, only erythrocytes infected by the parasite will show Hoechst fluorescence. Hoechst fluorescence is collected in channel 2 and is presented on the *X*-axis, while Fluo-4 fluorescence is collected in channel 3 and presented on the *Y*-axis. (**a**) Uninfected erythrocytes typically contain low levels of intracellular calcium. The majority will therefore exhibit weak Fluo-4 signal. However, senescence or injury can result in an influx of Ca^{2+} [18], resulting in a small population of uninfected erythrocytes with high Fluo-4 signal. (**b**) Some parasite strains do not maintain synchrony well. This set of data was obtained from a clinical isolate, showing a larger spread of parasite stages. Despite this, gating may still be performed to select the right stage for analysis. Although the starting parasitemia was 10 %, uninfected erythrocytes were only a small proportion of the cell population here; this is due to the exclusion of uninfected erythrocytes by the Cell Classifier option (mean intensity of Fluo-4 channel) during acquisition. (**c**) Representative bright-field images of trophozoite-stage parasites captured on the ImageStream. Note the prominent hemozoin pigment (denoted by *arrowheads*) used as an identifier of the trophozoite stage

each bin for image quality; start with the bin with the lowest Gradient RMS value. Create a linear region, R2 from the lowest acceptable Gradient RMS value to the maximum to gate for cells which are in focus (*see* **Note 10**).

5. Generate a scatter plot on region R2 with intensity of the Hoechst 33342 channel on the *X*-axis, and intensity of the Fluo-4 channel on the *Y*-axis (Fig. 4). Create a linear region, R3 along the *X*-axis to select for cells in the trophozoite stage (*see* **Note 11**).
6. Create a histogram on region R3 with the area of the Fluo-4 channel on the *X*-axis (Fig. 5). Click on the Statistics button and display the mean of the Fluo-4 area.
7. Save the analysis template (*.ast file) to apply to other samples.

a

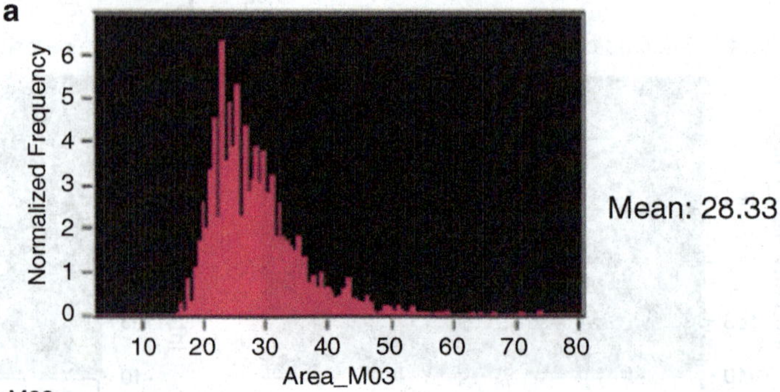

Area_M03

Population	Mean	Median	Std. Dev.
R3 & R2 single focused cell & R1 Single round cells	28.33	26.75	7.50

b

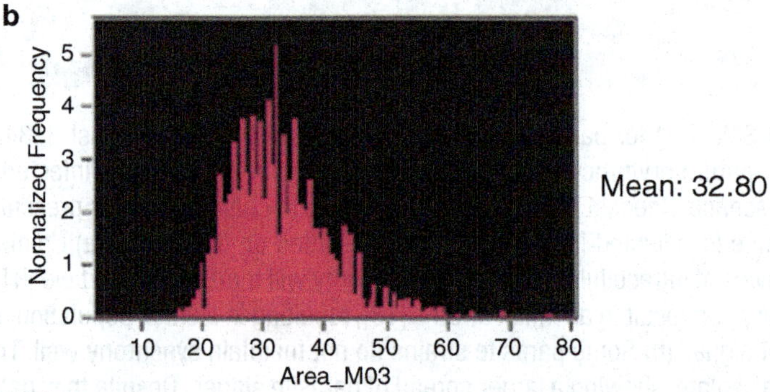

Area_M03

Population	Mean	Median	Std. Dev.
R3 & R2 single focused cell & R1 Single round cells	32.80	31.75	8.39

Fig. 5 Effect of compounds on mean Fluo-4 area in a clinical isolate. (**a**) PBS-treated negative control exhibited a small mean area of Fluo-4 signal. (**b**) Treatment of the parasites with 3 μM of chloroquine resulted in an increase in the mean Fluo-4 fluorescence area. Note that the difference in mean area was still detectable despite the poorer synchrony of the parasite culture (shown in Fig. 3)

8. Apply the compensation matrix and the analysis template to all your samples by batch processing, under the Tools menu. Generate a statistics report file of the mean area of the Fluo-4 channel for all your samples.
9. Perform at least two other replicate experiments. Statistically compare the mean Fluo-4 areas of the samples treated with the test compounds to that of the negative control by the paired samples *t*-test (*see* **Note 12**). Select hits that are statistically significant for a secondary screen at a lower test concentration (typically tenfold less than the primary screen) or validation by confocal microscopy.

4 Notes

1. It is easier to achieve a high parasitemia with cultures at a lower hematocrit; maintaining the culture in 2.5 or 1.25 % hematocrit reduces the risk of the culture crashing from overgrowth. Parasite strains differ in their sensitivity to high parasitemias; clinical isolates are typically less tolerant of culturing to higher parasitemia until a period of in vitro culture adaption has passed. This is mitigated by timing media changes, as described in [14].
2. Parasite growth rates differ across strains. The 36 hpi stage for the 3D7 strain may correspond with the 32 hpi stage in the K1 strain, for example. The growth rate has to be empirically determined and parasites should be at a stage morphologically similar to that of 32 hpi depicted in Fig. 2. Maintenance of synchrony also varies between strains, in that some strains remain synchronized for a longer duration than others. Some strains may therefore require multiple rounds of synchronization before use. Placing cultures in a shaking incubator may improve synchrony [15], but we have found that while shaking works well for 3D7, some strains such as K1 and Dd2 cannot tolerate much agitation.
3. If you have a large number of compounds to screen, it is advisable to include additional negative and positive control wells, to account for time-related effects on the results. Run the controls at the start and the end of the experiment to determine if this is a problem.
4. Longer treatment times are not recommended. The negative control parasites continue developing during this period, and the size of the digestive vacuole increases. This may confound later analysis as the relative effect size of a hit diminishes. Late-stage parasites are also reported to increase the uptake of Ca^{2+} [16].
5. The hypoxic chamber is also known as a microisolator [13].
6. Cells should never be placed on ice, as this results in the rapid diffusion of Fluo-4 fluorescence out of the digestive vacuole into the parasite cytosol. Fixation of cells should also be avoided as all Fluo-4 fluorescence will be lost. Imaging flow cytometry should therefore be conducted as promptly as possible after staining.
7. Ca^{2+}-bound Fluo-4 peak excitation/emission wavelengths are 494/506 nm. DNA-bound Hoechst 33342 peak excitation/emission wavelengths are 350/461 nm. Variations in equipment setup may require different channels to be used. The cell classifier of 50–500 for mean intensity of the Fluo-4 channel

was optimized with the ImageStream 100; this range may have to be re-optimized for newer models with a higher intensity resolution.

8. The area of the acquired image works in a similar way to the forward scatter in conventional flow cytometry, allowing us to exclude debris by size. Aspect ratio is the ratio of the minor axis to the major axis of an ellipse of best fit bounding the image. Doublet events fit a more elongated ellipse, and therefore have lower aspect ratios. Round cells fit a more circular ellipse, and their aspect ratios are therefore higher. A perfectly circular image would have an aspect ratio of 1, as the minor axis is equal to the major axis.
9. Check if the region has been correctly bounded by clicking on individual dots on the scatter plot, which will bring up the image of the object corresponding to that dot. Work your way along the borders of the region to verify that the region contains only single round cells, not doublets or debris.
10. Choosing the threshold for acceptable focus is a rather subjective exercise. We find that typically, the large majority of cells will be in good focus due to the autofocus function of the ImageStream, so this step has only a marginal effect on the analysis.
11. Infected erythrocytes, owing to the parasite DNA, will be stained with Hoechst 33342. This means that all cells in region R3 are infected. Excluding cells that are at the ring stage thus is a trivial task of gating cells with visible hemozoin pigment, seen as a black dot in the bright field. This sets the lower limit for region R3. Schizonts are excluded by selecting for cells without distinct multiple nuclei, visualized in the Hoechst 33342 channel; this is thus the upper limit for the region.
12. The large number of events acquired allows us to assume that the sample mean approximately follows a normal distribution; this is described by the central limit theorem [17].

References

1. World Health Organization (2013) World Malaria Report 2013. World Health Organization, Geneva, Switzerland
2. Fidock DA, Rosenthal PJ, Croft SL et al (2004) Antimalarial drug discovery: efficacy models for compound screening. Nat Rev Drug Discov 3:509–520. doi:10.1038/nrd1416
3. Bannister L, Hopkins J, Fowler R et al (2000) A brief illustrated guide to the ultrastructure of plasmodium falciparum asexual blood stages. Parasitol Today 16:427–433. doi:10.1016/S0169-4758(00)01755-5
4. Lew VL, Macdonald L, Ginsburg H et al (2004) Excess haemoglobin digestion by malaria parasites: a strategy to prevent premature host cell lysis. Blood Cells Mol Dis 32:353–359. doi:10.1016/j.bcmd.2004.01.006
5. Krugliak M, Zhang J, Ginsburg H (2002) Intraerythrocytic Plasmodium falciparum utilizes only a fraction of the amino acids

derived from the digestion of host cell cytosol for the biosynthesis of its proteins. Mol Biochem Parasitol 119:249–256. doi:10.1016/S0166-6851(01)00427-3

6. Francis SE, Sullivan DJ, Goldberg ED (1997) Hemoglobin metabolism in the malaria parasite Plasmodium falciparum. Annu Rev Microbiol 51:97–123. doi:10.1146/annurev.micro.51.1.97
7. Ch'ng J-H, Kotturi SR, Chong AG-L et al (2010) A programmed cell death pathway in the malaria parasite *Plasmodium falciparum* has general features of mammalian apoptosis but is mediated by clan CA cysteine proteases. Cell Death Dis 1:e26. doi:10.1038/cddis.2010.2
8. Ch'ng J-H, Liew K, Goh AS-P et al (2011) Drug-induced permeabilization of parasite's digestive vacuole is a key trigger of programmed cell death in *Plasmodium falciparum*. Cell Death Dis 2:e216. doi:10.1038/cddis.2011.97
9. Ch'ng J-H, Yeo S-P, Shyong-Wei Tan K (2012) Can a single "powerless" mitochondrion in the malaria parasite contribute to parasite programmed cell death in the asexual stages? Mitochondrion. doi:10.1016/j.mito.2012.10.014
10. Lee YQ, Goh ASP, Ch'ng JH et al (2014) A high-content phenotypic screen reveals the disruptive potency of quinacrine and 3′,4′-Dichlorobenzamil on the digestive vacuole of Plasmodium falciparum. Antimicrob Agents Chemother 58:550–558. doi:10.1128/AAC.01441-13
11. Biagini GA, Bray PG, Spiller DG et al (2003) The digestive food vacuole of the malaria parasite is a dynamic intracellular Ca2+ store. J Biol Chem 278:27910–27915. doi:10.1074/jbc.M304193200
12. Moll K, Ljungström I, Perlmann H et al (2008) Methods in malaria research, 5th edn. MR4 ATCC, BioMalPar, Manassas, VA
13. Maier AG, Rug M (2012) In vitro culturing Plasmodium falciparum erythrocytic stages. In: Ménard R (ed) Malaria, Methods and protocols. Humana, New York, pp 3–15
14. Radfar A, Méndez D, Moneriz C et al (2009) Synchronous culture of *Plasmodium falciparum* at high parasitemia levels. Nat Protoc 4:1899–1915. doi:10.1038/nprot.2009.198
15. Allen RJW, Kirk K (2010) *Plasmodium falciparum* culture: the benefits of shaking. Mol Biochem Parasitol 169:63–65. doi:10.1016/j.molbiopara.2009.09.005
16. Glushakova S, Lizunov V, Blank PS et al (2013) Cytoplasmic free Ca^{2+} is essential for multiple steps in malaria parasite egress from infected erythrocytes. Malar J 12:41. doi:10.1186/1475-2875-12-41
17. Fischer H (2011) History of the central limit theorem: from classical to modern probability theory. Springer, New York
18. Lang F, Lang E, Föller M (2012) Physiology and pathophysiology of eryptosis. Transfus Med Hemoth 39:308–314. doi:10.1159/000342534

Part IV

IFC in Studies of Cell Death and Apoptosis

Chapter 15

Accurate Assessment of Cell Death by Imaging Flow Cytometry

Aja M. Rieger and Daniel R. Barreda

Abstract

The number of investigators using cell death analysis applications has greatly expanded since the introduction of flow cytometry. The Annexin V/propidium iodide (PI) method is among the most commonly used procedures and allows users to determine if cells are viable, apoptotic, or necrotic, based on changes in membrane lipid composition, integrity, and permeability. Unfortunately, PI can intercalate into RNA, in addition to DNA, which contributes to a large number of events showing PI staining within the cytoplasmic compartment. We show that this occurs across a broad range of animal primary cells and commonly used cell lines, and is most prevalent in large cells (nuclear:cytoplasmic ratio <0.5). Any cellular system where RNA levels change throughout an experiment will be particularly affected, such as those that utilize virally infected cells. As two examples, we highlight our recent work on cells infected with vesicular stomatitis virus (VSV), an RNA virus, and herpes simplex virus-1 (HSV-1), a DNA virus. Similarly, these issues are relevant to experimental systems where cells have increased RNA content such as during genotoxic stress, following exposure to cell cycle arrest drugs such as thymidine or hydroxyurea, or where developmental progression promotes discrete changes in cellular RNA synthesis. This chapter outlines a modified Annexin V/PI method that addresses cytoplasmic RNA staining issues to allow for accurate assessment of cell death. This protocol takes advantage of an additional cellular permeability caused by fixation to promote RNase A entry into the cell. Based on our observations, cell morphological parameters are well maintained and less than 5 % of total cellular events exhibit cytoplasmic PI staining under this protocol.

Key words Apoptosis, Cell death, Propidium iodide, Annexin V, Necrosis, RNA staining

1 Introduction

Analysis of cell death is central to many fields, and has become increasingly used since the advent of flow cytometry. While there are now multiple identified forms of cell death, this chapter focuses on the detection of apoptosis and necrosis [1]. Apoptosis is a programmed form of cell death, triggered by a number of physiological signals including DNA damage, cell damage, or immunological signals [2]. Necrosis, on the other hand, is an uncontrolled, premature cell death that involves autolysis [2]. Necrotic cells are also characterized by

Natasha S. Barteneva and Ivan A. Vorobjev (eds.), *Imaging Flow Cytometry: Methods and Protocols*, Methods in Molecular Biology, vol. 1389, DOI 10.1007/978-1-4939-3302-0_15, © Springer Science+Business Media New York 2016

mitochondrial dysfunction which promotes disruption of nuclear membrane integrity [3]. This process can result in damage and loss of surrounding cells and activation of the immune system.

Apoptosis is characterized by key morphological features [4]. Changes in membrane composition are one of the first characteristics that can be measured—namely by the presence of phosphatidylserine on the extracellular membrane, which can be detected by Annexin V binding [5]. In living cells, phosphatidylserine is generally sequestered on the inner leaflet, preventing extracellular Annexin V from binding. As the apoptotic cascade continues, additional morphological features develop, including the condensation of the cytoplasm and nucleus, cleavage of proteins, and cleavage of DNA [6].

Necrosis is readily detected by membrane impermeant DNA dyes. Of these, propidium iodide (PI) is among the most cost effective and commonly utilized. When used in combination with Annexin V, users are able to distinguish between live, apoptotic, and necrotic cells based on differences in membrane permeability [5, 7]. Live cells are identified by a lack of staining with Annexin V and PI ($AnxV^-$/PI^-); apoptotic cells are stained with Annexin V by lack PI staining ($AnxV^+$/PI^-); necrosis is identified by PI staining with no Annexin V staining ($AnxV^-$/PI^+); late apoptotic/secondary necrotic cell death is marked by staining with both Annexin V and PI ($AnxV^+$/PI^+) [5, 7–9]. During late apoptotic/secondary necrotic cell death, there is increased permeability of nuclear pores, which allows large proteins and complexes to enter [9]. Beyond the analysis of cell death, PI has been used in measurements of cell cycling and tumor ploidy.

Early studies that predate flow cytometry-based assays showed that PI, in addition to binding DNA, could also intercalate into double-stranded regions of RNA [10]. This led to inclusion of RNase in microscopy-based PI procedures to increase the specificity of PI for DNA [10, 11]. Unfortunately, RNase treatment has not been included in widely used flow cytometry-associated protocols and in protocols from major developers including Molecular Probes, BD Biosciences, eBioscience, Abcam, MBL international, Santa Cruz, Miltenyi Biotech, and Roche Applied Science. As a result, since the publication of the first Annexin V/PI protocol in 1995 [5], only 3 of 1019 publications include an RNase step [12]. The loss of this important step may stem from perceived changes in staining due to fixation or because RNase does not readily enter live cells.

Fig. 1 (continued) (**a**) Representative analysis shows procedure for classification of primary kidney macrophage cells into true positive (R4) and false positive (R5) populations. Using a mask that highlighted the nuclear area, based on the pixel intensity of the DRAQ5 stain, the degree of colocalization between DRAQ5 and PI was determined. (**b**) All cell types tested exhibited false-positive PI staining to varying degrees. To allow for easier visual determination of PI and DRAQ5 colocalization, the DRAQ5 stain was given a green pseudocolor. Areas where PI and DRAQ5 colocalize appear *yellow*. Cells with a high degree of colocalization were characterized as true positive. Cells with a low degree of colocalization were characterized as false positive. Chicken blastodermal cells, $n=5$; goldfish PKM, $n=3$; porcine lung and spleen macrophages, $n=3$; murine splenocytes and bone marrow (BM) macrophages, $n=2$; Jurkat T-cells, $n=3$; RAW macrophages, $n=3$. Modified figure from Rieger et al. (2010) [15]

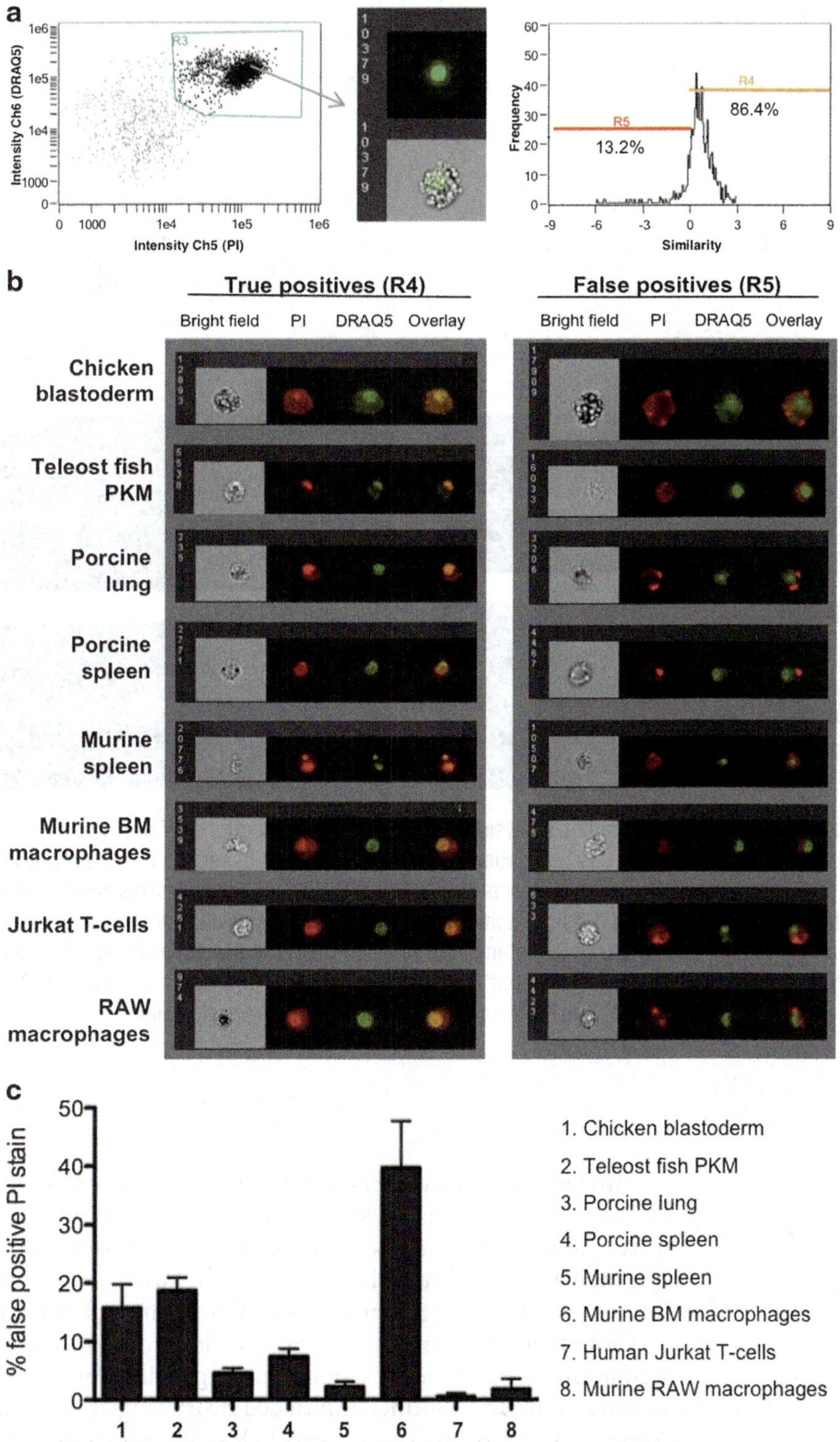

Fig. 1 Primary cells and commonly used cell lines exhibit false-positive PI staining. Primary cells and cell lines were harvested and stained with propidium iodide and DRAQ5. Primary cells included chicken blastodermal cells, goldfish primary kidney macrophages (PKM), porcine lung macrophages, porcine spleen macrophages, murine splenocytes, and bone marrow (BM) macrophages. Cell lines included human Jurkat T-cells and murine RAW macrophages. Imagery was acquired by an ImageStream multi-spectral imaging flow cytometer.

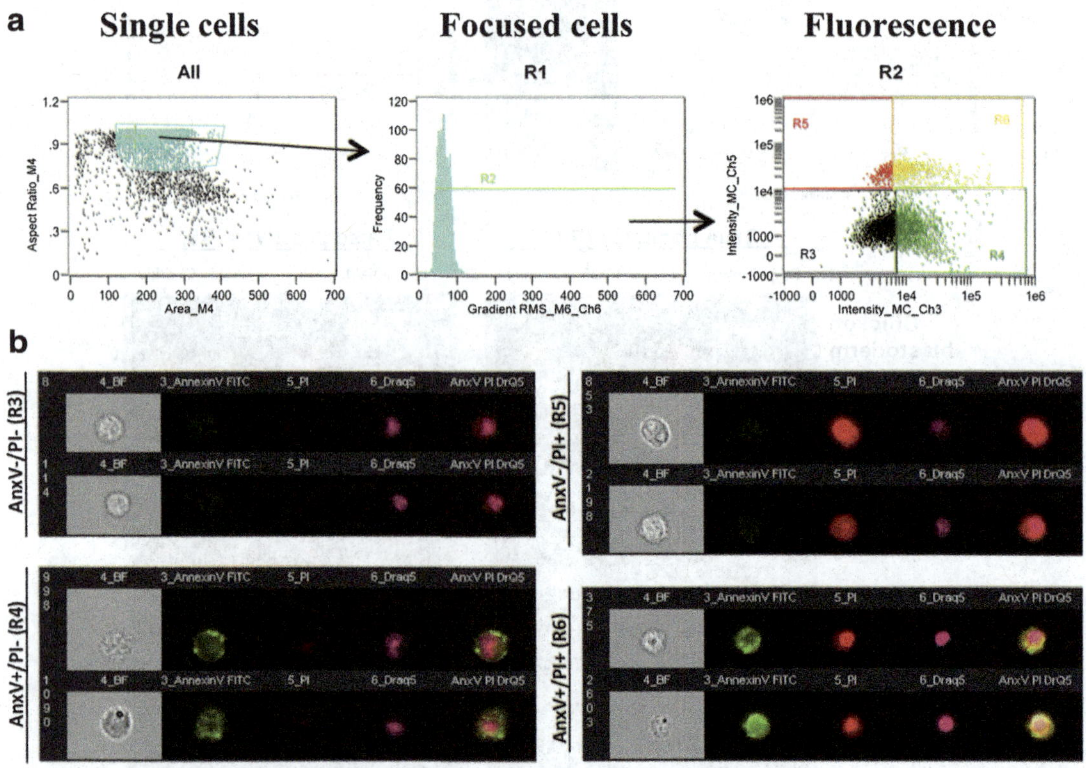

Fig. 2 Gating strategy for ImageStream analysis. (**a**) To analyze cell death, cells are first gated to remove clumps and debris (*far left*). Using this population, cells are then gated based on focus using gradient RMS (*middle*). To create this gate, progressively selects bins to find identify focused cells. This population represents the single, focused cell population. This population is used for all downstream analyses. To determine cell death, a *dot plot* denotes the fluorescence intensity of Annexin V versus the fluorescence intensity of PI. (**b**) Representative images correspond to cells that are Annexin V$^-$/PI$^-$ (healthy cells), Annexin V$^+$/PI$^-$ (apoptotic cells), Annexin V$^-$/PI$^+$ (necrotic cells), and Annexin V$^+$/PI$^+$ (late apoptotic/necrotic cells)

Imaging flow cytometry is a relatively new advance in the field of flow cytometry that is starting to hit its stride with regards to novel applications and complementarity to conventional flow cytometry [13]. In this chapter, we use an ImageStream multi-spectral flow cytometer (Amnis Corporation, EMD Millipore) to show that conventional Annexin V/PI protocols contribute to significant cytoplasmic PI staining among cellular events, thus preventing accurate quantification of cell death (Fig. 1). The modified Annexin V/PI protocol outlined below incorporates fixation and RNase treatment to remove cytoplasmic PI staining (Figs. 2 and 3). These modifications greatly reduce the number of events showing cytoplasmic PI staining in a range of cell types while having no

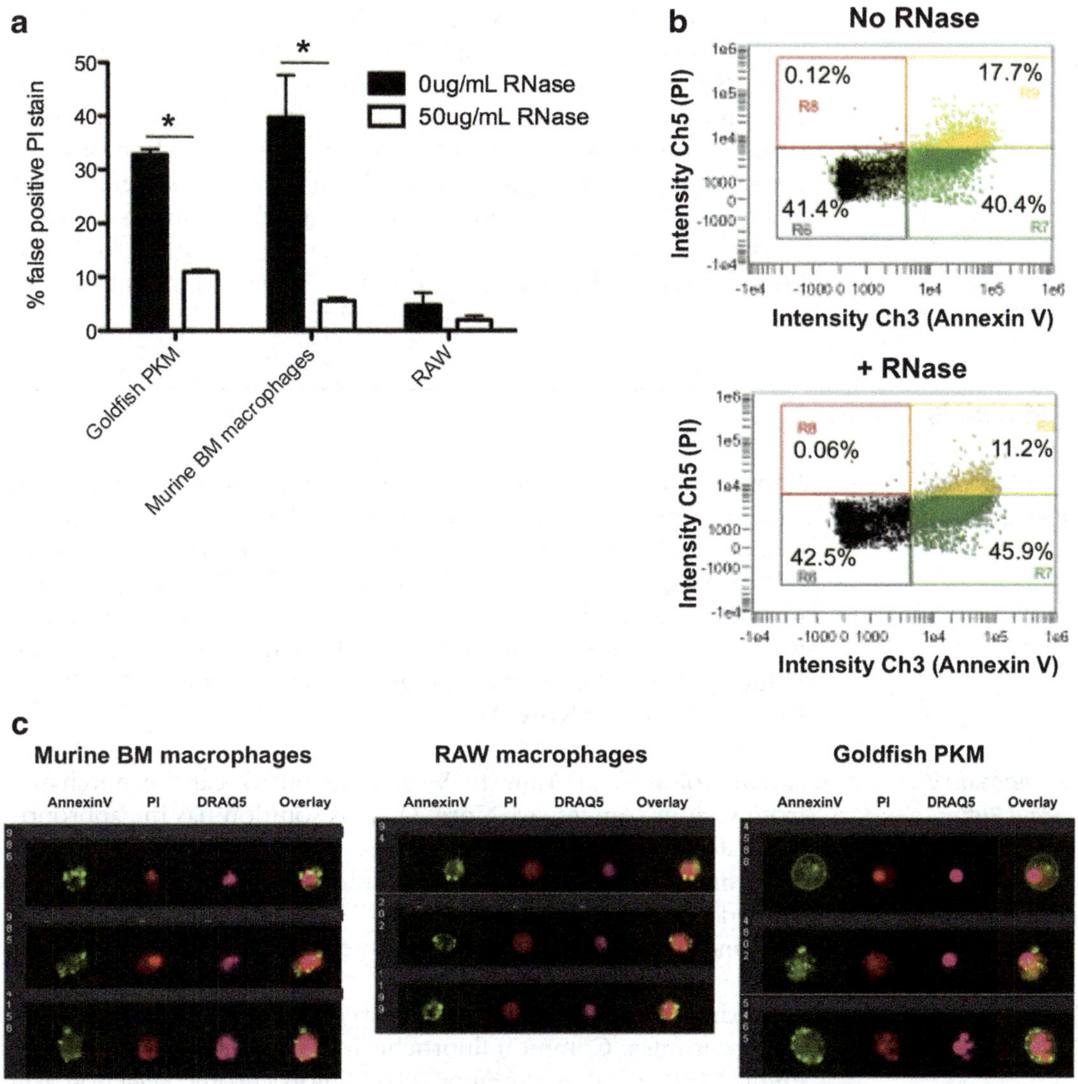

Fig. 3 Improving the Annexin V/PI based assessment of cell death across a range of experimental models using imaging flow cytometry. (**a**) Cells were fixed with 1 % formaldehyde and treated ±50 μg/mL DNase-free RNase for 15 min, according to our modified protocol. A conventional staining protocol leads to marked PI cytoplasmic staining, whereas modification of the protocol and inclusion of RNase A decreases cytoplasmic RNA staining. (**b**) Scatterplots depicting Annexin V/PI stain of goldfish primary macrophages (PKM). Following RNase treatment, there was a decrease in the percentage of PI$^+$ cells due to removal of false-positive events. Note: the decrease in false positive events is shown in the context of the whole population of cells (PI$^+$ and PI$^-$). (**c**) Representative images of goldfish PKM, murine bone marrow macrophages, and RAW cells stained using the modified Annexin V/PI protocol. Using this protocol, cytoplasmic PI staining is reduced while intensity/pattern of Annexin V stain is unchanged. Modified figure from Rieger et al. (2010) [15]

detectable impact on the staining of Annexin V or nuclear PI. This chapter focuses largely on application of this protocol under the ImageStream platform; however, this protocol is also amenable to conventional flow cytometry (Fig. 4b), microscopy, and microplate-based approaches. Importantly, while these modifications are essential to accurately quantify cell death in normal systems, these changes are especially important in biological systems that display marked changes in RNA content. As one example we highlight how changes in RNA content during viral infections can impact accurate measurement of cell death (Fig. 4).

2 Materials

2.1 Phosphate-Buffered Saline

Phosphate-buffered saline (PBS) is an isotonic solution made with sodium phosphate and sodium chloride. The formulation we use lacks calcium and magnesium. A 10× solution is prepared with: KCl–2.00 g, KH_2PO_4–2.00 g, NaCl–80.00 g, $Na_2HPO_4.7H_2O$–21.60 g, top to 1 L with ultrapure water (prepared by purifying deionized water to attain a sensitivity of 18.2 MΩ cm at 25 °C) (*see* **Note 1**). Diluting 10× solution with ultrapure water creates a 1× $PBS^{-/-}$ stock solution (*see* **Note 2**).

2.2 Annexin V Binding Buffer

A 10× solution of Annexin V binding buffer can be purchased from various sources (*see* **Note 3**). This solution has the appropriate calcium and magnesium concentrations for staining with Annexin V (*see* **Note 4**). Prior to staining, a 10× solution should be diluted to 1× with ultrapure water. This solution should be used cold and made fresh for each experiment.

2.3 Annexin V

Annexin V can be purchased complexed with a number of different fluorochromes. Common fluorochromes include FITC, PE, APC, eFluor®450, or PerCP-eFluor®710. Fluorochrome selection will depend on the laser configuration of the flow cytometer or microscope platform used, and the fluorochromes of any other stains being used in the experiment.

2.4 Propidium Iodide

PI intercalates into double-stranded nucleic acids (RNA or DNA). This dye cannot penetrate living membranes, so it is only incorporated into cells with disrupted membranes. PI can be purchased from many vendors.

2.5 Formaldehyde

Formaldehyde is a commonly used tissue fixative and embalming agent. Many vendors supply formaldehyde at a 37 % solution. To create an isotonic 2 % stock solution, add 54 mL 37 % formaldehyde to 100 mL 10× $PBS^{-/-}$ and top to 1 L with ultrapure water. This solution can be stored a room temperature for a number of months.

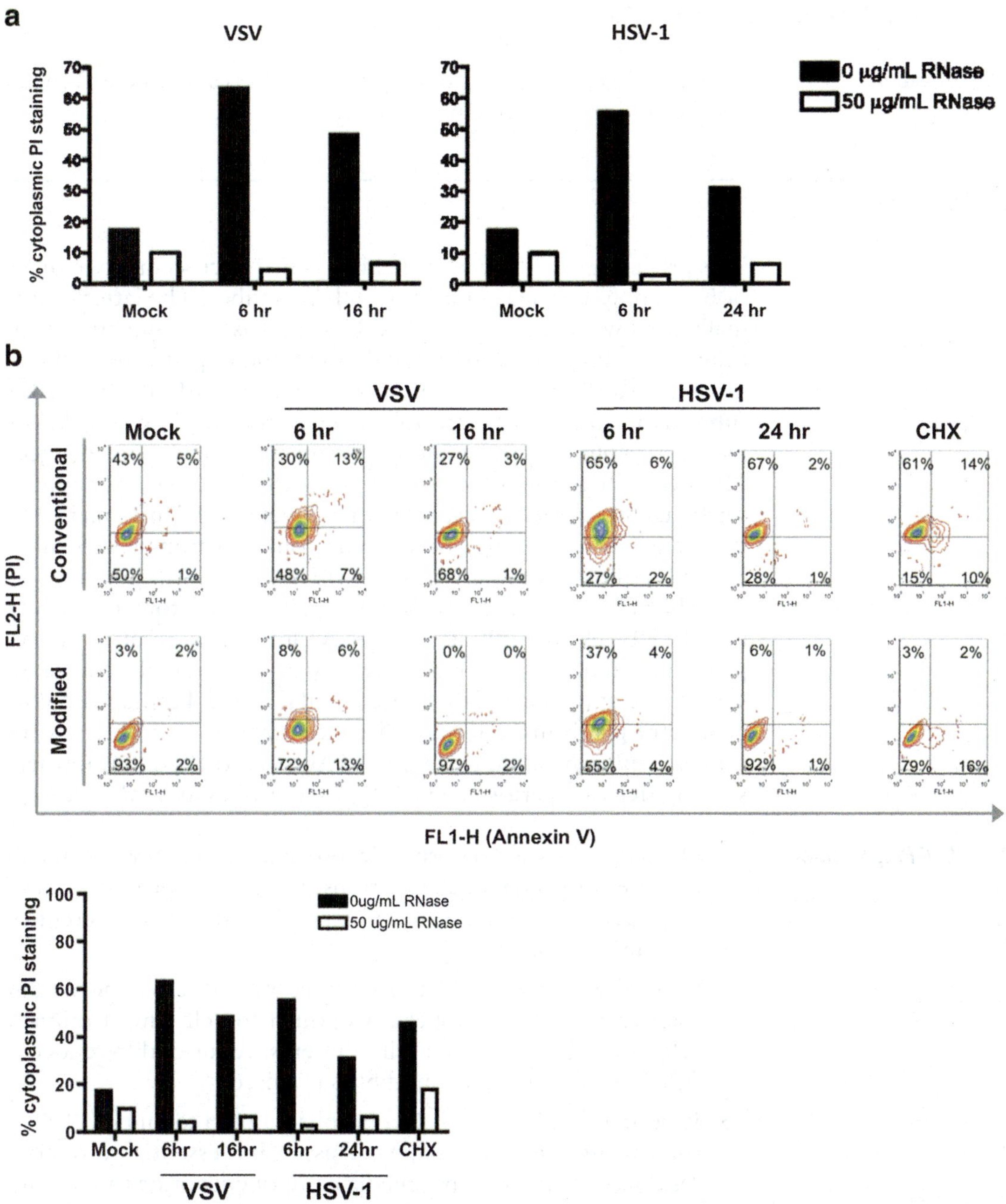

Fig. 4 Application of modified Annexin V/PI protocol for examination of cell death in virally infected cells. (**a**) Application of modified Annexin V/PI protocol to virally infected cells using an ImageStream platform. Vero cells were infected with vesicular stomatitis virus (VSV) or herpes simplex virus (HSV-1) with a multiplicity of infection of five. Cells were harvested at the indicated time points and stained using either the conventional Annexin V/PI protocol (no fixation, no RNase) or with the modified Annexin V/PI protocol (with fixation and RNase addition). Viral infection increases the number of false positive PI events detected. Treatment with RNase greatly reduces the percentage of cells with cytoplasmic PI staining. (**b**) Application of modified Annexin V/PI protocol to virally infected cells using conventional flow cytometry platform. Vero cells were infected with VSV or HSV1 and stained using the conventional Annexin V/PI protocol or the modified Annexin V/PI protocol, as described above. Cell death was analyzed on a BD FACSCanto II. Conventional flow cytometry analysis also shows that RNase treatment reduces cytoplasmic PI staining associated with increased RNA content from viral infection

2.6 RNase A

RNase A was purchased from Sigma-Aldrich (catalogue # R4642) and derived from bovine pancreas (*see* **Note 5**).

3 Methods

This procedure can be performed in 1.5 mL microcentrifuge tubes or 5 mL polystyrene round-bottom FACS tubes. The latter is normally used when carrying out viability analysis in conjunction with other procedures or for standard flow cytometry platforms. All volumes provided are for 5 mL polystyrene round-bottom FACS tubes. For 1.5 mL microcentrifuge tubes, reduce all wash volumes by 1/5; volumes of staining parameters remain unchanged. Microcentrifuge tubes are used for ImageStream analysis. Staining for ImageStream analysis can be done directly in 1.5 mL microcentrifuge tubes or in 5 mL polystyrene round-bottom FACS tubes and subsequently transferred to 1.5 mL microcentrifuge tubes.

The optimum concentration for imaging flow cytometry analysis is $2–4 \times 10^6$ cells per 200 μL volume. Cell loss may result from this procedure and thus we recommend that each sample consist of 4×10^6 cells at the start of the procedure if possible. For standard flow cytometry platforms, a lower cell concentration is possible. 1×10^6 cells are sufficient when using a BD FACSCanto II platform, where staining steps are performed in 100 μl and washes in 2 mL volumes.

3.1 Cell Preparation

1. Harvest cells—use lab-specific procedures for corresponding cell lines or primary cell isolations. Harvesting procedures will vary greatly depending on cell type (e.g., if cell is adherent or non-adherent).
2. Centrifuge samples at 311 ×*g* for 10 min at 4 °C and decant the supernatant. This speed is optimal to pellet most primary cells and cell lines. If the cell of interest requires different centrifugation parameters, modify as required.
3. Resuspend cells in 2 mL 1× phosphate buffered saline $(PBS)^{-/-}$ (no calcium, no magnesium). This wash is essential to remove residual serum and other media components prior to staining.
4. Centrifuge samples at 311 ×*g* for 10 min at 4 °C and decant the supernatant.
5. Resuspend cells in 1 mL 1× Annexin V binding buffer. This wash is essential to set magnesium and calcium concentrations to appropriate levels for Annexin V staining.
6. Centrifuge samples at 311 ×*g* for 10 min at 4 °C and decant the supernatant.
7. Resuspend cell pellet in 100 μL 1× Annexin V binding buffer.

3.2 Application of Annexin V/PI Stain

1. Add Annexin V according to the manufacturer's recommendations The volume will vary depending on the supplier.
2. Incubate tubes in the dark for 15 min at room temperature.
3. Add 100 μL of 1× Annexin V binding buffer to each reaction tube. There should be approximately 200 μL in each tube.
4. Add 4 μL of PI (Sigma, Cat# P-4864-10 ML) that has been diluted 1:10 in 1× Annexin V binding buffer (i.e., 1 μL PI with 9 μL 1× Annexin V binding buffer). This will yield a final PI concentration of 2 μg/mL in each sample. If a different stock solution of PI is being used, adjust addition volume to get a final concentration of 2 μg/mL in the 200 μL staining volume.
5. Incubate tubes in the dark for 15 min at room temperature.
6. Add 2 mL 1× Annexin V binding buffer to wash the cells.
7. Centrifuge samples at 311 ×*g* for 10 min at 4 °C and decant the supernatant.
8. Resuspend cells in 500 μL 1× Annexin V binding buffer and 500 μL 2 % formaldehyde to create a 1 % formaldehyde (fixative) solution. Mix tubes by gentle inversion.
9. Fix samples on ice for 10 min. Alternatively, samples can be stored overnight at 4 °C in the dark. If the latter method is chosen, make sure to be consistent across all tubes labeled with Annexin V/PI (including compensation controls).
10. Add 1 mL 1× $PBS^{-/-}$ to each sample and mix gently by flicking.
11. Centrifuge tubes at 393 ×*g* for 10 min at 4 °C and decant the supernatant. The increased speed is required to pellet fixed cells.
12. Repeat **steps 10** and **11**.
13. Resuspend pellet by flicking the tube.
14. Add 16 μL of 1:100 diluted RNase A (Sigma, R4642) to give a final concentration of 50 μg/mL. Incubate for 15 min at 37 °C.
15. Add 1 mL 1× $PBS^{-/-}$ and mix gently by flicking.
16. Centrifuge tubes at 393 ×*g* for 10 min at 4 °C.
17. Samples are now ready to be analyzed. Alternatively, samples can be used for subsequent staining steps if Annexin/PI staining is being performed in parallel with other procedures.

3.3 Acquisition and Analysis: ImageStream Platform

The analysis procedure described below is based on the ImageStream platform (Amnis Corporation, EMD Millipore). Other imaging flow cytometry platforms may utilize different features but the basics of the analysis will remain similar.

1. During acquisition, set lasers to balance signals from Annexin V and PI so that no over exposure occurs.
2. Set classifiers to eliminate debris and cell clumps based on size and aspect ratio of the bright-field channel [14].
3. After acquisition, a compensation matrix will need to be applied to all data to correct for spectral overlap. This can be done using the wizard that comes standard with IDEAS® (Image Data Exploration and Analysis Software, Amnis Corporation, EMD Millipore) software. Ensure that there is no over- or under-compensation of fluorochromes, as this can greatly affect results.
4. All analyses must be completed on a population of spectrally compensated, single, focused cells. A new IDEAS® wizard has recently been developed to study cell death. This wizard simplifies the analysis steps. The wizard will aid in identification of single, focused cells and set up a graph the fluorescence intensity of Annexin V vs. PI. To analyze cell death, gate cells based on staining pattern: Annexin V$^-$/PI$^-$ (healthy cells), Annexin V$^+$/ PI$^-$ (early apoptosis), Annexin V$^-$/PI$^+$ (necrotic), and Annexin V$^+$/PI$^+$ (late apoptotic/necrotic).
5. To complete analysis manually first graph bright field size vs. aspect ratio. This will allow identification and gating of single cells, and eliminate clumps and debris from subsequent analyses. In the next step, graph "Brightfield Gradient RMS" and gate on cells that have a high degree of focus. Using this population, graph the fluorescence intensity of Annexin V vs. PI. To analyze cells death, gate cells based on staining pattern: Annexin V$^-$/PI$^-$ (healthy cells), Annexin V$^+$/ PI$^-$ (early apoptosis), Annexin V$^-$/PI$^+$ (necrotic), and Annexin V$^+$/PI$^+$ (late apoptotic/necrotic). Gating strategy is shown in Fig. 2.

3.4 Acquisition and Analysis: Flow Cytometry

The analysis procedure described below is based on acquisition with a BD FACSCanto II (Becton Dickson) and analyzed with FACSDiva or FCS Express v3. Other imaging flow cytometry platforms may utilize different features but the basics of the analysis will remain similar.

1. During acquisition, set lasers to balance signals from Annexin V and PI and perform compensation.
2. Set forward scatter and side scatter parameters to eliminate debris.
3. To analyze, gate on cell population based on forward and side scatter intensities. Be aware that cell populations tend to shift in cells that are dying.
4. All analyses must be completed on a population of spectrally compensated cells. Graph the fluorescence intensity of Annexin

V vs. PI. To analyze cell death, gate cells based on staining pattern: Annexin V^{-}/PI^{-} (healthy cells), Annexin V^{+}/PI^{-} (early apoptosis), Annexin V^{-}/PI^{+} (necrotic), and Annexin V^{+}/PI^{+} (late apoptotic/necrotic). This plot is present in Fig. 4.

4 Notes

1. Add salts to approximately 500 mL ultrapure water on a stir plate. Once all salts are added, top solution up to 1 L. Check pH; generally solution has a pH of 7.2–7.4. If 10× solution is outside of this range, pH solution to this level. The 10× solution is then filter sterilized using a 0.2 μm filter. This solution can then be stored at room temperature. Storing at 4 °C will cause salts to precipitate out of solution. If this occurs, warm solution and remix. Salts will return into solution.
2. To dilute to a 1× $PBS^{-/-}$ solution, add one part of 10× $PBS^{-/-}$ solution to nine parts filter sterilized ultrapure water. This solution should then be stored at 4 °C.
3. Common vendors are BD Biosciences (catalogue #556454), eBioscience (catalogue # BMS500BB), or Life Technologies (catalogue #PNN1001).
4. Buffers with different magnesium or calcium concentrations from 1× Annexin V binding buffer will result in a lack of staining with Annexin V. Without the correct concentration of these essential salts, Annexin V is incapable of binding externalized phosphatidylserine. If other buffers, such as 1× $PBS^{-/-}$, are used during staining procedures no Annexin V staining will be detected. However, once cells are fixed later in the protocol, Annexin V binding buffer is no longer required to maintain Annexin V binding to the cells.
5. The fixation step is critical for RNase A to pass through the cell membrane. Without fixation, RNase A is excluded from cells and is therefore not able to degrade intracellular RNA and will fail to reduce false-positive events.

Acknowledgments

This work was supported by Natural Sciences and Engineering Council of Canada (NSERC) research grant and an Alberta Agriculture Funding Consortium grant to D.R.B. A.M.R. is supported through an NSERC Vanier Canadian Graduate Scholarship and the University of Alberta (graduate teaching assistantship, Queen Elizabeth II graduate scholarship, and Dissertation Fellowship).

References

1. Muppidi J, Porter M, Siegel RM (2004) Measurement of apoptosis and other forms of cell death. Curr Protoc Immunol Chapter 3, Unit 3. 17
2. Green DR (2011) Means to an end: apoptosis and other cell death mechanisms. Cold Spring Harbor Laboratory Press, Cold Spring Harbor
3. Kroemer G, Dallaporta B, Resche-Rigon M (1998) The mitochondrial death/life regulator in apoptosis and necrosis. Annu Rev Physiol 60:619–642
4. Overbeeke R, Steffens-Nakken H, Vermes I, Reutelingsperger C, Haanen C (1998) Early features of apoptosis detected by four different flow cytometry assays. Apoptosis 3:115–121
5. Vermes I, Haanen C, Steffens-Nakken H, Reutelingsperger C (1995) A novel assay for apoptosis. Flow cytometric detection of phosphatidylserine expression on early apoptotic cells using fluorescein labelled Annexin V. J Immunol Methods 184:39–51
6. Telford WG, Komoriya A, Packard BZ (2004) Multiparametric analysis of apoptosis by flow and image cytometry. Methods Mol Biol 263:141–160
7. Vermes I, Haanen C, Reutelingsperger C (2000) Flow cytometry of apoptotic cell death. J Immunol Methods 243:167–190
8. Darzynkiewicz Z, Bruno S, Del Bino G, Gorczyca W, Hotz MA, Lassota P, Traganos F (1992) Features of apoptotic cells measured by flow cytometry. Cytometry 13:795–808
9. Faleiro L, Lazebnik Y (2000) Caspases disrupt the nuclear-cytoplasmic barrier. J Cell Biol 151:951–959
10. Deitch AD, Law H, deVere WR (1982) A stable propidium iodide staining procedure for flow cytometry. J Histochem Cytochem 30:967–972
11. Fried J, Perez AG, Clarkson BD (1976) Flow cytofluorometric analysis of cell cycle distributions using propidium iodide. Properties of the method and mathematical analysis of the data. J Cell Biol 71:172–181
12. http://apps.isiknowledge.com/summary.do?qid=17&product=UA&SID=3EdJcO897oK2Gfd6EH4&search_mode=Refine. (2009) Web of knowledge search
13. Zuba-Surma EK, Kucia M, Abdel-Latif A, Lillard JW Jr, Ratajczak MZ (2007) The ImageStream system: a key step to a new era in imaging. Folia Histochem Cytobiol 45:279–290
14. Henery S, George T, Hall B, Basiji D, Ortyn W, Morrissey P (2008) Quantitative image based apoptotic index measurement using multispectral imaging flow cytometry: a comparison with standard photometric methods. Apoptosis 13:1054–1063. doi:10.1007/s10495-008-0227-4
15. Rieger AM, Hall BE, Luong LT, Schang LM, Barreda DR (2010) Conventional apoptosis assays using propidium iodide generate a significant number of false positives that prevent accurate assessment of cell death. J Immunol Methods 358:81–92. doi:10.1016/j.jim.2010.03.019

Chapter 16

Temporal Heterogeneity in Apoptosis Determined by Imaging Flow Cytometry

Ivan A. Vorobjev and Natasha S. Barteneva

Abstract

Apoptotic process is highly heterogeneous, and a long-standing question is how many parameters define time and reversibility of the apoptotic response at a population and single-cell levels. Cell death analysis applications have greatly expanded since the introduction of flow cytometry. Classical approach for evaluation of apoptosis is en masse analysis of cells treated with different stimuli, but these methods cannot demonstrate heterogeneity in the population. Single-cell heterogeneity is now usually assessed by multicolor fluorescence microscopy; however obtaining reasonable statistics is time consuming and laborious. Therefore we combined flow cytometry, imaging flow cytometry, and fluorescent microscopy to characterize at a single-cell and population level sequence of apoptotic events induced by a variety of treatments (Vorobjev, Barteneva, J Histochem Cytochem 63:494–510, 2015). We show that simultaneous use of membrane potential dye TMRE, caspases 3/7 sensor, Annexin V and nuclear staining along with morphological parameters demonstrate heterogeneity of the whole process and is a valuable method for quantitative study of the apoptosis execution. Imaging flow cytometry allowed us to analyze correlation between TMRE, caspases 3/7, and Annexin V staining and morphological characteristics providing valuable information on the process of apoptotic execution. Importantly, comparisons of different data sets obtained by three methods allowed us to achieve temporal resolution of the whole process superior to that had been obtained by only one method.

Key words Apoptosis, Cell death, Caspase 3/7, Annexin V, Mitochondria, TMRE, Cell volume

1 Introduction

Apoptosis is a programmed form of cell death, triggered by a variety of signals including TNF receptors, p53 activation, mitochondria damage, oxidative stress, DNA damage, etc. Analysis of apoptosis is essential to many fields, and is usually performed on the population level. The key events in the progression and execution of apoptosis include apoptotic volume decrease (AVD), cell shrinkage, phosphatidylserine (PS) externalization on the plasma membrane, nuclear condensation, blebbing, and fragmentation of dying cell. These features are orchestrated by the activation of cysteine

Natasha S. Barteneva and Ivan A. Vorobjev (eds.), *Imaging Flow Cytometry: Methods and Protocols*, Methods in Molecular Biology, vol. 1389, DOI 10.1007/978-1-4939-3302-0_16, © Springer Science+Business Media New York 2016

proteases, namely caspases [1]. Usually morphological changes of apoptotic cells are preceded or supplemented with mitochondria outer membrane permeabilization (MOMP) and occur along with the executioner caspases activation [2–5]. However, the relationship between AVD, MOMP, caspase activation, PS externalization, and morphologically distinguishable events is poorly investigated.

Flow cytometry studies of apoptosis are often limited by use of PI (or other DNA dyes) to study permeability of the plasma membrane and Annexin V staining as a probe for PS externalization (see Rieger and Barreda, this volume for details), while other characteristics like MOMP and AVD are usually studied by microscopy and caspase activation is usually studied by immunoblotting and related methods. Recently we showed that TMRE staining could be successfully used in flow cytometry (with 561 nm excitation laser) and loss of TMRE staining is tightly coordinated with apoptosis execution [6, 7]. Microscopic studies of apoptosis are mainly devoted to the study of MOMP [1], apoptotic volume decrease [8, 9], and rarely—to the activation of caspases [3, 10].

Microscopic studies of apoptosis performed at a single-cell level had been using caspase FRET sensors demonstrated that caspase 3/7 activation can occur immediately after MOMP [3, 11, 12]; however limited statistics was obtained.

To the best of our knowledge no microscopic study of apoptosis evaluated PS externalization along with other events. Thus characterization of apoptosis performed using different methods (microscopy, flow cytometry, immunoblotting) is focused on different markers of the process and thus results obtained are not easy to put into one comprehensive picture.

Because of the technical problems with Annexin V staining—it requires high concentration of calcium and dissociation constant for Annexin V from PS is two orders of magnitude higher than for antibodies—intensive microscopic studies of Annexin V-labeled cells are difficult.

Imaging flow cytometry is a relatively new method allowing to narrow the gap between standard flow cytometry with the lack of information about a given cell and classical microscopy, where obtaining quantitative parameters for cell populations is tedious and time consuming.

In this chapter, we use an ImageStream multi-spectral flow cytometer (Amnis Corporation, EMD Millipore) parallel with conventional flow cytometry and time-lapse fluorescent microscopy to show that multicolor staining of tissue culture cells undergoing apoptosis using caspase 3/7 sensor, TMRE as mitochondria potential probe and Annexin V, conjugated with Alexa 647 as a marker of PS externalization (Fig. 1) allow detailed analysis of apoptosis execution on the population level and analysis of image galleries allows one to make conclusions on the events might or might not happen simultaneously. We show that PS externalization happens always after MOMP, yet before caspase 3/7 activation. Intensity of

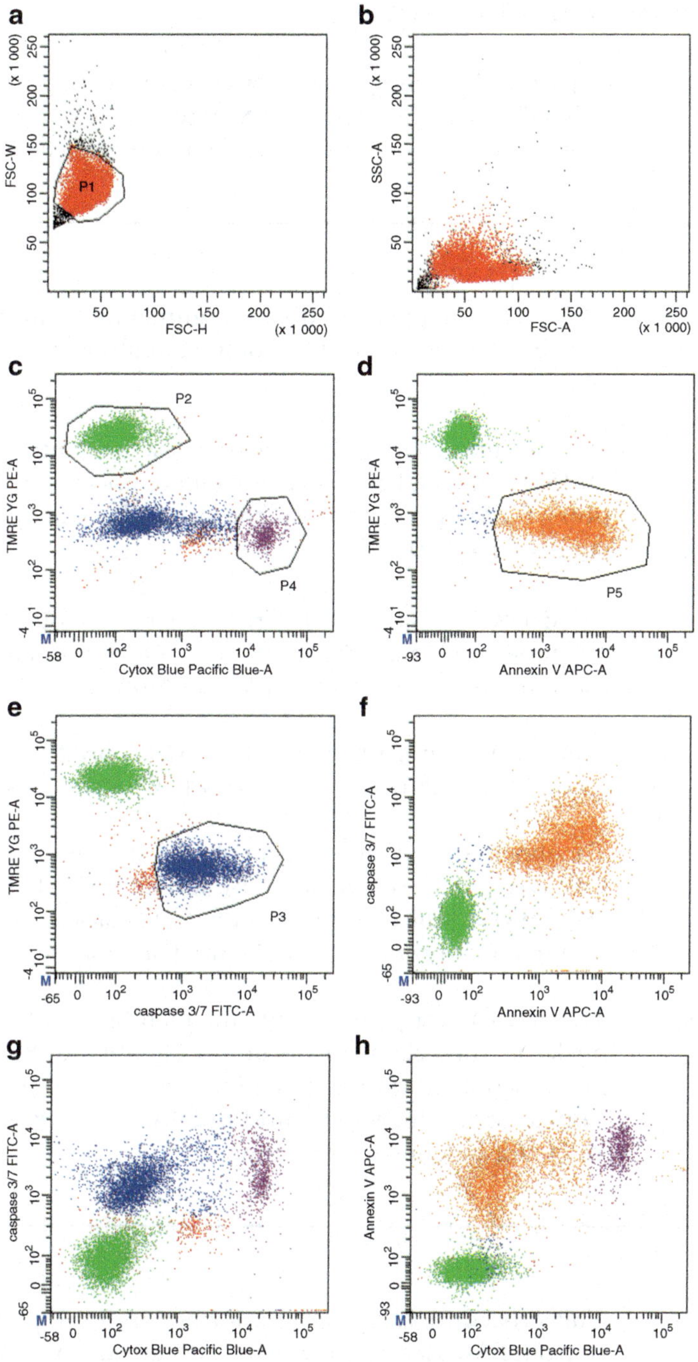

Experiment Name: Experiment_001

Population	#Events	%Parent	caspase 3/... Mean	TMRE YG P... Mean
P1	9 450	94,5	1 117	1 1 422
P2	4 840	51,2	94	21 634
P3	4 282	45,3	2 374	588
P4	796	8,4	3 650	413
P5	4 498	47,6	2 229	580

Fig. 1 Dotplots from FACSAria II. THP1 cells treated with TRAIL, 100 ng/ml for 20 h. (**a**) FSC-W versus FSC-H, region *P1* denotes singlet events. (**b**) FSC/SSC plot. (**c**) TMRE/Sytox blue plot, *P2* TMRE positive cells, *P4* Sytox Blue-positive (dead) cells. (**d**) TMRE/Annexin V plot, *P5* Annexin V-positive cells. (**e**) TMRE/caspase 3/7 plot, *P3* caspase-positive cells. (**f**) Annexin V/caspase plot. (**g**) Sytox Blue/caspase 3/7 plot. (**h**) Sytox Blue/Annexin V plot

caspase 3/7 staining does not correlate with Annexin V staining indicating that process of PS externalization and caspase 3/7 activation are largely independent at a single-cell level.

2 Materials

2.1 Tissue Culture Medium

RPMI-164 supplemented with L-glutamine, antibiotics, and 10 % fetal bovine serum (FBS). Complete culture medium is stored at +4 °C.

2.2 TMRE (Tetramethyl-rhodamine, Ethyl Ester) Solution

Stock solution should be prepared at a concentration of 50 μM in DMSO and could be stored at room temperature. Final concentration of TMRE depends on the particular cell type and is in the range of 10–100 nM.

2.3 Caspase 3/7 Sensor

CellEvent™ Caspase 3/7 Green reagent is purchased from Life Technologies Inc. (Grand Island, NY, USA) and stored at the room temperature (in the dark) (*see* **Note 1**).

2.4 Annexin V

Annexin V conjugated with Alexa-647 is used for the current protocol. Stock solution is stored at +4 °C in the dark (*see* **Note 2**).

2.5 Annexin V Binding Medium

To obtain strong signal from Annexin V high concentration of Ca^{++} is required. It is achieved by adding $CaCl_2$ from stock solution (500 mM) to the culture medium. $CaCl_2$ stock solution is prepared in sterile deionized water and stored at room temperature.

2.6 Sytox Blue (SB)

This dye cannot penetrate plasma membrane so it is only staining nuclei in the cells with disrupted membrane. Sytox Blue can be purchased from different vendors. For our purposes, we used SB from a stock solution of 1 mM in DMSO. Stock solution is kept at +4 °C for indefinitely long time.

3 Methods

This procedure is typically performed in 5 mL polystyrene round-bottom FACS sterile tubes with caps using 0.5–1 ml of cell suspension. For ImageStream analysis 50–70 μl of stained suspension is taken. The optimum concentration for imaging flow cytometry analysis is $\sim 5 \times 10^6$ cells per 1 mL volume. If cell concentration is significantly less, larger volume might be recommended. When large files obtained by imaging flow cytometry are necessary, cells are concentrated two- to fivefold immediately before ImageStream analysis. Microcentrifuge tubes are used for ImageStream analysis.

Special staining for ImageStream in 1.5 mL microcentrifuge tubes should be avoided, since small volume does not guarantee cell viability for long enough time (*see* **Note 3**).

3.1 Cell Preparation

1. Harvest cells—use lab-specific procedures for corresponding cell lines or primary cell isolations. Harvesting procedures will vary greatly depending on cell type (e.g., if cell is adherent or non-adherent). Cell suspension is finally prepared in culture medium.
2. Centrifuge samples at $400 \times g$ for 4–5 min at room temperature and decant the supernatant. This speed is sufficient to pellet most suspension cell lines. If the cell of interest requires different centrifugation parameters, modify as required (*see* **Note 4**).
3. Resuspend cells in ~0.5 mL of fresh culture medium to achieve concentration of $1–2 \times 10^6$ cells per ml.

3.2 TMRE and Caspase 3/7 Staining

1. Add TMRE from stock solution at a final concentration 50–100 nM. Typical stock solution is 50–100 μM in DMSO.
2. Add caspase 3/7 reagent—one drop per 0.5 ml is recommended, proper concentration could be titrated.
3. Place vial with cells into CO_2 incubator (37 °C) for 30 min.

3.3 Application of Annexin V Stain

1. Add $CaCl_2$ from the stock solution to the sample to achieve final concentration of 3 mM (*see* **Note 5**).
2. Add Annexin V according to the manufacturer's recommendations (e.g., 5 μL Annexin V-Alexa Fluor 647 if using from Life Technologies). The volume might vary depending on the supplier.
3. Incubate tubes in the dark for 15 min at room temperature (not less than 23 °C) or in CO_2 incubator (37 °C).

3.4 Application of Sytox Blue Stain

1. Add ~0.5 μl Sytox Blue to the sample and take sample for analysis within 2 min.

3.5 Acquisition and Analysis: Flow Cytometry Platform

The analysis procedure described below is based on acquisition with a FACS Aria cells sorter (Becton Dickinson) and analyzed with FACSDiva software. It could be performed in the similar way with any instrument equipped with 405/407, 488, 561, and 635/640 lasers. Consequently caspase 3/7 is viewed in FITC channel (bandpass 530/30), TMRE in PE channel (yellow–green laser; bandpass 575/25), Sytox Blue in Pacific Blue channel (bandpass 450/50), and Annexin V-Alexa 647—in APC channel (bandpass 660/20) (*see* **Note 6**).

1. During acquisition, set lasers to balance signals from Annexin V, caspase, TMRE, and Sytox Blue so that no overexposure occurs. Usually minimal compensation is required when using 4-laser platform (*see* **Notes 7–9**).
2. Set forward scatter and side scatter parameters to visualize debris. To analyze, gate on cell population based on forward scatter width and height parameters to exclude debris and cell doublets.
3. Data analysis is performed in a standard way for flow cytometry. It is important to notice that intensity of staining (particularly for TMRE and caspase substrate) might be heterogeneous in the entire population and populations with intermediate intensity of staining represent certain stages of apoptotic execution (*see* **Note 10**).

3.6 Acquisition and Analysis Using ImageStream Platform

The analysis procedure described below is based on the ImageStream X Mark II platform and IDEAS software (Amnis Corporation, EMD Millipore). Original (uncompensated) images acquired with an ImageStream™ are stored in a raw image file (*.rif). Single-stained controls are used to produce a compensation matrix (stored as *.ctm file) and, subsequently, compensated image files (*.cif) are generated. Data analysis is performed using *.cif file and saved as *.daf file.

1. Before acquisition, set lasers to balance signals from Annexin V-Alexa 647, caspase 3/7 sensor, TMRE, and Sytox Blue so that no over exposure occurs (*see* **Notes 11** and **12**).
2. Switch off bright-field channel and adjust laser power to achieve a bright signal without saturated pixels for each channel. Acquire at least 200 bright images for each compensation control and create compensation matrix (*.ctm file). Compensation matrixes can be stored and applied to other data files employing the same probes and acquired with same lasers power (*see* **Notes 11** and **13**).
3. Acquire your samples. At least 5000 events should be acquired (10–20 K events is preferable). Recording files containing more than 20,000 pictures should be avoided, since post-acquisition processing of large files is rather slow.
4. After acquisition of the datasets, a compensation matrix has to be applied to correct for spectral overlap. The compensation is performed independently for each of the thousands pixels per image rather than just once for one cellular event in conventional cytometry. This can be done using the wizard that comes standard with IDEAS® (Image Data Exploration and Analysis Software, Amnis Corporation, EMD Millipore) software. The multispectral image compensation in ImageStream system includes nine steps as follows: (1) dark current pixel offset measurement; (2) pixel gain measurement; (3) inter-image

spatial offset measurement; (4) application of dark current corrections; (5) application of spatial alignment corrections; (6) fluorescence crosstalk measurement; (7) bright-field cross talk measurement; (8) application of compensation matrix; (9) application of bright-field pixel gain compensation [13].

5. To start analysis of the data first graph bright field size vs. aspect ratio. Set classifiers to eliminate debris and cell aggregates based on size and "Aspect ratio_M2" of the bright-field channel using linear scale (*see* **Note 14**). This will allow identification of single cells, and eliminate debris from subsequent analyses.
6. In the next step, graph "Brightfield Gradient RMS" and gate on cells that are in focus. Using this population, graph the fluorescence intensity in each channel. It is useful to make a new *.cif and *.daf files including only gated events at this step.
7. Build two-dimensional dotplots in *.cif file and check for the compensation. Whether compensation obtained using IDEAS compensation wizard is not correct, manual compensation could be applied (*see* **Note 15**).
8. To analyze correlation between different staining patterns generate dotplots in Annexin V/TMRE, TMRE/caspase and AnnV/caspase axes in *.daf file.
9. Selection of negative and positive events is performed on the base of image analysis. To put reasonable threshold first gate positive and negative populations on the dot plots. Then look through image galleries generated after gating in each channel. Whether cells with suspicious staining pattern are seen as positive, minimal threshold value should be increased. Useful tips for image analysis are narrowing intensity histogram and enhancing gamma. After selecting the unambiguous threshold value it is applied for statistical analysis.
10. For presentation image galleries, dotplots and histograms are exported and mounted into ppt slides, then scaled and transferred into tiff files using Adobe Photoshop, Paint, or ImageJ software. Example of the output montage is presented in Fig. 2.

3.7 Acquisition and Analysis by Fluorescent Microscopy

The analysis procedure described below is based on acquisition with an Axiovert 200 fluorescent microscope (Zeiss) equipped with following filter cubes: DAPI, FITC, TRITC/Cy-3, Cy5, and SlideBook software. Similar analysis could be done on any other fluorescent microscope.

1. Protocol for attached cells: Subculture cells into 96-well glass-bottom dish at a density of 10,000–20,000 cells per well (in

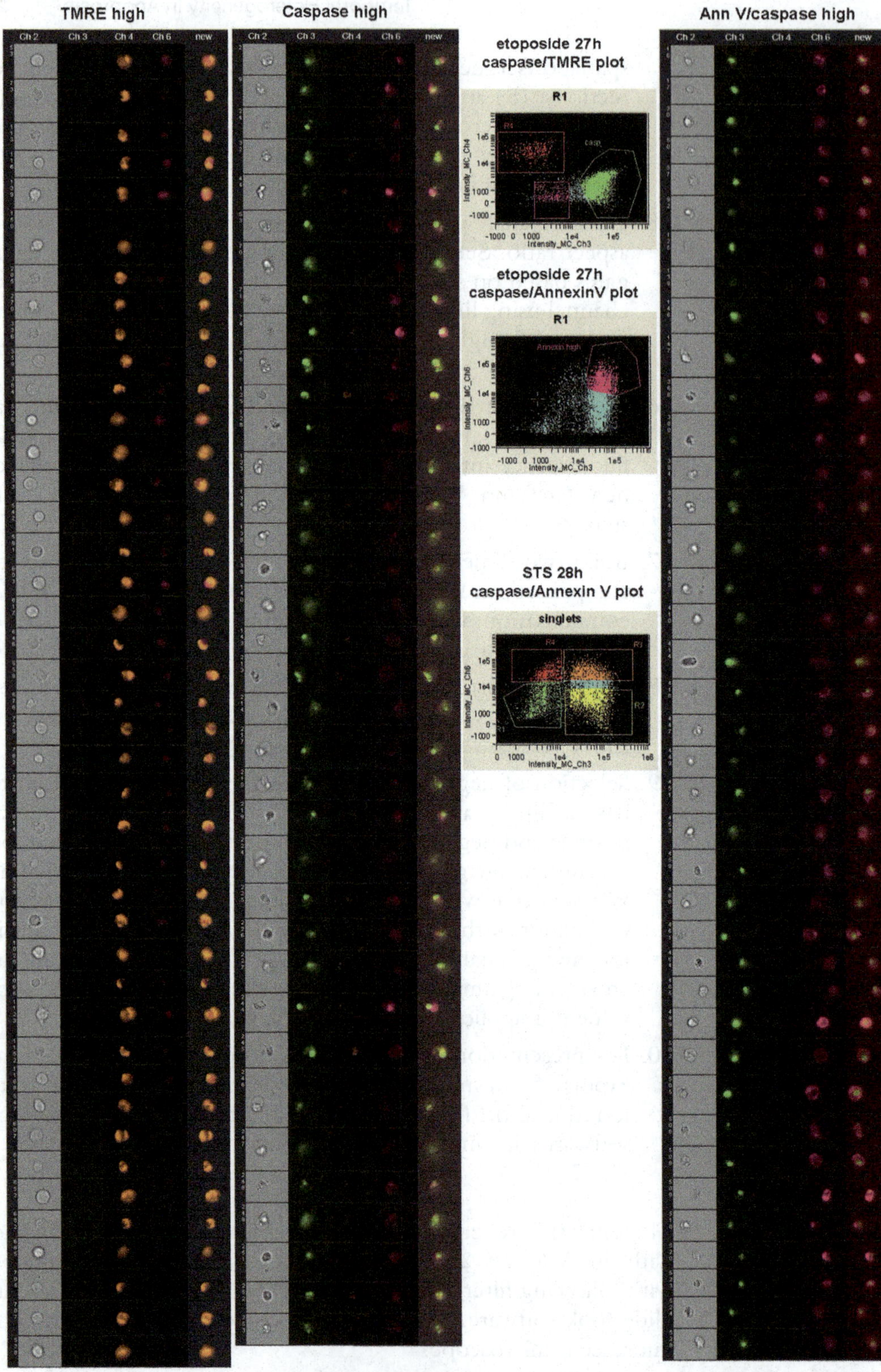

Fig. 2 ImageStream output (Jurkat cells treated with etoposide, 50 μM for 27 h). Images correspond to cells that are TMRE positive (*left column*); caspase positive/Annexin dim (*middle column*), and Annexin V positive (*right column*). Regions are taken from corresponding dotplots (caspase sensor—Intensity Ch3, TMRE—Intensity Ch4, Annexin V intensity–Ch6)

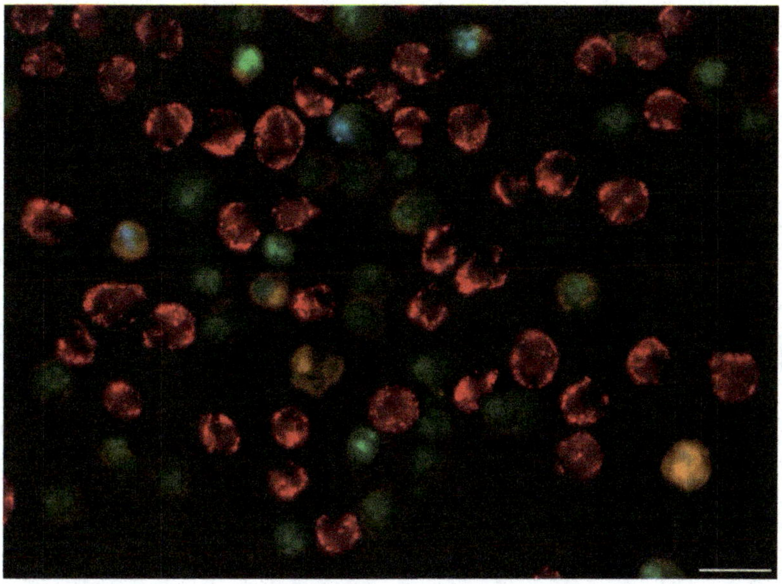

Fig. 3 THP-1 cells treated with TRAIL, 100 ng/ml, 24 h. Quadruple staining, microscopy, objective lens ×40/1.3. *Red*—TMRE staining, *green*—caspase 3/7 staining, *yellow*—Annexin V-Alexa 647 staining, *blue*—Sytox Blue staining. Scale bar—20 μm

0.3 ml volume). Experiment could be started next day after plating. Staining is performed in the following sequence: TMRE, caspase 3/7 substrate, then Annexin V (with additional $CaCl_2$), and finally Sytox Blue.

2. Pictures are taken using DIC and each fluorescent channel. Exposure time should be determined to achieve signal for the positive cells at least 20 % of the dynamic range of camera (*see* **Note 16**).
3. Perform compensation by subtracting images from neighboring channels in the same way as it is recommended for the flow cytometry. Prepare color images using multichannel option and subtracted channels (under View knob—User-defined colors) in SlideBook software. Colors have to be selected from the palette. Example of such image is given on Fig. 3.
4. Protocol for suspended cells. Plate cells into 96 well glass bottom dish at a density of 20,000–50,000 cells per well (in 0.3 ml volume). Staining is performed as described for the flow cytometry. Alternatively, cells already stained with TMRE and caspase sensor could be plated into the dish from the tubes.
5. Allow cells to sediment (it takes not less than 20 min) in CO_2 incubator.

6. Stain cells either in appliance with complete protocol (Subheadings 3.2–3.4) or if TMRE/caspase staining were performed in the tube with Annexin V and then with Sytox Blue as described in Subheadings 3.3–3.4.
7. Pictures are taken 2–3 min after addition of Sytox Blue using DIC and each fluorescent channel. Exposure time should be determined to achieve signal for the positive cells at least 20 % of the dynamic range of camera (*see* **Note 17**).

4 Notes

1. For caspase sensor there is a single vendor—Life Technologies (catalogue # R37111). The big advantage is that caspase substrate can be stored at room temperature. In our hands proper dilution was 5 μL of caspase sensor for 300 μL of cell suspension.
2. Other fluorescent conjugates of Annexin V could be used (with APC etc.), however according to our experience the best results are obtained with Alexa-647 conjugate purchased from Life Technologies. Stock solution of Annexin V conjugate cannot be stored for prolonged period—according to our experience it should be used within 1–2 months upon purchase. After this time intensity of staining of apoptotic cells and reproducibility of results go down.
3. Cell loss may result from staining procedure and thus we recommend that each sample consist of $2–4 \times 10^6$ cells at the start of the procedure. For standard flow cytometry platforms, a lower cell concentration is possible. In our experience 1×10^6 cells is enough when using a BD FACSAria II platform.
4. Apoptotic cells often become fragile, thus centrifuge acceleration should be minimized not to damage cells at late stages of apoptosis. After centrifugation cells should be resuspended gently.
5. Buffers and culture media have different calcium concentrations. Low concentration of calcium might result in a lack of staining with Annexin V of externalized phosphatidylserine or staining is short-living. To our experience when complete culture media like DMEM with calf serum is used Annexin V staining is detected ambiguously. However, once cells are stained and further kept in the presence of >3 mM Ca^{2+}, Annexin V binding to externalized phosphatidylserine is strong and unambiguous.
6. If using three laser platform (488; 407 and 640 nm) excitation of TMRE is rather low and compensation between FITC and

PE channels should be performed using single-stain controls of untreated cells stained with TMRE and cells at late apoptotic stages stained with caspase 3/7 sensor.

7. For FITC and Pacific Blue channels strong autofluorescence might be present (depends on the cell type)—unstained controls are essential.
8. Relative intensity of TMRE, caspase sensor, Annexin V and Sytox Blue staining usually is different from standard signals obtained with conjugated antibodies viewed in the same channels. PMT voltage and compensation table needs to be modified to optimize signals for positive and negative populations in all four channels. For proper compensation single stained probes are required, since emission spectra of these TMRE and caspase sensor differ from standard control beads.
9. It is important to notice that positive staining for caspase and Annexin V could be obtained only on the cells in late apoptotic stages, while TMRE staining is stronger in the untreated cells.
10. On our experience relatively large TMRE intermediate population often appears after staurosporine treatment and smaller populations after other treatments. Intensity of caspase staining could vary in the range of two decades. Dead cells are determined by high intensity of SytoxBlue staining, SytoxBlue intermediate cells might be alive.
11. Since $TMRE^{bright}$ cells dominate in control specimen, while caspase $3/7^{bright}$ and $SytoxBlue^{bright}$ cells are found at the late stages of apoptosis, at least two types of specimens—control (untreated) and late apoptotic probes—should be used for calibration of laser settings.
12. Intensity of TMRE staining depends on the cell type and this dye should be titrated prior to experiment. Usually TMRE gives very strong signal compared to the other fluorophores and 488 nm laser should be set to a minimal intensity (~20 mW). Whether TMRE signal is saturating less concentration of TMRE should be used. Using 561 nm laser for TMRE excitation might give excessive signal even at minimal power.
13. The problems in spectral compensation matrixes can happen because of a number of reasons such as high autofluorescence background and inclusion of saturated pixels. The compensation matrix can be manually adjusted and edited by manually changing crosstalk matrix values in increments and using Preview option. Detailed information is provided in IDEAS manual. The post-acquisition compensation is critical and is one of major obstacles for development of cell sorting option for imaging flow cytometer, which would require a spectral compensation of images available during a sorting.

14. Elimination of cell clamps is better performed manually, since narrowing aspect ratio will exclude a lot of single cell events from subsequent analysis.
15. Our experience is that data obtained through IDEAS compensation wizard (*.ctm file) are usually undercompensated. Manual compensation should be done cautiously by gradually increasing coefficients in compensation matrix not to obtain overcompensated data. During compensation emphasis should be put on the brightest events for each channel. After building proper compensation it is useful to make another *.ctm file that can be used in analysis of other files in the current and similar experiments.
16. To obtain reasonable statistics high-magnification objectives are not useful. Our experience is that the best objective to be used is PlanApo 20×/0.8. Analysis of the pictures taken with 10× objective is possible, but often time consuming.
17. Since cells are not attached to the substrate Brownian motion will blur the image. Thus minimal exposure will give sharper pictures. To obtain reasonable statistics high-magnification objectives are not useful. Our experience is that the best objectives to be used are PlanApo 20×/0.8 (preferable) or PlanApo 40/1.3, but not dry ×40 objective. Analysis of the pictures taken with 10× objective is possible; however objective with high numerical aperture and the same magnification is preferable.

Acknowledgments

The work was supported in part by NIH S10 RR023459 grant, Harvard Pilot grant, the Russian Foundation for Basic Research grants 11-04-01749a and 13-04-40189-H.

References

1. Green DR (2005) Apoptotic pathways: ten minutes to dead. Cell 121:671–674. doi:10.1016/j.cell.2005.05.019
2. Kroemer G, Petit P, Zamzami N, Vayssiere JL, Mignotte B (1995) The biochemistry of programmed cell death. FASEB J 9:1277–1287
3. Albeck JG, Burke JM, Aldridge BB, Zhang M, Lauffenburger DA, Sorger PK (2008) Quantitative analysis of pathways controlling extrinsic apoptosis in single cells. Mol Cell 30:11–25. doi:10.1016/j.molcel.2008.02.012
4. Tait SW, Green DR (2010) Mitochondria and cell death: outer membrane permeabilization and beyond. Nat Rev Mol Cell Biol 11:621–632. doi:10.1038/nrm2952
5. Tait SW, Green DR (2013) Mitochondrial regulation of cell death. Cold Spring Harb Perspect Biol 5(9). doi: 10.1101/cshperspect.a008706
6. Barteneva NS, Ponomarev E, Tsytsykova A, Armant M, Vorobjev IA (2014) Mitochondrial staining allows robust elimination of apoptotic

and damaged cells during cell sorting. J Histochem Cytochem 62:265–275. doi:10.1369/0022155413520404

7. Vorobjev IA, Barteneva NS (2015) Temporal heterogeneity metrics in apoptosis induced by anticancer drugs. J Histochem Cytochem 63: 494–510. doi:10.1369/0022155415583534
8. Bortner CD, Cidlowski JA (2003) Uncoupling cell shrinkage from apoptosis reveals that Na^+ influx is required for volume loss during programmed cell death. J Biol Chem 278:39176–39184
9. Bortner CD, Cidlowski JA (2011) Life and death of lymphocytes: a volume regulation affair. Cell Physiol Biochem 28:1079–1088. doi:10.1159/000335864
10. Spencer SL, Gaudet S, Albeck JG, Burke JM, Sorger PK (2009) Non-genetic origins of cell-to-cell variability in TRAIL-induced apoptosis. Nature 459:428–432. doi:10.1038/nature08012
11. Rehm M, Dussmann H, Janicke RU, Tavare JM, Kogel D, Prehn JH (2002) Single-cell fluorescence resonance energy transfer analysis demonstrates that caspase activation during apoptosis is a rapid process. Role of caspase-3. J Biol Chem 277:24506–24514
12. Spencer SL, Sorger PK (2011) Measuring and modeling apoptosis in single cells. Cell 144:926–939. doi:10.1016/j.cell.2011.03.002
13. Ortyn WE, Hall BE, George TC, Frost K, Basiji DA, Perry DJ, Zimmerman CA, Coder D, Morrissey PJ (2006) Sensitivity measurement and compensation in spectral imaging. Cytometry A 69:852–862

Part V

Imaging Cytometry in Cell Classification and Morphological Analysis

Chapter 17

FlowCam: Quantification and Classification of Phytoplankton by Imaging Flow Cytometry

Nicole J. Poulton

Abstract

The ability to enumerate, classify, and determine biomass of phytoplankton from environmental samples is essential for determining ecosystem function and their role in the aquatic community and microbial food web. Traditional micro-phytoplankton quantification methods using microscopic techniques require preservation and are slow, tedious and very laborious. The availability of more automated imaging microscopy platforms has revolutionized the way particles and cells are detected within their natural environment. The ability to examine cells unaltered and without preservation is key to providing more accurate cell concentration estimates and overall phytoplankton biomass. The FlowCam® is an imaging cytometry tool that was originally developed for use in aquatic sciences and provides a more rapid and unbiased method for enumerating and classifying phytoplankton within diverse aquatic environments.

Key words Phytoplankton, Flow cytometry, FlowCam, Chlorophyll fluorescence, Image analysis

1 Introduction

The Flow Cytometer and Microscope (FlowCam) is an imaging flow cytometer developed for rapid particle detection and identification. It uses a combination of imaging and laser light to detect particles within a fluid stream, enabling the capture of particle images for downstream image analysis and particle identification [1]. The FlowCam, originally developed to assist in water quality analysis and in the identification and enumeration of phytoplankton, is now widely used as an instrument for general particle detection in aquatic environments, public drinking water treatment, pharmaceuticals industry, petrochemical industry, and the food and beverage industry. Using the instrument's VisualSpreadsheet® Software (ViSp), the FlowCam can capture images and provide real-time image analysis, providing over 40 different parameters, such as length, width, equivalent spherical diameter, and fluorescence values for each particle imaged. The instrument has the capability of detecting two fluorescence values depending on the laser

Natasha S. Barteneva and Ivan A. Vorobjev (eds.), *Imaging Flow Cytometry: Methods and Protocols*, Methods in Molecular Biology, vol. 1389, DOI 10.1007/978-1-4939-3302-0_17,

and the fluorescence application required. For phytoplankton detection and water quality monitoring the FlowCam is unique as it can rapidly detect, enumerate, and classify different phytoplankton species within natural communities. Traditional microscopic methods, such as the Utermöhl method, are very slow and tedious and also require sample preservation. The Utermöhl method has been used to enumerate both nano- and micro-plankton and is considered the most accurate for identification and enumeration [2]. Automated techniques, such as the FlowCam, have been developed more recently and are able to process, enumerate [3–5], and size plankton rapidly [1, 6]. The FlowCam can be used in the laboratory and the field and has the ability to process samples in real time without preservation.

In addition, the FlowCam is very useful in determining the size structure of the planktonic community [7–10]. Automatic classification by the ViSp software can provide reliable data for most ecological studies; however, traditional microscopic techniques are able to identify phytoplankton to the species level, whereas, the FlowCam and other automated techniques can usually classify cells to the level of genus [9]. The FlowCam methodology complements traditional microscope techniques, it is useful for automated size structure analysis and can obtain a vast amount of information in a very short period of time. Classification by the FlowCam is limited to low taxonomic resolution when compared to microscopy, however, as automated classification algorithms continue to improve the taxonomic resolution will improve. The use of the FlowCam to enumerate and classify a phytoplankton community will depend on different operational factors that will be discussed in this chapter these include the density and variation in the plankton assemblage, the time available for analysis, the dependence on the operator for classification determination, and the need to analyze unpreserved samples.

2 Materials

2.1 Instrumentation and Accessories

Depending on the requirements of the operator, different FlowCam instrumentation options are available for phytoplankton quantification. The FlowCam is available as a traditional laboratory benchtop unit (Fig. 1a) and as a portable unit for field use (Fig. 1b). Two different benchtop units are available with and without a laser, FlowCam VS-4 and FlowCam VS-1, respectively. Each FlowCam is available with numerous options, including black and white or color digital cameras, green (532 nm) or blue excitation (488 nm) lasers, and four different objectives (2×, 4×, 10×, and 20×) for appropriate cell magnification and detection. Depending on the laser used, three different fluorescence emission filters are available and a interchangeable for detecting both naturally occurring pigments (chlorophyll and phycoerythrin) and other fluorophores

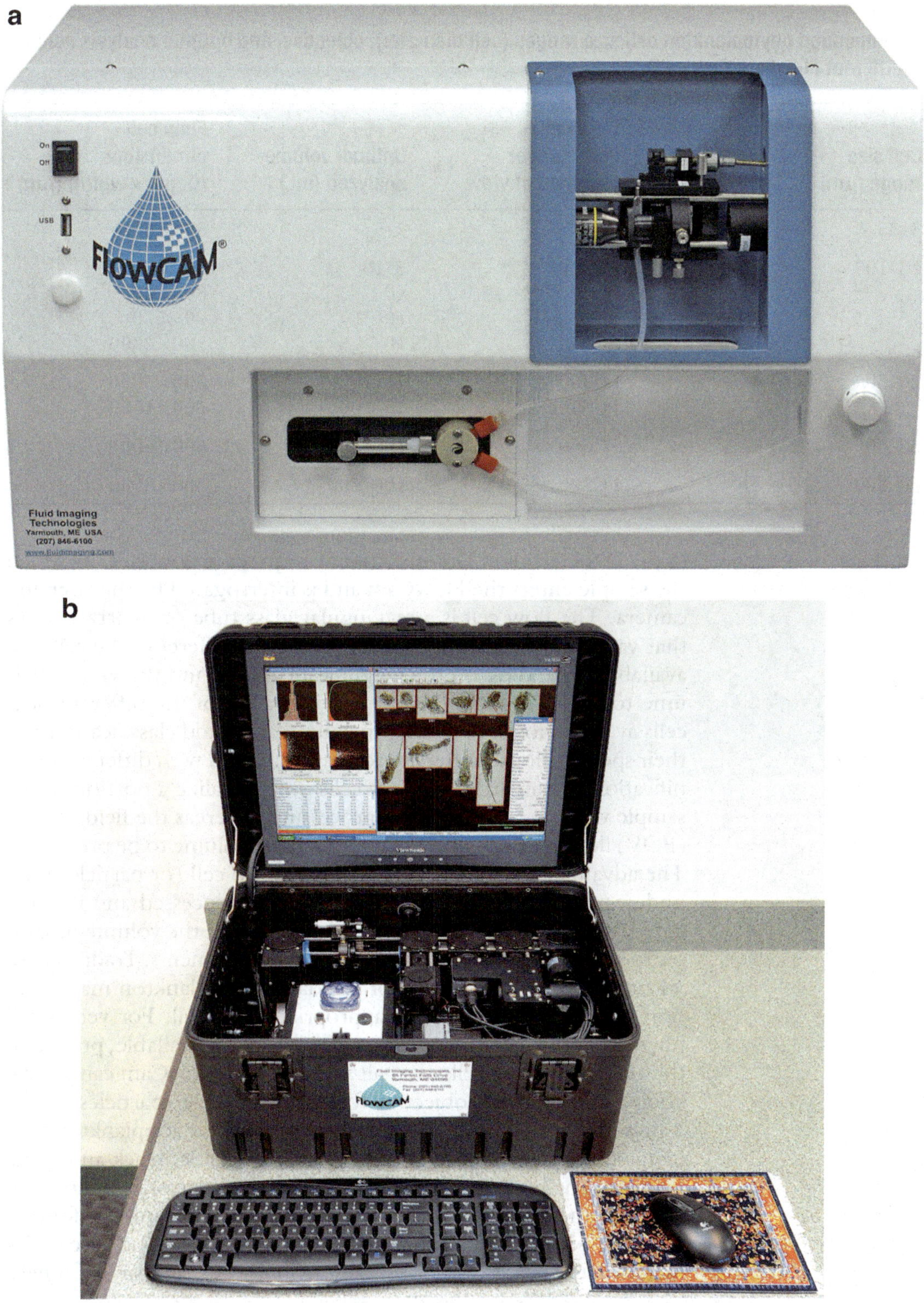

Fig. 1 FlowCam instrumentation. (**a**) Laboratory benchtop unit. (**b**) Portable unit

Table 1
Recommended phytoplankton cell size ranges (cell diameter), objective, and optimal analysis volume for different FlowCam flow cells

Cell size range (μm)	Objective	Flow cell Part number FV = field of view	Optimal volume analyzed (mL)	Flow cell dimensions (Depth × width) (μm)
5–35	20×	FC50	2–5	50 × 1000
10–100	10×	FC80FV	5–10	80 × 570
	10×	FC80-7FV	10	80 × 700
	10×	FC90-FV	10	90 × 570
	10×	FC100	10	100 × 2000
20–300	4×	FC300	25–150	300 × 3000
	4×	FC300FV	25–100	300 × 1500
60–600	4×	FC600	100–200	600 × 6000
60–800	4×	FC800	100–300	800 × 8000

(e.g., DNA stains). The instrument requires a flow cell, in which the sample enters the FlowCam and is interrogated by the laser and camera. The flow cell is a rectangular glass tube or quartz cuvette that varies in size (depth and width). Many different flow cells are available depending on the magnification used and the sample volume to be processed. Table 1 provides a list of the different flow cells available for phytoplankton enumeration and classification and their specifications. Different flow cells are used with different magnifications. Standard (FC) flow cells only visualize a portion of the sample volume during sample processing, whereas the field of view (FOV) flow cells visualize the entire sample volume to be processed. The advantage of a FOV flow cell is that no cell (or particle) goes undetected, however, much less volume is processed and interrogated. The type of flow cell utilized depends on the volume of sample to be processed and the cell detection requirements. Traditionally, ×4 or ×10 objective is used for standard phytoplankton magnification in conjunction with an appropriate flow cell. For very small phytoplankton cells or cultures, ×20 objective is available, providing the ability to image cells as small as 2 μm. The FlowCam can also be equipped with a 2× objective for use with larger particles, up to 2 mm in size such as colonial phytoplankton and zooplankton. For image capture, two different cameras are available, black and white or color. The black and white camera has higher resolution however a color camera adds six additional image analysis parameters for particle analysis and differentiation. Other consumables and accessories include latex beads of various sizes, nylon sieves, and small funnels.

2.2 Sample Preparation

Environmental samples need to be unpreserved and analyzed within a couple of hours of sampling. Samples should be kept in the dark and cool (4–6 °C). A sterile filtered diluent needs to be prepared for cleaning the sample tubing and flow cell prior to analysis and for diluting samples that are too concentrated. Typically, 0.2 μm filtered fresh or salt water is used depending on the type of samples collected. If the only option is to preserve samples prior to processing, the samples should be preserved using a solution of 10 % buffered formalin solution (*see* **Note 1**), such that the final concentration is 1 or 2 %. Preserved samples need to be stored at 4 °C, in the dark and processed as soon as possible, a the algal pigment fluorescence within the cells decreases with prolonged sample storage. Auto Image Mode is recommended for best particle detection when working with preserved samples. For different environments the effects of preservation should be tested and compared to unpreserved samples. While formalin is the preferred method for phytoplankton preservation, the FlowCam can handle samples preserved by any method—Utermöhls, methanol, etc. (*see* **Note 2**).

3 Methods

3.1 Instrument Setup

The FlowCam is primarily used for nano- and micro-phytoplankton detection (cell size range 15–300 μm), and is ideally suited for analysis of both cultures and natural samples from diverse aquatic environments. It can also be used to both quantify phytoplankton and assist in the identification of phytoplankton to the genus level and in some cases to the species level.

1. The instrument and laser should warm up for a minimum of 20 min, especially in older units.
2. To detect standard fluorescent phytoplankton pigments (chlorophyll and phycoerythrin), the use of either a green or a blue laser instrument is required.
3. The ViSp software is "set up" and threshold values for Trigger Mode detection (fluorescence detection) and particle cell size requirements need to be set within the "Context menu" of the ViSp software. Context settings are defined by the operator, and will be based on the phytoplankton cell size, sample type, location, and depth (*see* **Note 3**).
4. Using Table 1, install the appropriate flow cell vertically within the instrument. Flow cell selection depends on the size range of the phytoplankton cells to be enumerated and the sample volume analyzed (*see* **Note 4**).
5. The FlowCam needs to be focused prior to initiating a run, this can be done manually for older instruments, or automatically using the ViSp software (purchased as an option with new instruments). Fluorescent or non-fluorescent latex beads can be used to focus the instrument prior to sample analysis.
6. The FlowCam pump (peristaltic or syringe) needs to be primed with the appropriate sample diluent (fresh or salt water).

3.2 Modes of Operation

The FlowCam VS-4 is equipped with two primary modes of operation, Trigger Mode and Auto Image Mode. Trigger Mode uses a laser as the primary tool for detecting particles via fluorescence or scatter detection. For determining the presence of algae in a natural sample the algal pigments (chlorophyll and/or phycoerythrin) are detected by auto-fluorescence using laser light allowing for image capture. In the case of Trigger Mode the density of the phytoplankton needs to be ~1–1.5 cells per camera field of view. As the cells are imaged and archived by the ViSp software the particles per image are calculated and updated.

If dense or concentrated phytoplankton samples need to be analyzed, such as phytoplankton cultures, Auto Image Mode is the preferred mode of detection. Auto Image Mode does not use the laser for cell or particle detection. In this mode the camera is set at a specified frame rate ranging from 0 to 20 frames per second. When the camera shutters, images of all particles or cells within the camera's field of view are captured and saved throughout the sample run. In Auto Image Mode the number of camera images captured is determined and compared to the volume of sample analyzed and the ViSp Software determines the concentration of the cells in the sample.

3.3 Sample Processing

1. To determine the abundance of phytoplankton in a sample, the sample will require thorough mixing by gentle rotation in order to resuspend particles and cells that may have settled (*see* **Note 5**). If the sample was cooled (4–6 °C), allow the sample to warm up briefly prior to processing (*see* **Note 6**).
2. Using the FlowCam fluidics system the pump will need to be primed with the sample diluent prior to sample analysis. Priming the tubing will prevent air bubbles from entering the flow cell (*see* **Note 7**). A syringe pump is recommended for FlowCam fluidics control, as it can measure the precise volume analyzed and is calibrated and controlled using the ViSp software. If using a peristaltic pump for fluidics control (as found in older instruments) the sample volume needs to be measured using a volumetric flask or cylinder prior to the sample run (*see* **Note 8**).
3. Trigger Mode is used within the ViSp software to start the sample run, and the pump flow is initiated. For most environmental samples 10–100 mL are processed. As the sample is analyzed the camera will only capture cell images associated with a fluorescent event due the to laser excitation of the phytoplankton pigments (chlorophyll and phycoerythrin).
4. During sample processing, the ViSp software, uses a segmentation algorithm based on operator specifications and particle/cell images are captured and archived as a collage of images (Fig. 2).

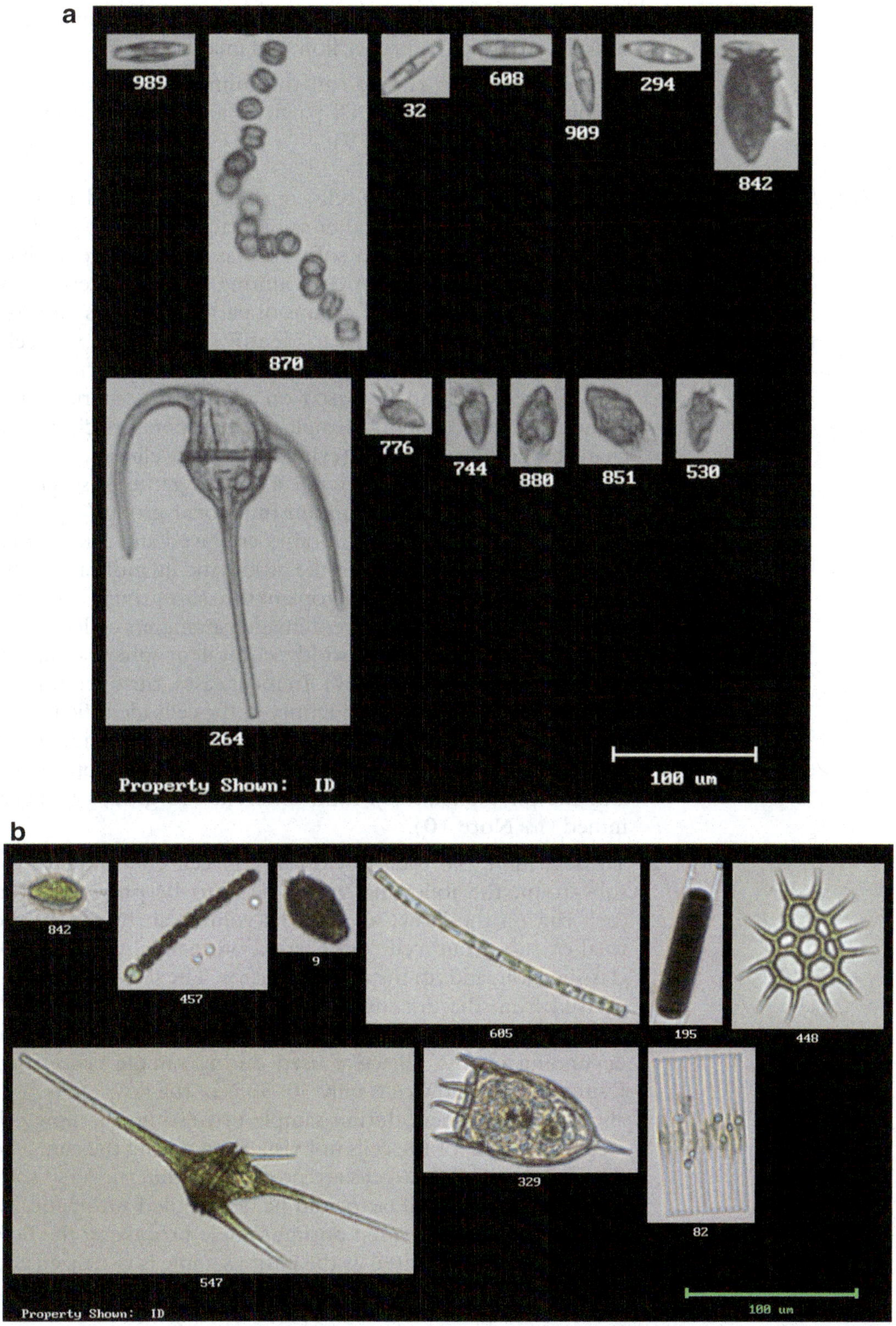

Fig. 2 FlowCam collage images from a marine sample (**a**) containing a diverse group of phytoplankton, including different types of diatoms and dinoflagellates and a freshwater sample using a color camera (**b**). The scale bar = 100 μm

As the run continues the operator will monitor the sample occasionally, ensuring the proper flow and image capture.

5. At the end of the sample run the funnel is rinsed with the sample diluent to ensure all particles and cells are removed from the funnel (*see* **Note 9**).

3.4 Sample Analysis

1. Analysis of phytoplankton cells/images acquired and archived can be done immediately after sample acquisition or at a later time. The ViSp FlowCam software has many image analysis features that allow for particle counting and characterization. Initially, image libraries or groups of particles (classes) are created by the operator in order to identify specific groups of cells with similar properties. Image libraries and/or cell classes are user defined and can be based on any image criteria found within the environmental sample. Some operators choose to characterize cells by shape (cylinder, sphere, ellipsoid) or can choose to identify or characterize a specific genus or species of phytoplankton or phytoplankton functional group (e.g., diatoms, dinoflagellates). The libraries are saved and can be used on future samples to assist in the automatic quantification and classification of specific phytoplankton sub-groups within a sample, using the 40+ different image parameters collected for each particle (e.g., length, width, equivalent spherical diameter, edge gradient, circularity). In most cases, the operator will need to manually verify the identity of the cells identified using automatic classification. After the cells have been identified and classified the concentration of the total phytoplankton and the phytoplankton within different sub-groups is determined (*see* **Note 10**).
2. To determine the total phytoplankton cell concentration or sub-group, the following values need to be provided from each run on the FlowCam: (a) the volume analyzed, (b) the total or sub-group cell count based on manual or automatic classification, and (c) the flow cell factor. The flow cell factor is derived from the dimensions of the Standard flow cell (FC) used during each run and the camera's micron per pixel value depending on the objective used during sample processing. Using standard flow cells only 30–50 % of the flow cell is visualized by the camera during sample processing, the flow cell factor accounts for the cells not visualized during the run. The dimensions of all flow cells are provided within the ViSp software and the flow cell factor can be determined automatically. If FOV flow cells are used during sample processing the flow cell factor is not required as the entire sample is visualized and enumerated.

4 Notes

1. Preparation of 10 % buffered formalin solution: Combine 900 mL distilled water and 100 g paraformaldehyde powder in a beaker on a stirring/hot plate within a fume hood and mix with a stir bar. Heat the solution to approximately 60 °C, and do not to let the solution boil. Continue to stir for 1 h, and remove from heat. Add 100 μl 1 N Sodium Hydroxide to "clear" the solution and cool to room temperature. Note: not all the paraformaldehyde powder will go into solution. Add 100 mL of phosphate buffered solution (PBS) or 0.2 μm filtered seawater to "buffer" the formalin solution. The use of PBS or filtered seawater will depend on the type of environmental samples to be preserved (freshwater or marine, respectively). Filter the buffered formalin solution through a GF/F glass fiber filter to remove the precipitate. Test the pH of the buffered formalin solution, if lower than 7.4–8.0, add sodium hydroxide drop by drop. It is best to prepare this solution fresh, prior to sample collection in the field.
2. Iodine-based preservatives, such as Utermohl's and Lugol's iodine, are not ideal for use with the FlowCam. Using these preservatives causes the natural pigment auto-fluorescence to be lost and the iodine staining causes the cells to be very high in contrast, therefore making it very difficult to identify for classification and in some cases difficult to separate from sample debris. However, iodine preserved samples have been used previously and evaluated using auto image mode [11].
3. Preparation of the context file is key for determining how the FlowCam will evaluate a sample and capture images. Within the context file the following are defined: camera settings (light, color balance, etc.), image segmentation settings and thresholds, particle cell size range to be collected, calibration factors, sample stop conditions, camera settings, and a location for inserting metadata. Understanding how the context file is generated is required and provided with basic instrument training.
4. The phytoplankton concentration and diversity can vary widely from sample to sample, depending on the ecosystem or environment of collection (e.g., coastal vs. open ocean). Since each FlowCam objective and flow cell is limited to a specific cell size range, it is possible that two FlowCam runs will be required to best quantify and classify the majority of the phytoplankton within a specific environment. In some cases, higher magnification will be necessary to quantify and classify smaller and dense phytoplankton. Initial testing and examination

of the FlowCam data may be required prior to running a series of samples from one ecosystem. In other cases, the phytoplankton density can be very dilute, as in oligotrophic regions of the ocean. In this case, particle concentration will be required using reverse filtration. In this method the sample is concentrated gently using either 10 or 20 μm nylon mesh minimizing damage to the cells [12].

5. As a sample is processed, the operator should occasionally mix the sample, to prevent cells from settling. If the original environmental sample was not initially screened with a Nitex/nylon mesh (100 or 300 μm) prior to analysis, the operator should check the flow cell periodically for clogs or partial blockages at the flow cell entrance. When the flow cell is clogged or a particle will not pass through the flow cell, the run should be stopped and restarted with a clean flow cell. To prevent or minimize clogging the operator could choose to prescreen all samples with Nitex/nylon mesh. Disposable and sterile 40, 70, and 100 μm Nitex meshes are available from Becton Dickinson and from Fluid Imaging Technologies.
6. In humid environments, especially on board boats when working in the field it is important that the phytoplankton sample is warmed up few degrees or to room temperature prior to processing. This will prevent condensation from forming on the flow cell, if condensation forms, the camera will be unable to capture images and the run will need to be stopped and flow cell will need to be cleaned.
7. It is essential to prevent air from entering the flow cell or allowing the flow cell to run dry. This will prevent air bubbles from forming within the sample line. Bubbles are highly fluorescent and will be detected as particles and imaged. If captured, bubbles and other detrital particles will need to be removed from the image dataset prior to determining the total phytoplankton cell abundance.
8. The size of sample funnel will depend on the volume to be processed and analyzed. If necessary, a large ring stand may be required for sample volumes over 100 mL.
9. Depending on the type of flow cell used, cleaning between samples may be required or after a day a sample processing. Only the quartz cuvette FOV flow cell requires cleaning, as the standard (FC) flow cells are considered disposable. In some cases these FOV flow cell will require soaking overnight in a diluted laboratory surfactant (e.g., Micro90, Contrad70, Simple Green). If necessary an overnight soak in 10 % hydrochloric acid may be required.
10. In some cases the operator will only need to determine the total phytoplankton count and concentration; therefore,

phytoplankton classification using image libraries is not required. In these cases, the operator needs to scan and identify any detritus, bubbles, and zooplankton that were captured during sample processing. Libraries can also be created to identify bubbles and sample debris fairly rapidly using specific image analysis criteria. Once these particles are removed from the image dataset, the total phytoplankton concentration can be determined accurately and quickly.

Acknowledgements

Support for this work was provided by Bigelow Laboratory for Ocean Sciences and the J.J. MacIsaac Facility for Aquatic Cytometry. I would like to thank Harry Nelson from Fluid Imaging Technologies for providing color images for the manuscript.

References

1. Sieracki CK, Sieracki M, Yentsch CS (1998) An imaging-in-flow system for automated analysis of marine microplankton. Mar Ecol Prog Ser 168:285–296
2. Utermöhl H (1958) Zur vervollkommnung der quantitativen phytoplankton-methodik. Mitt Int Ver Theor Angew Limnol 9:1–38
3. Buskey E, Hyatt CJ (2006) Use of the FlowCAM for semi-automated recognition and enumeration of red tide cells (*Karenia brevis*) in natural plankton samples. Harmful Algae 5:685–692
4. See JH, Campbell L, Richardson TL et al (2005) Combining new technologies for determination of phytoplankton community structure in the northern Gulf of Mexico. J Phycol 41:305–310
5. Zarauz L, Irigoien X, Urtizberea A et al (2007) Mapping plankton distribution in the Bay of Biscay during three consecutive spring surveys. Mar Ecol Prog Ser 345:37–39
6. Spaulding SA, Jewson DH, Bixby RJ et al (2012) Automated measurement of diatom size. Limnol Oceanogr Meth 10:882–890
7. Álvarez E, Moyano M, López-Urrutia A et al (2013) Routine determination of plankton community composition and size structure: a comparison between FlowCAM and light microscopy. J Plankton Res. doi:10.1093/plankt/fbt069
8. Zarauz L, Irigoien X, Fernandas JA (2009) Changes in plankton size structure and composition, during the generation of a phytoplankton bloom, in the central Canatabrian Sea. J Plankton Res 31:193–207
9. Alvarez E, Lopez-Urrutia A, Nogueira E et al (2011) How to effectively sample the plankton size spectrum? A case study using FlowCAM. J Plankton Res 33:1119–1133
10. Jakobsen HH, Carstensen J (2011) FlowCAM: sizing cells and understanding the impact of size distributions on biovolume of planktonic community structure. Aquat Microb Ecol 65:75–87
11. Zarauz L, Irigoien X (2008) Effects of Lugol's fixation on the size structure of nano-microplankton analyzed by means of an automated counting method. J Plankton Res 30: 1297–1303
12. Dodson AN, Thomas WH (1978) Reverse filtration. In: Sournia A (ed) Phytoplankton manual. UNESCO, Paris, pp 104–107

Chapter 18

Detection and Characterization of Rare Circulating Endothelial Cells by Imaging Flow Cytometry

Leigh Samsel and J. Philip McCoy Jr.

Abstract

Circulating endothelial cells (CECs) are angiogenic cells that appear in increased numbers in the peripheral circulation either as a result of vascular injury or in response to angiogenic stimuli. Elevated levels of CECs have been correlated with various disease states, indicating the use of CECs as a biomarker of disease. Flow cytometry is a widely accepted method for detecting and quantitating CECs. Flow cytometry provides statistical information on large numbers of cells but no information on morphological characteristics. Imaging flow cytometry combines traditional flow cytometry and microscopy, providing a streamlined, multiparameter approach to characterize the biological properties and morphology of large numbers of cells, and is particularly amenable for rare event analysis such as CECs. This approach for identifying and characterizing CECs allows the morphological characterization of large numbers of live, nucleated, single CECs, and alleviates the need for prior enrichment.

Key words Circulating endothelial cells, CECs, Imaging flow cytometry, Rare event flow cytometry, CD146

1 Introduction

The endothelium is a layer of cells providing the inner lining of blood vessels, forming a barrier between the blood and the outer portion of vessels. Direct assessment of endothelial damage or repair is difficult to accomplish with unobtrusive techniques, thus circulating endothelial cells (CECs) have been proposed as surrogate biomarkers of endothelial health in various diseases. CECs have been reported to be an indicator of vascular injury in many conditions such as sickle cell disease, SLE, pulmonary hypertension, vasculitis, rickettsial infection, cardiovascular disease and percutaneous coronary intervention, renal transplantation, and others. Levels of CECs have frequently been reported to be elevated in many types of cancers such as breast, lymphoma, prostate, pancreatic, renal, lung, and rectal cancer, among others. The extreme rarity of these cells necessitates the need for accurate resolution of

Natasha S. Barteneva and Ivan A. Vorobjev (eds.), *Imaging Flow Cytometry: Methods and Protocols*, Methods in Molecular Biology, vol. 1389, DOI 10.1007/978-1-4939-3302-0_18, © Springer Science+Business Media New York 2016

one CEC among a background of millions of leukocytic and other non-endothelial cells and particles in the peripheral blood.

Several different approaches have been developed for detection and enumeration of CECs using combinations of surface markers. A problem common to many of the markers used to identify CECs is the lack of complete specificity for endothelial cells. CD34, for example, is well known to be expressed on endothelial cells [1, 2] but in the hematology and cytometry communities is more regarded as a marker of hematopoietic stem cells [3, 4]. CD31, also called PECAM-1, can mark endothelial cells but is also found on a wide variety of leukocytes [5, 6]. CD146, also known as MCAM, had been described as an endothelial-specific marker [7–10] but is now known to be expressed on a small percentage of circulating T lymphocytes whose expression can be upregulated by activation [11, 12].

Flow cytometric immunophenotyping is ideally suited for detection and quantification of CECs, as it can rapidly analyze tens of thousands of cells per second using multiple markers to identify cells of interest. Flow cytometry can accurately quantify the number of positive events, and, if desired, can physically separate these cells for further studies. Imaging flow cytometry offers many of the same benefits for identification of CECs as does regular flow cytometry (except for cell sorting), and in addition offers the ability to visualize these cells. Visualization not only offers morphologic information regarding these cells, but can also provide data concerning antigen co-localization, antigen capping, cell health, and so forth.

Millions or tens of millions of events from peripheral blood need to be analyzed in order to acquire a sufficient number of the rare CECs to assure, or at least suggest, statistical significance. Several special steps need to be taken in flow cytometric detection of rare events. Of great importance is the need for the cytometer to be as clean as possible, thus lessening the amount of debris or background material ending up in the data files collected. It is recommended to clean the cytometer after running the single color compensation controls and before running the *N*−1 control, because the latter control is the most critical for assessing background events. Crucial components of this assay in addition to thorough cleaning of the cytometer include: collection of a large number of events, use of a blocking serum to help minimize nonspecific binding, exclusion of dead cells, inclusion of mononuclear cells, exclusion of doublets and aggregates, use of a nuclear stain to ensure that cells rather than clumps of particles are examined, time-resolved gating to exclude system variation, CD45 and CD3 to exclude leukocytes, use of an *N*−1 control to help identify background events, ensure accurate compensation, and set gates, use of the brightest fluorochrome for the rare event marker, and the use of CD146 to identify endothelial cells.

As alluded to above, setting up this assay for CECs requires the use of multiple fluorochrome-conjugated antibodies for staining the cells. A generally accepted principle in designing polychromatic flow cytometry assays is to use the fluorochrome with the greatest quantum yield (i.e., the "brightest") with the marker that is the least dense (i.e., the "dimmest") in order to achieve acceptably bright staining of the marker in question [13]. In rare event analysis not only does this apply, but also does the desire to have the most rare marker stain as brightly as possible in order to best separate stained cells from background. Markers such as CD146 are generally found as a continuum of intensities in peripheral blood rather than as a discrete population of events. Thus the use a bright fluorochrome with CD146 is highly desirable. It should be emphasized that the goal of imaging flow cytometry of CECs is the visualization and characterization of these cells, not their quantification. Imaging flow cytometry has much slower acquisition rates and creates much larger file sizes, making this technique difficult to replace routine flow cytometry as the method of choice for enumeration. However, imaging flow cytometry is clearly a preferred method for visualization of CECs as it is much quicker than manual analysis, is capable of using multiple fluorochromes, and yields data which can be gated in a Boolean manner and statically analyzed.

2 Materials

1. 8.5 ml Vacutainer tubes containing ACD solution A (BD Biosciences, San Jose, CA).
2. 50 and 15 ml conical polypropylene tubes.
3. ACK Lysing Buffer (Quality Biologicals, Gaithersburg, MD): A commercial product containing ammonium chloride and potassium designed to lyse red blood cells but not leukocytes; is utilized to lyse red blood cells in whole blood, leaving white blood cells intact.
4. FACS buffer: PBS pH 7.2, 0.5 mM EDTA, 0.2 % BSA. 0.5 mM EDTA, and 0.2 % BSA were added to the PBS to help prevent cells from aggregating, while pH 7.2 better enabled Hoechst labeling. Store FACS buffer at 4 °C.
5. Normal mouse serum.
6. Antibodies (*see* **Note 1**):

   CD146 PE (clone P1H12).

   CD3 AlexaFluor 647 (clone UCHT1).

   CD45 APC-Cy7 (clone 2D1).
7. Hoechst 33342.
8. 7-AAD.

9. 70 μ nylon mesh strainers.
10. ImageStream X or MKII imaging flow cytometer equipped with 405, 488, 561 (optional), 642 (or 658), and 785 nm lasers and two cameras/12 channels; INSPIRE acquisition software and IDEAS analysis software (Amnis, Seattle, WA).

3 Methods

3.1 Preparation of Sample for Immunophenotyping

1. Prior to beginning this procedure, all antibodies should be titrated according to McLaughlin, et al. [14] (*see* **Note 2**).
2. Collect whole blood (*see* **Note 3**) by venipuncture in vacutainer tubes containing ACD.
3. Centrifuge blood in the original vacutainer tubes for 10 min at 400 ×*g* at room temperature.
4. Remove the top plasma layer by vacuum aspiration (or similar method) and discard. For more complete lysis of blood, remove as much of plasma layer as possible.
5. Transfer remaining blood cell pack (~3–4 ml) from vacutainer tube to a 50 ml polypropylene conical tube. Use polypropylene tubes for all subsequent steps to minimize cell loss. Use one 50 ml conical tube for each vacutainer used.
6. Dilute the blood cell pack 1:10 with ACK Lysing Buffer.
7. Lyse blood for 10 min at room temperature with gentle rocking.
8. Centrifuge the lysed blood for 10 min at 400 ×*g* at room temperature.
9. Carefully remove the supernatant by vacuum aspiration (or similar method) and discard.
10. Resuspend cell pellet in 5 ml ACK Lysing Buffer. Cells can be combined into one 50 ml conical tube at this step.
11. Lyse cells a second time for 10 min at room temperature with gentle rocking.
12. Centrifuge the lysed blood for 10 min at 400 ×*g* at room temperature.
13. Carefully remove the supernatant by vacuum aspiration (or similar method) and discard.
14. Resuspend the cell pellet in 10 ml FACS buffer to wash the cells.
15. Centrifuge for 10 min at 400 ×*g* at room temperature.
16. Carefully remove the supernatant by vacuum aspiration (or similar method) and discard.

17. Resuspend the cell pellet in 800 μl FACS buffer.
18. Add 200 μl normal mouse serum.
19. Incubate for 10 min at room temperature.
20. Perform a cell count and determine cell concentration.
21. Label 1.5 ml microcentrifuge tubes for the following (*see* **Note 4**):
    (a) Unstained control.
    (b) CD146 PE-only control.
    (c) CD3 AF647-only control.
    (d) CD45 APC-Cy7-only control.
    (e) Hoechst 33342-only control.
    (f) 7-AAD-only control.
    (g) *N*− 1 control.
22. Transfer 1×10^6 cells to each control tube. Bring the volume in each control tube up to 100 μl with FACS buffer. Use remaining cells in 50 ml conical tube for CEC sample staining.
23. Add the appropriate amount of antibody, as determined in **step 1** of Subheading 3.1, to the tubes for controls and CEC sample. The *N*− 1 control will contain CD3 AlexaFluor 647 and CD45 APC-Cy7, but should NOT contain CD146 PE. The CEC sample will contain CD146 PE, CD3 AlexaFluor 647, and CD45 APC-Cy7.
24. Incubate at room temperature for 30 min in the dark.

    (Prior to completion of sample preparation, the ImageStream should be started up, fluidics should be initialized, and all ASSIST tests and calibrations should be run. This can be done during the incubation steps.)
25. Add 1 ml FACS buffer to the control tubes and 10 ml FACS buffer to CEC sample tubes to wash the cells.
26. Centrifuge for 10 min at $400 \times g$ at room temperature.
27. Carefully remove the supernatant by vacuum aspiration (or similar method) and discard.
28. Resuspend the single-color controls and *N*− 1 control in 50 μl FACS buffer, and the CEC cell pellet in FACS buffer to a concentration of ~ $3–5 \times 10^6$ cells in 50 μl.
29. Add Hoechst 33342 to the Hoechst only control, *N*− 1 control, and CEC sample tubes to a final concentration of 0.5 μM per million cells and incubate at 37 °C for 30 min. For example, since the Hoechst-only control and *N*− 1 control

have 1×10^6 cells in 50 μl, add 1.5 μl of 0.1 mM Hoechst 33342. For the CEC sample, multiply the number of millions of cells by 1.5 μl to determine how much 0.1 mM Hoechst 33342 to add. Do not wash cells after incubation.

30. Add 1.5 μl of a 50 mg/ml stock solution of 7-AAD per million cells to the 7-AAD-only control, *N*– 1 control, and CEC sample tubes.

3.2 Imaging Flow Cytometry

3.2.1 Instrument Setup

Prior to completion of sample preparation, the ImageStream should be started up, fluidics should be initialized, and all ASSIST tests and calibrations should be run. This can be done during the incubation steps of the preparation of sample for immunephenotyping. All acquisition is performed in INSPIRE software.

This experiment was performed on a two camera ImageStream MKII equipped with 405, 488, 561, and 642 lasers. ImageStreams may have different lasers and varying powers. Modifications concerning which fluorochromes to use can be made based on instrument configuration. Instrument settings, laser powers, and acquisition criteria may be different on an ImageStream^X and on ImageStreams with varying laser powers.

1. For the panel described above, the following channels and lasers are used:
   (a) Bright-field imagery is collected in Channels 1 and 9 at an intensity of 800.
   (b) SSC imagery is collected in Channel 6 (745–800 nm filter) off of the 785 nm laser.
   (c) PE imagery is collected in Channel 3 (560–595 nm filter) off of the 488 and/or 561 nm laser.
   (d) 7-AAD imagery is collected in Channel 5 (660–745 nm filter) off of the 488 and/or 561 nm laser.
   (e) Hoechst imagery is collected in Channel 7 (430-505 nm filter) off of the 405 nm laser.
   (f) AlexaFluor 647 imagery is collected in Channel 11 (660–745 nm filter) off of the 642 or 658 nm laser.
   (g) APC-Cy7 imagery is collected in Channel 12 (745-800 nm filter) off of the 642 or 658 nm laser.
2. Set the desired magnification. In Samsel et al. [15], all data were collected at 60×. The collection speed will vary according to instrument model and magnification.
3. Laser powers for all lasers should be determined such that the events off a given laser are as bright as possible but are not saturating the cameras in those channels. This can be accomplished by running each single-color control sample, plotting the Raw Max Pixel (RMxP) feature (non-background-subtracted

pixel intensity) for that channel, and ensuring that events are not falling at or above a RMxP value of 4094. Increase or decrease laser power as needed. Return the sample, which will not be acquired until all laser powers have been determined. Repeat this process for each single-color control until all optimal laser powers have been established.

4. Set lasers at the optimal powers determined.

3.2.2 Sample Acquisition and Enrichment by Gating

Setting acquisition gates or cell classifiers in INSPIRE acquisition software (Amnis Corporation, Seattle, WA) designates which events will be acquired into the data file, and which events will be eliminated. This provides a powerful tool for rare event cytometry, in allowing for enrichment of the cell of interest at the time of acquisition. Due to the extremely rare nature of CECs and the large number of cells that must be imaged in order to find CECs, a single raw image file (.rif) data file from the ImageStream containing sufficient events for detection and analysis of CECs would need to contain several million events and the file size would be hundreds of gigabytes. The analyses of files of such size are not practical using standard desktop computer systems; therefore it is necessary to acquire ImageStream data differently than is usually done for flow cytometry. To minimize file size, data should be collected using an acquisition gate or cell classifier (live acquisition gate) so that the majority of CD146 PE negative events are not collected, thus enriching for CD146$^+$ events in the acquired data file. All CECs are CD146$^+$, but only a small percentage (2–3 %) of other mononuclear cells express this marker. Using this approach, a .rif data file containing 10,000 events could be collected, having a file size of roughly 1–4 GB. Even using this enrichment by gating approach, it is necessary to collect multiple 10,000 event .rif files in order to analyze the entire aliquot of any given sample. Each .rif is then analyzed individually.

This gating by enrichment or cell classifying can be accomplished several ways. One way would be to use the Intensity feature of the CD146 PE channel (Channel 3) to identify CD146 negative and positive events. However, to minimize the number of different acquisition plots used, we used the Raw Max Pixel_CD146 PE (Channel 3) feature, which we also used to eliminate saturating events. Whichever features you choose, however, extreme caution should be taken and these gates should be set generously so that cells of interest are not inadvertently omitted.

Acquisition gates for an ImageStream MKII were set as follows (Fig. 1) (*see* **Note 5** for setting acquisition gates for an ImageStream^X).

1. Load the unstained sample.
   (a) Create a two-parameter dot plot of the Area_Channel 1 feature versus the Aspect Ratio_Channel 1 feature and

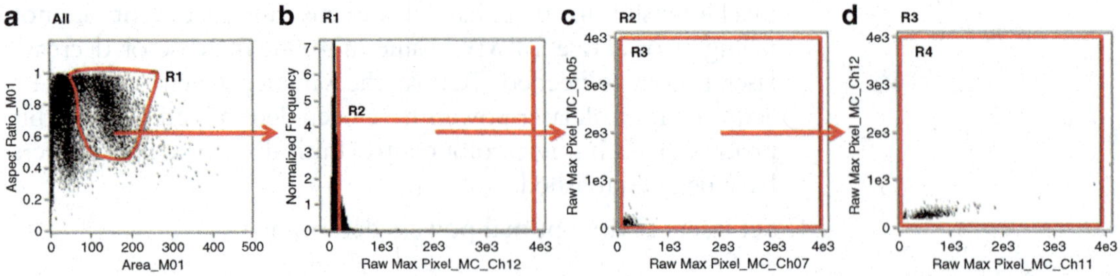

Fig. 1 Gating strategy for enrichment of CECs during acquisition on an ImageStream MKII. Debris and large aggregates were eliminated with the R1 gate (**a**). The majority of CD146 PE-negative events were eliminated with the R2 gate to enrich for CD146-positive cells. Great care must be taken to ensure all PE-positive events are included in the R2 gate. The left edge of the R2 gate should be set generously. Events which may have saturated the camera in the PE channel were also eliminated in the R2 gate (**b**). 7-AAD or Hoechst fluorescence which may have saturated the camera was gated out (**c**), as was any saturating CD3 AlexaFluor 647 or CD45 APC-Cy7 fluorescence (**d**)

draw a generous amorphous gate ("R1") around the single-cell population to eliminate the acquisition of debris and aggregates into the acquired data file (Fig. 1a).

(b) Create a histogram, gated on "R1," of the Raw Max Pixel_ Channel 3 feature. Create a region ("R2") whose left edge excludes the majority of the unstained cells and whose right edge is set at a RMxP_Ch 3 value of 4090 to exclude events which were saturating the camera (Fig. 1b). Great care must be taken to ensure that dim CD146 PE positive events are included in this gate. The gate should be set generously. Stricter gating can be done in subsequent analysis after compensation has been performed. Exact values for the lower RMxP_Ch 3 setting are not given here, as the actual lower RMxP_Ch 3 setting may vary from instrument to instrument, with laser powers, and with staining conditions.

The following steps were only used to eliminate saturating events. These are not strictly necessary, but are highly valuable in gating out events which are saturating the camera, and thus could have incorrectly calculated feature values.

(c) Create a two-parameter dot plot, gated on the CD146 positive gate "R2", of the Raw Max Pixel_Channel 7 feature (Hoechst) versus Raw Max Pixel_Channel 5 feature (7-AAD). Create a large rectangular gate ("R3"), which includes all events except those containing saturating fluorescence (Fig. 1c). Set the upper left and lower right vertices to a raw max pixel value of 4090.

(d) Create a two parameter dot plot, gated on the "R3" gate, of the Raw Max Pixel_Channel 11 feature (CD3-AF647) versus the Raw Max Pixel_Channel 12 feature (CD45 APC-Cy7). Create a large rectangular gate ("R4") (Fig. 1d) which includes all events except those containing saturating fluorescence. Set the upper left and lower right vertices to a raw max pixel value of 4090. Setting the gate in this manner allowed the study of Th17 phenotype identified by CD146 expression on $CD3^+$ cells [16].

2. Set the destination folder, enter the file name, and number of events to collect.
3. Designate the R1 acquisition gate in the File Acquisition area in INSPIRE. It is necessary to use the R1 gate at this step, as the R2 gate is for acquiring PE-positive events, which should not occur in the unstained sample. Acquire 10,000 events of the unstained sample.
4. Load and acquire 10,000–20,000 events of the $N-1$ sample using the R1 acquisition gate. Similarly as for the unstained sample, it is necessary to use the R1 gate at this step, since the R2 gate is set for acquiring PE-positive events, which should not occur in the $N-1$ sample. Any PE positive events that do occur in the acquired file will have arisen as a result of nonspecific binding or dead cells, etc., and will help with eliminating false PE positive events from the final CEC gate.
5. Load the CEC sample and ensure that the acquisition gate acquisition gate in the File Acquisition area in INSPIRE is set ("R4"). The cells which are being currently viewed (top left pull-down menu in INSPIRE) can also be set to R4 to visualize only the events of interest as they are being acquired.
6. Acquire multiple 10,000–20,000 event files.

3.2.3 Acquiring Single-Color Compensation Controls

It is absolutely essential that compensation controls be acquired properly and that the compensation matrix is accurate (see Subheading 3.3.1 for creating the compensation matrix). Single-color controls for compensation must be run with bright field and the 785 nm laser turned off, and with all other lasers set to the power used while collecting the CEC samples. The acquisition gates will be different for the single-color controls than they were for acquiring the unstained, $N-1$, or CEC sample because turning off bright field invalidates the R1 acquisition gate (as no events will fall in the R1 bright-field gate). In fact, the acquisition gates will be different for each single color control, corresponding with which fluorescence channel is currently being used. Further, extremely bright events which saturate the camera should be eliminated from the acquisition file using the Raw Max Pixel feature, to ensure proper compensation. Since the CEC sample was acquired uncompensated, the single-color controls can be acquired after the samples.

Acquisition gates for the single-color controls were set as follows:

(a) Turn off both bright field and the 785 laser. Files acquired with bright field turned off will be annotated with "_NoBF.rif". These files should be used when creating the compensation matrix in IDEAS following acquisition.

(b) Create a histogram gated on All events using the Raw Max Pixel_Channel 3 (CD146 PE) feature and draw a region so that events above 4090 are eliminated from the acquired file. Use this gate while acquiring the CD146 single-color control.

(c) Acquire 10,000 events.

(d) Repeat **steps (b)** and **(c)** above for each fluorescence channel used, changing the Raw Max Pixel feature channel as appropriate for each single-color control. A new acquisition gate, which eliminates the saturating fluorescence for the channel currently being used, will need to be created for each channel used, and set as the acquisition gate.

3.3 Data Analysis

All data analysis is performed in IDEAS analysis software.

3.3.1 Compensation Matrix

1. Prior to creating the compensation matrix, each single color control file can be separately analyzed to identify true positively stained cells, and a new "_NoBF" rif can be created which contains only true positively stained cells, eliminating debris, particles staining nonspecifically for the dye, or autofluorescence (*see* **Note 6**).
2. Follow the steps in the compensation wizard to create a compensation matrix utilizing the "_NoBF" single color control rif files.
3. View the created matrix and ensure that there are values for each channel used. It is absolutely imperative that all compensation values in the matrix be verified and modified if necessary. All single-color controls should be checked individually against every other single color control used in the panel with the calculated compensation matrix to ensure that appropriate values have been determined. If values are over- or under-compensated, the matrix should be edited until compensation is accurate.

3.3.2 Gating Strategy

A hierarchical gating strategy is created to identify CECs (Fig. 2).

1. Open a CEC rif file and apply the optimized compensation matrix.
2. Create a Brightfield_Area versus Intensity_Side Scatter plot to identify and create a gate on low side scatter mononuclear cells (Fig. 2a).

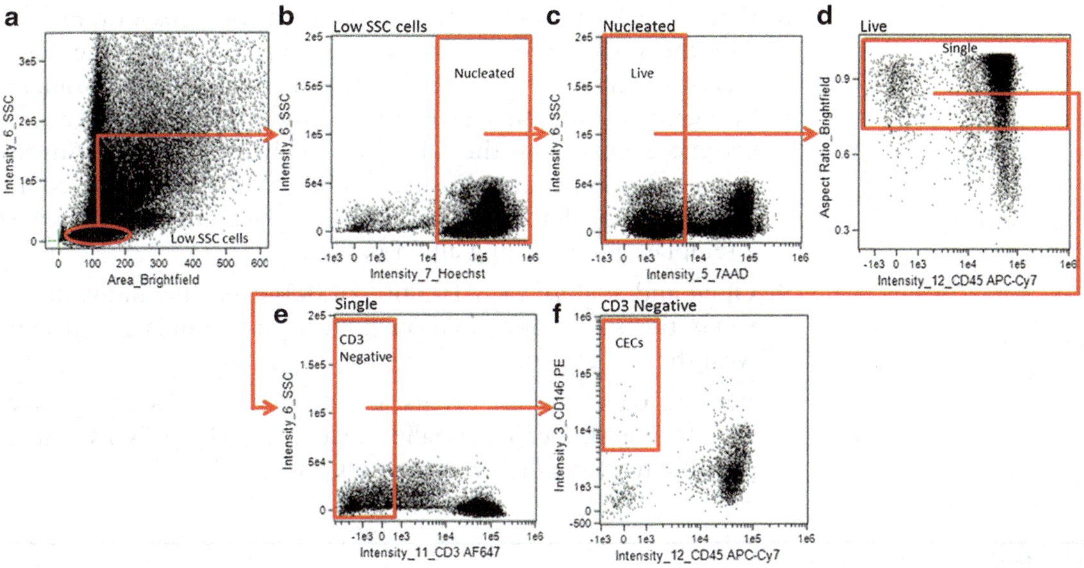

Fig. 2 Analysis gating strategy for identifying CECs from Samsel, et al. [15]. CECs were identified as low SSC (**a**), Hoechst positive (nucleated) (**b**), 7-AAD negative (live) (**c**), aspect ratio high (single cells) (**d**), CD3 negative (non-T-cells) (**e**), CD45 negative (non-leukocytic) CD146-positive cells (**f**)

3. Create an Intensity_Hoechst (Channel 7) versus Intensity_Side Scatter (Channel 6) plot, gated on the low side scatter mononuclear cells, to identify and gate on Hoechst positive (nucleated) cells (Fig. 2b).
4. Create an Intensity_7AAD (Channel 5) versus Intensity_Side Scatter (Channel 6) plot, gated on nucleated low side scatter cells, to identify and gate on Live (7AAD negative) cells (Fig. 2c).
5. Create an Intensity_CD45 APC-Cy7 (Channel 12) versus Aspect Ratio_Brightfield plot, gated on Live, Nucleated, low side scatter cells, to identify and gate on single cells (Fig. 2d). Single, round cells have a higher aspect ratio while elongated cells, aggregates, or more than one cell in a frame have a lower aspect ratio.
6. Create an Intensity_CD3 AlexaFluor 647 (Channel 11) versus Intensity_Side Scatter plot, gated on Live, nucleated, single cells with low side scatter, to identify and gate on CD3 negative cells (Fig. 2e).
7. Create an Intensity_CD45 APC-Cy7 (Channel 12) versus Intensity_CD146 PE (Channel 3) plot, gated on Live, nucleated, single, CD3-negative cells with low side scatter, to identify and gate on CD146+ CD45− cells (Fig. 2f). Save this file and use it as a template to verify the location of all gates in the *N*−1 control file.

8. Open the *N*− 1 control file with the final compensation matrix and the CEC template created in **step 7**. Adjust the final CD146+ CD45− CEC gate location as necessary to eliminate background and non-specific events, and minimize those background events from the CD146+ CD45− CEC gate. Adjust other gates as needed. Ensure that all data are properly compensated and adjust matrix if necessary. Save this file and use as a template for analyzing all CEC files.
9. Open and analyze each 10,000–20,000 event file individually using the final compensation matrix and template adjusted with the *N*− 1 control.
10. Because these are rare events, the number of CECs may vary greatly in each file. To visualize a larger number of CECs, it is possible to merge files containing CECs.

4 Notes

1. Panel design: There are many guidelines and considerations which should be followed for choosing antibodies and fluorochromes when designing a panel for rare event analysis.
   (a) Instrument configuration: Instrument configurations vary depending on the wavelengths and number of lasers and channels available. Panels should be designed to perform optimally with the configuration of the instrument being used.
   (b) Antibody and fluorochrome selection, minimizing spectral overlap: The brightest fluorochromes in a panel should be utilized for the antibody specific to the rare event marker, or marker with lowest antigen density [17]. For this reason, we used phycoerythrin (PE), which has a high staining index, for CD146, the marker used for detecting CECs. In addition, care should be taken to minimize the spectral overlap between the fluorochromes used in the panel. Spectral overlap and fluorochromes being excited by more than one laser can both lead to data spread which can negatively affect sensitivity [18]. For example, even though PE-Cy5 has a higher staining index than PE, we chose not to use tandems due to spillover optics. All lasers available in the instrument configuration should be used to spread fluorochromes out across channels and reduce spectral overlap.
   (c) Specificity and exclusion gating: Several markers have been described for identification of CECs, but none are specific only to endothelial cells, and are found on other cell types [7–10]. One such marker, CD146, also known

as MelCAM and MCAM, was originally thought to be an endothelial-specific marker. However, CD146 was subsequently found to be present on a subset of activated T-lymphocytes [11, 12, 16]. It is important to include additional markers in the panel which will exclude non-CEC events that also express the antigen of interest (in the case of CECs, CD146) and ensure specificity. Elimination of events not meeting your criteria is referred to as exclusion gating, or a "dump" channel. For example, the pan-leukocyte marker CD45 is expressed on all leukocytes in the peripheral circulation. As such, CD45-positive events are commonly excluded from CEC detection so that the vast majority of non-CEC events will be eliminated from the analysis. For this reason, we included CD3 and CD45 antibodies into our panel to exclude T-lymphocytes which also express CD146. By excluding or "dumping" CD3 positive and CD45 positive cells out of further analysis, only CD146-positive cells which are negative for expression of CD3 and CD45 could be considered as a CEC.

(d) Viability and nuclear staining: It is important to remove dead cells from the analysis as dead cells may bind antibodies nonspecifically, thus giving rise to background and false-positive events. Additionally, use of a nuclear dye can be used to ensure that events being detected as CECs are single cells, rather than clumps or particles. Because of the desire to spread fluorochromes across all lasers, the collection channels still available, and because the cells were not fixed, Hoechst 33342 was chosen as the nuclear dye and 7-AAD was chosen as a viability dye in this panel.

2. Antibody titration: In order to optimize the staining of cells with fluorochrome-conjugated antibodies, the antibodies were first titered to determine the concentration that maximally distinguishes cells staining positive for the respective antigen from the cells which do not bind to the antibody. This was accomplished by testing dilutions (in this case from 1:2 to 1:100) of each antibody against one million PBMCs from a healthy donor. The concentrations of the initial stock solutions of each antibody were determined by the vendors, thus the titers used will vary from vendor to vendor and even from lot to lot of antibody. 10 μl of the antibody dilutions was added to 100 μl of PBS containing the 10^6 PBMCs and incubated for 30 min at room temperature. The cells were then washed once, centrifuged, and analyzed immediately. The cells were analyzed on a Beckman Coulter MoFlo and the staining index (SI) calculated for each antibody. The SI is the mean channel number of the positive staining divided

by the mean channel number of the negative staining. The dilution yielding the highest SI was selected as the concentration to be used for all subsequent staining.

3. Quantity of whole blood needed: Because the number of CECs is extremely low in normal human peripheral blood and varies in different diseases and disease states, the volume of whole blood needed may vary [19]. We utilized two 8.5 ml vacutainer tubes for basic immunophenotyping and detection of CECs by imaging flow cytometry. A higher volume of blood may be desired to obtain a higher "*N*" of detected CECs, for further characterization of CECs with additional markers, for validation of immunophenotype, or if subject type is normal control blood. If a higher number of detected CECs are desired, or if the subject type is normal control, scale up the amount of whole blood used. Validation of this immunophenotypic imaging flow cytometry method for CEC detection was accomplished by sorting a portion of the sample which was used for imaging, and performing Q-RT PCR for endothelial genes [15]. Because enough CECs were needed for both sorting followed by Q-RT PCR *and* detection of CECs by imaging flow cytometry, and because CECs are a rare population, we started with three 8.5 ml vacutainer tubes when performing both sorting and imaging simultaneously. Further, if using an ImageStream^X, only approximately 50 % of the sample loaded can be imaged due to priming of the pumps, while using an ImageStream MKII, approximately 95 % of the sample loaded can be imaged. If an ImageStream^X is being used, starting with 2–3 tubes of whole blood can be advantageous.
4. Controls: Controls are needed for performing spectral compensation and proper setting of gates. To ensure proper correction of spectral overlap, single color controls are run to perform compensation. Each single color control should contain only one antibody/dye. The Hoechst single color control will need to be incubated at 37 °C for 20 min after addition of Hoechst. The $N-1$ control should contain CD3 AlexaFluor 647, CD45 APC-Cy7, Hoechst, and 7AAD, but should NOT contain CD146 PE. For the $N-1$ control, antibody labeling should be performed first per the protocol, cells should be washed with FACS buffer, the supernatant should be removed, cells should be resuspended in FACS buffer, and incubated with Hoechst at 37 °C for 20 min. After Hoechst incubation, the cells should not be washed. 7-AAD is added last.
5. Setting cell classifiers for acquisition on ISX: Images were acquired on a 2 camera, 12-channel ImageStream^X (Amnis Corporation, Seattle, WA) utilizing 405, 488, 658, and 785 nm lasers. Bright field was collected in Channels 1 and 9 at an intensity of 800, SSC was collected in Channel 6 at a 785 nm

power of ~2 mW, Hoechst was detected in Channel 7 (430-505 nm filter), PE and 7AAD were detected in Channels 3 (560–595 nm filter) and 5 (660–745 nm filter), respectively, and AlexaFluor 647 and APC-Cy7 were detected in channels 11 (660–745 nm filter) and 12 (745–800 nm filter), respectively. Cell classifiers were set for channel 1 area lower limit of 25 to eliminate collection of debris, channel 3 raw max pixel lower limit of 20 (non-background-subtracted pixel intensity of the PE channel) such that the majority of CD146 PE-negative events were not collected, and a raw max pixel upper limit of 4094 for all channels used so that events containing saturating fluorescence were not collected. Care must be taken to ensure that you are eliminating any PE-positive events.

6. Creating compensation matrix: Because accurate compensation is critical for this procedure, it can be helpful to analyze each single color control file before creating the compensation matrix. In doing this, new *.rif files can be created which contain only true single positive cells, and events which appear to be positive for the channel of fluorescence but may be debris, particles staining non-specifically for the dye, or autofluorescence can be eliminated. To do this, open each single color *.rif without applying any compensation. Plot of intensity value of the single-color channel currently being used against an adjacent channel. For example plot the Intensity of Channel 3 for PE against the Intensity of Channel 2. The true PE-positive cells should be identifiable by clustering together and can be verified by clicking on the events to visualize the imagery. A region can be drawn around the true positively stained cells, thus creating a new population. A new *.rif file can be created for this true positive staining population by clicking on "Tools," "Create new data file from population," and selecting the true positive staining population. This can be done for each single color control prior to calculating the compensation matrix, and will help with obtaining a more accurate compensation matrix.

References

1. Anthony PP, Ramani P (1991) Endothelial markers in malignant vascular tumours of the liver: superiority of QB-END/10 over von Willebrand factor and Ulex europaeus agglutinin 1. J Clin Pathol 44:29–32
2. Nickoloff BJ (1991) The human progenitor cell antigen (CD34) is localized on endothelial cells, dermal dendritic cells, and perifollicular cells in formalin-fixed normal skin, and on proliferating endothelial cells and stromal spindle-shaped cells in Kaposi's sarcoma. Arch Dermatol 127:523–529
3. Srour EF, Brandt JE, Briddell RA, Leemhuis T, van Besien K, Hoffman R (1991) Human CD34^{+} HLA-DR^{-} bone marrow cells contain progenitor cells capable of self-renewal, multilineage differentiation, and long-term in vitro hematopoiesis. Blood Cells 17:287–295
4. Krause DS, Fackler MJ, Civin CI, May WS (1996) CD34: structure, biology, and clinical utility. Blood 87:1–13
5. Woodfin A, Voisin MB, Nourshargh S (2007) PECAM-1: a multi-functional molecule in

inflammation and vascular biology. Arterioscler Thromb Vasc Biol 27:2514–2523. doi:10.1161/ATVBAHA.107.151456

6. Privratsky JR, Newman DK, Newman PJ (2010) PECAM-1: conflicts of interest in inflammation. Life Sci 87:69–82. doi:10.1016/j.lfs.2010.06.001
7. Woywodt A, Streiber F, de Groot K, Regelsberger H, Haller H, Haubitz M (2003) Circulating endothelial cells as markers for ANCA-associated small-vessel vasculitis. Lancet 361:206–210. doi:10.1016/S0140-6736(03)12269-6
8. Bull TM, Golpon H, Hebbel RP, Solovey A, Cool CD, Tuder RM, Geraci MW, Voelkel NF (2003) Circulating endothelial cells in pulmonary hypertension. Thromb Haemost 90:698–703
9. Beerepoot LV, Mehra N, Vermaat JS, Zonnenberg BA, Gebbink MF, Voest EE (2004) Increased levels of viable circulating endothelial cells are an indicator of progressive disease in cancer patients. Ann Oncol 15:139–145. doi:10.1093/annonc/mdh017
10. Solovey A, Lin Y, Browne P, Choong S, Wayner E, Hebbel RP (1997) Circulating activated endothelial cells in sickle cell anemia. N Engl J Med 337:1584–1590
11. Elshal MF, Khan SS, Takahashi Y, Solomon MA, McCoy JP, Jr. (2005) CD146 (Mel-CAM), an adhesion marker of endothelial cells, is a novel marker of lymphocyte subset activation in normal peripheral blood. Blood 106: 2923-2924. doi: http://dx.doi.org.ezp-prod1.hul.harvard.edu/10.1182/blood-2005-06-2307
12. Elshal MF, Khan SS, Raghavachari N, Takahashi Y, Barb J, Bailey JJ, Munson PJ, Solomon MA, Danner RL, McCoy JP Jr (2007) A unique population of effector memory lymphocytes identified by CD146 having a distinct immunophenotypic and genomic profile. BMC Immunol 8:29. doi:10.1186/1471-2172-8-29
13. Mahnke YD, Roederer M (2007) Optimizing a multicolor immunophenotyping assay. Clin Lab Med 27:469–485. doi:10.1016/j.cll.2007.05.002
14. McLaughlin BE, Baumgarth N, Bigos M, Roederer M, De Rosa SC, Altman JD, Nixon DF, Ottinger J, Oxford C, Evans TG, Asmuth DM (2008) Nine-color flow cytometry for accurate measurement of T cell subsets and cytokine responses. Part I: Panel design by an empiric approach. Cytometry A 73:400–410
15. Samsel L, Dagur PK, Raghavachari N, Seamon C, Kato GJ, McCoy JP Jr (2013) Imaging flow cytometry for morphologic and phenotypic characterization of rare circulating endothelial cells. Cytometry B Clin Cytom 84:379–389. doi:10.1002/cyto.b.21088
16. Dagur P, Tatlici G, Gourley M, Samsel L, Raghavachari N, Liu P, Liu D, McCoy JP Jr (2010) CD146+ T lymphocytes are increased in both the peripheral circulation and in the synovial effusions of patients with various musculoskeletal diseases and display pro-inflammatory gene profiles. Cytometry B Clin Cytom 78:88–95. doi:10.1002/cyto.b.20502
17. Biancotto A, McCoy JP Jr (2014) Studying the human immunome: the complexity of comprehensive leukocyte immunophenotyping. Curr Top Microbiol Immunol 377:23–60
18. Maecker HT, Frey T, Nomura LE, Trotter J (2004) Selecting fluorochrome conjugates for maximum sensitivity. Cytomety A 62A:169–173
19. Khan S, Solomon MA, McCoy JP Jr (2005) Detection of circulating endothelial cells and endothelial progenitor cells by flow cytometry. Cytometry B Clin Cytom 64:1–8

Chapter 19

Imaging Flow Cytometric Analysis of Primary Bone Marrow Megakaryocytes

Lisa M. Niswander, James Palis, and Kathleen E. McGrath

Abstract

In light of the indispensible role of platelets in the maintenance of hemostasis, understanding the biology of platelet production from bone marrow megakaryocytes (MKs) may uncover new therapeutic strategies for thrombocytopenia. While there has been much recent interest in optimizing culture systems to facilitate the study of the morphologically unique MK lineage, these systems lack the intricacy of in vivo megakaryopoiesis. Given the limitations of many common techniques for the in vivo study of MKs, in this chapter we describe a method to quantify and analyze primary murine bone marrow megakaryocytes utilizing imaging flow cytometry.

Key words Megakaryocyte, Imaging flow cytometry, Bone marrow, Megakaryopoiesis, Ploidy

1 Introduction

Platelets are small, anucleate, blood cells that are critical components of hemostasis and constitute the second most abundant type of circulating blood cell. Platelets have a life-span of only 4–5 days in rodents and 7–10 days in humans, which necessitates a robust synthesis system for platelet production [1, 2]. Decreased numbers of platelets, or thrombocytopenia, can have severe consequences for human health due to the enhanced risk for bleeding. Accordingly, platelet synthesis, which occurs from megakaryocytes (MKs) in the bone marrow, is a topic of active research. However, some of the maturational characteristics of MKs complicate their study, particularly in primary, in vivo models.

MKs are large, polyploid cells that develop as a rare cell population in the bone marrow from lineage-committed MK progenitor cells (MKPs) [3, 4]. MK maturation is characterized by a progressive increase in cell size and DNA content through abortive endomitotic cell cycles [5, 6]. Ultimately, following cytoplasmic maturation, MKs release circulating platelets from their cytoplasm [7, 8].

Natasha S. Barteneva and Ivan A. Vorobjev (eds.), *Imaging Flow Cytometry: Methods and Protocols*, Methods in Molecular Biology, vol. 1389, DOI 10.1007/978-1-4939-3302-0_19, © Springer Science+Business Media New York 2016

MKs have long been defined by their large size and increased DNA content with histochemical staining of marrow aspirates or sections. More recently, MKs have been characterized by specific surface immunohistochemical markers [9]. However, the low frequency of MKs in the marrow creates a technical hurdle for the collection of sufficient data with these methods. While the ability to sample many cells very quickly with conventional flow cytometry is ideal for the analysis of rare cell populations, the maturational characteristics of the MK lineage create challenges for accurate quantification and analysis with flow cytometry. These challenges include the large continuum of cell size and DNA content as well as a shared surface immunophenotype with their abundant platelet progeny. While the in vitro derivation of MKs from upstream hematopoietic progenitors is an attractive alternative for a relatively pure, enriched population of MKs, these culture systems do not fully recapitulate the complexity of in vivo megakaryopoiesis and thrombopoiesis.

This chapter describes a method for the quantification and characterization of uncultured bone marrow murine megakaryocytes utilizing imaging flow cytometry (IFC), on the Amnis/EMD Millipore ImageStreamX platform. This technology is perfectly poised to facilitate the study of the rare, morphologically unique MK population as it combines the ability to analyze many cells rapidly using flow cytometry with the morphometric capabilities and localization of immunophenotypic markers afforded by microscopy. While the techniques presented here are aimed at accurately determining the frequency of MKs from flushed marrow samples, they can also easily be expanded to facilitate the study of other aspects of the maturation and biology of primary MKs in the murine model [10].

2 Materials

2.1 Isolation of Bone Marrow

1. 7–9-week-old C57BL/6 J mice.
2. PB2 + heparin: Dulbecco's phosphate-buffered saline, 0.3 % BSA, 0.1 % glucose supplemented with 25 μg/ml heparin (*see* **Note 1**).
3. 5 ml syringes attached to 22-G (length = 25 mm) and 20-G (length = 40 mm) needles.
4. Single-edged razor blade.
5. Dissecting scissors and forceps.
6. 15 ml conical tubes.
7. 40 μm nylon cell strainer on a 5 ml round-bottom tube (*see* **Note 2**).
8. Hemocytometer and microscope.

2.2 Cell Staining and Preparation for IFC

1. PB2: Dulbecco's phosphate-buffered saline, 0.3 % BSA, 0.1 % glucose.
2. Blocking buffer: 25 % normal rat serum prepared in PBS (*see* **Note 3**).
3. 5 ml round-bottom FACS tubes.
4. Swinging bucket centrifuge.
5. Anti-mouse Fluorescein isothiocyanate (FITC)-CD9, Phycoerythrin (PE)-Cy7-Kit, and eFluor450-CD41 antibodies (all eBioscience) (*see* **Notes 4–6**).
6. 1.7 ml microfuge tubes.
7. DRAQ5 (eBioscience) (*see* **Note 7**).

2.3 IFC Data Acquisition and Analysis

1. Imaging flow cytometer (*see* **Note 8**).
2. IDEAS analysis software (Amnis Corporation).

3 Methods

3.1 Isolation of Bone Marrow

1. Sacrifice mouse with preferred method (*see* **Note 9**).
2. Remove hind limb. Dissect and clean femur.
3. Open marrow cavity on both ends with the razor just distal to the femoral head and greater trochanter and immediately proximal to the patellar groove.
4. Flush in a distal to proximal direction (knee to hip) with 4 ml of PB2 + heparin in 5 ml syringe with 22-G needle by placing needle into distal/hip epiphysis of the femur (*see* **Note 10**).
5. Triturate marrow by drawing suspension in and out of a 5 ml syringe with 20-G needle (*see* **Note 11**).
6. Pass cell suspension through cell strainer.
7. Count single cell suspension by hemocytometer and record total cells per flushed femur. The number of total cells per femur will be used to determine total MKs per femur following IFC analysis.

3.2 Cell Staining and Preparation for IFC

1. Pellet 5×10^6 bone marrow cells in a 1.7 ml microfuge tube for 5 min at 200 relative centrifugation force (RCF), 4 °C.
2. Remove all but 100 μl of the supernatant of the 5×10^6 cell pellet and resuspend. Store unstained sample on ice. This sample will be stained with DRAQ5 only and used to determine the ratio of nucleated to enucleated cells in the marrow (*see* **Note 12**).
3. Pellet 10^7 bone marrow cells in a 5 ml round-bottom tube for 5 min at 200 RCF, 4 °C.

4. Gently resuspend 10^7 cell pellet in 50 μl of blocking buffer.
5. Incubate on ice for ≥20 min.
6. Add 50 μl of MK panel antibodies prepared in PB2 (1:50 FITC-CD9, 1:50 PE-Cy7-Kit, 1:50 eFluor450-CD41) (*see* **Notes 13** and **14**).
7. Incubate on ice in the dark for ≥20 min.
8. Wash cells with 4 ml of PB2 and pellet for 5 min at 200 RCF, 4 °C.
9. Remove all but ~100 μl of wash supernatant, gently resuspend pellet, and transfer to 1.7 ml microfuge tube.
10. Add DRAQ5 (final concentration 2.5 μM) to each sample (including unstained sample from **steps 1** and **2**) 5–10 min prior to data collection on an imaging flow cytometer. Keep samples on ice and in the dark (*see* **Notes 15** and **16**).

3.3 IFC Setup and Acquisition

1. Channel setup on 2-camera ImageStreamX (*see* **Notes 17** and **18**):

   Channel 1 (430–480 nm): Brightfield.

   Channel 2 (480–650 nm): FITC-CD9.

   Channel 6 (740–800 nm): PE-Cy7-Kit.

   Channel 7 (420–505 nm): eFluor450-CD41.

   Channel 9 (570–595 nm): Brightfield 2.

   Channel 11 (660–740 nm): DRAQ5.
2. Additional settings:

   Excitation lasers with typical intensity settings: 405 nm (80 mW), 488 nm (100 mW), 658 (40 mW) (*see* **Note 19**).

   40× objective lens.

   Cell classifier: Area lower limit 25 in channel 1 (Brightfield) (*see* **Note 20**).
3. Collect two to five 50,000 event files for each MK panel sample and one file for DRAQ5-stained control (*see* **Notes 21** and **22**).
4. Collect single-stained controls for spectral compensation (*see* **Note 23**).

3.4 IFC Gating and Analysis

1. Following spectral compensation with IDEAS software, use DRAQ5 signal intensity to determine the number of DRAQ5$^+$ nucleated cells in each 50,000-event MK panel file (Fig. 1a). Additionally, use the same strategy to determine the frequency of nucleated cells in the unmanipulated DRAQ5-only file for each biologic sample. Record both the number of nucleated

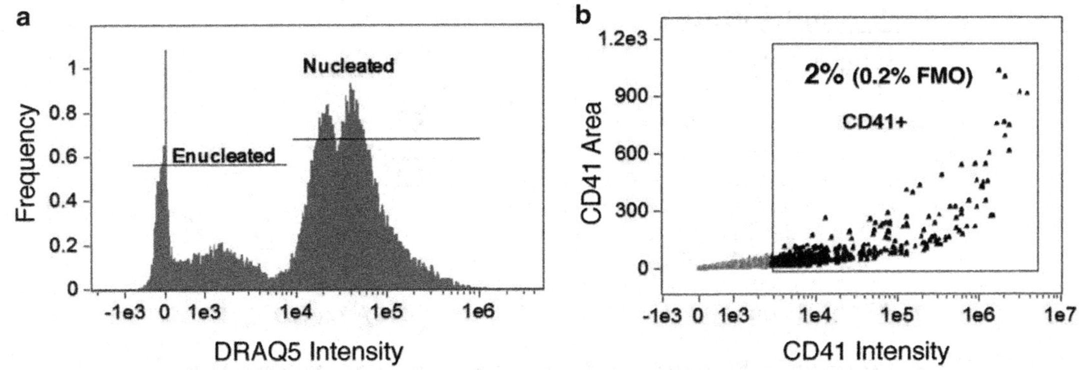

Fig. 1 Initial processing of IFC MK panel files and merging of CD41+ events. (**a**) Determination of enucleated red blood cells versus nucleated cells by DRAQ5 signal intensity. This strategy is used to record the number of nucleated cells in each MK panel file as well as to record the frequency of nucleated cells in the unmanipulated DRAQ5-only file for each biologic sample. This population contains all collected events, including multicellular events. For more distinct 2N and 4N ploidy peaks, single cells can be gated first with the IDEAS "single-cell" building block. (**b**) Events containing CD41 positivity are gated based on CD41 signal intensity. For this sample, 2 % of the 50,000 collected events fall into the CD41+ gate, which is set based on a combined isotype/flour minus one (FMO) control

cells in each MK panel file and the frequency of nucleated cells in the DRAQ5-only file (nucleated/[nucleated + enucleated]) (*see* **Note 24**).

2. For each 50,000-event MK panel file, gate events with CD41 positivity (Fig. 1b) and create a CIF file from this $CD41^+$ population.
3. Merge the $CD41^+$ CIF files from all MK panel files for a given sample into a single $CD41^+$ file. Sum the nucleated cells for the original 50,000-event files that contributed to the merged file (from **step 3.4.1**) to obtain total nucleated cells.
4. Figure 2 details the sequential gating strategy to determine the number of MKs from the merged $CD41^+$ file. Overall, the gates are set based on MK yield rather than purity, and purity is achieved with nested gating. The percentage of parent population cells that fall within each gate for this sample is indicated in Fig. 2; however, it is important to visualize the cells at gate boundaries and to adjust these gates to include all MKs (Fig. 3) (*see* **Notes 25** and **26**).
5. The final MK population is defined in a Boolean population of "Round" (Fig. 2 #5) and "Reclaimed from oddly shaped" (Fig. 2 #6). Examples of MKs are provided in Fig. 3. Record the final number of MKs in the merged $CD41^+$ file.

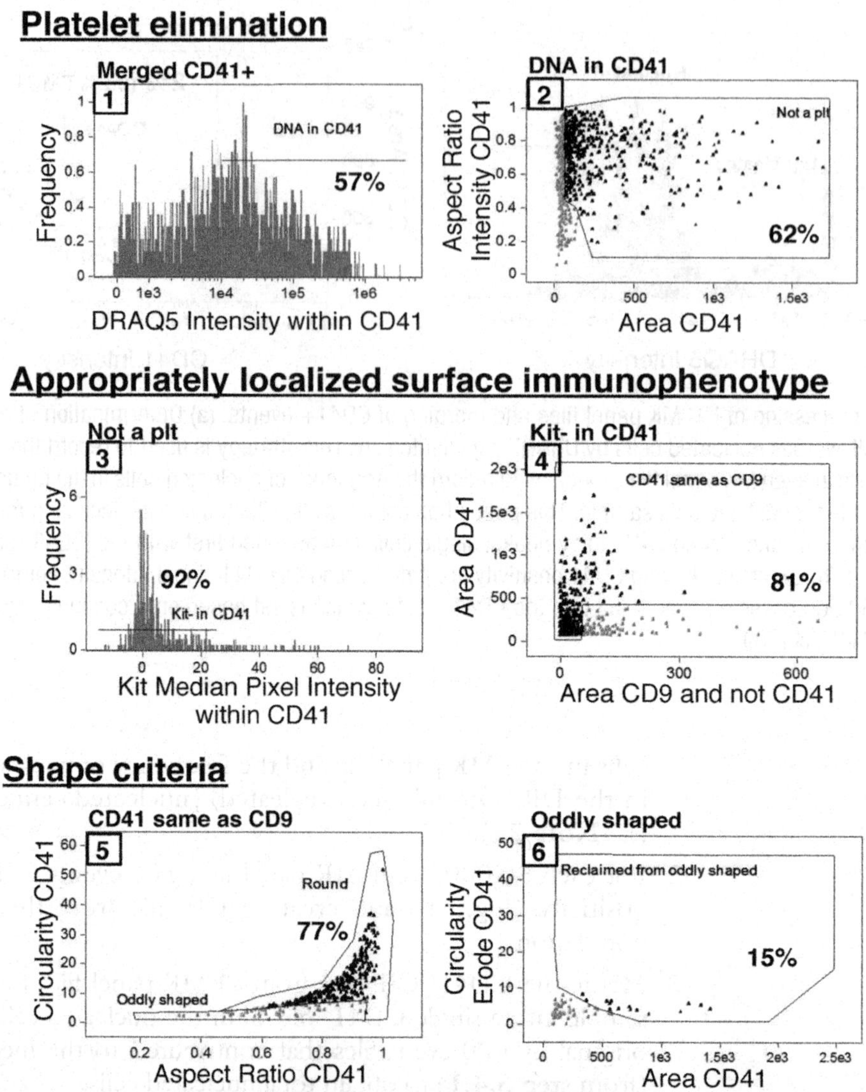

Fig. 2 Nested IFC gating strategy for MKs beginning with merged CD41+ populations. Gates are numbered sequentially and the percentage of the parent population that falls within a given gate is indicated for this sample. First, CD41+ platelets attached to non-MK nucleated cells are eliminated from analysis based on a lack of DRAQ5 signal intensity within the CD41 mask (1) as well as small area and irregular shape (2). The MK population is further refined by both surface immunophenotype, and appropriate localization of immunophenotypic markers with CD41 within an event. MKs are negative for Kit signal within the CD41 mask (3) and contain overlapping CD9 and CD41 signal areas (4). Additionally, cells positive for Kit signal within the CD41 mask can be further immunophenotypically refined with the strategy by Pronk et al. [13] to analyze MKPs. Finally, events that make it through the four above-described gates must fulfill shape-based criteria, which eliminate cytoplasmic fragment debris. The final MK population consists of mostly cells in the "Round" gate (5) with a few MKs with cytoplasmic extensions that are originally classified as "oddly shaped" (5) but are reclaimed based on the circularity of an eroded CD41 mask (6). The axis names have been changed for clarity; however, full descriptive names of features and masks can be found in Table 1

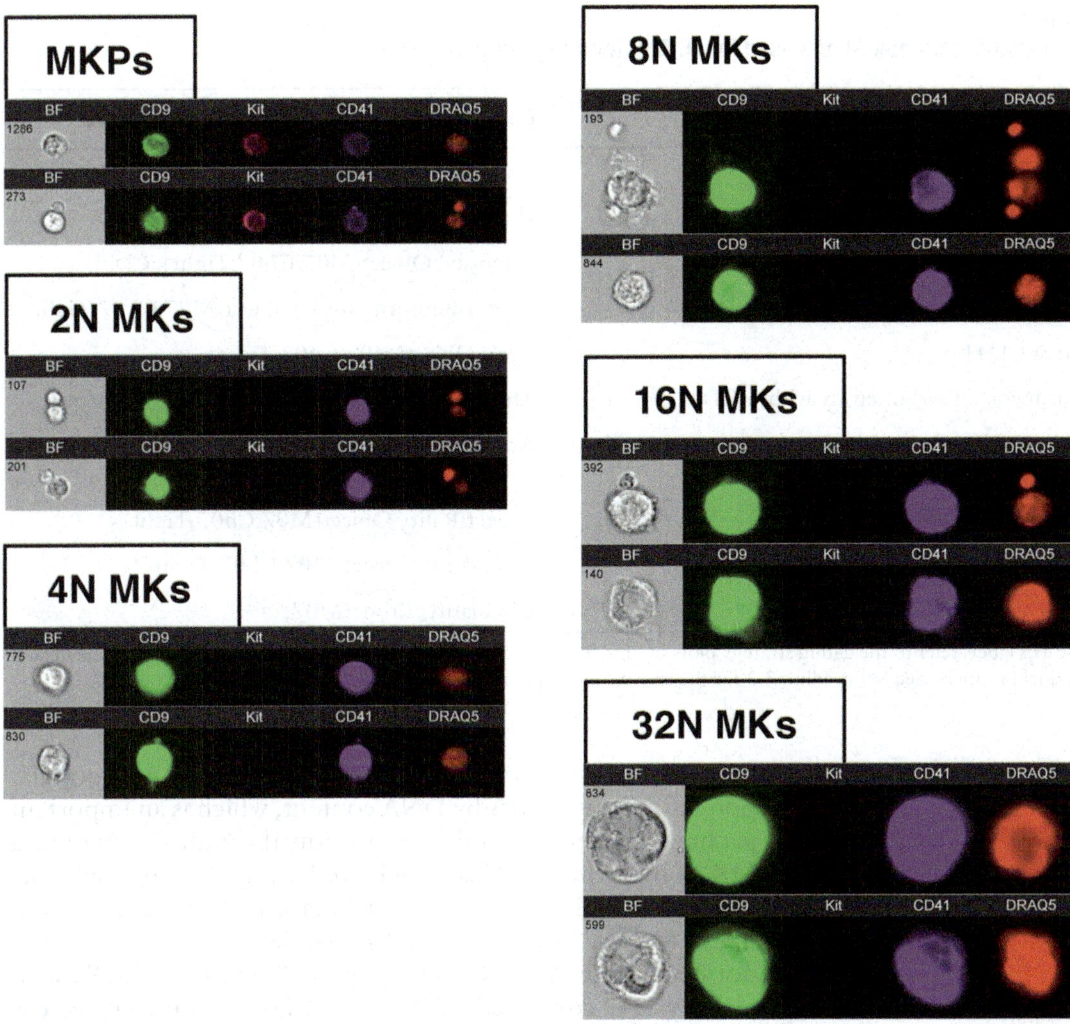

Fig. 3 Examples of MK lineage cells by IFC. Immunophenotypically defined MK progenitors can be analyzed from the Kit+ population (Fig. 2, plot 2, ungated population). Examples of MKs in ploidy classes ranging from 2N to 32N are shown. Images from *left* to *right* for each event row are Brightfield (BF), CD9, Kit, CD41, and DRAQ5. All MK images in this figure were obtained with identical display settings, with the exception of MKP images which have been adjusted to optimize viewing. Note that localization of signal measurements to the CD41 mask in the gating strategy presented in Fig. 2 allows for the analysis of MKs even when physically associated with other nucleated cells in an event (top 8N and 16N event examples)

6. Calculate MKs/femur: [MKs/merged $CD41^+$ file (**step 3.4.5**)] × [sum of total nucleated cells/merged $CD41^+$ file (**steps 3.4.1** and **3**)] × [frequency of nucleated cells (**step 3.4.1** DRAQ5-only file)] × [total cells/femur (**step 3.1.7**)] (*see* **Note 27**).
7. In addition to the quantification of MK frequency in murine bone marrow, this gating strategy generates a primary MK population that can be further studied by restricting analyses to

Table 1
Mask/feature combinations used in IFC MK lineage gating strategy

Axis label	Default IDEAS mask/feature nomenclature
DRAQ5 intensity	Intensity_Morphology(M11,Ch11)_Ch11
CD41 intensity	Intensity_Object(M07,Ch07,Tight)_Ch07
DRAQ5 intensity within CD41	Intensity_Object(M07,Ch07,Tight)_Ch11
Aspect ratio intensity CD41	Aspect Ratio Intensity_Object(M07,Ch07,Tight)
Area CD41	Area_Object(M07,Ch07,Tight)
Kit median pixel intensity within CD41	Median Pixel_Object(M07,Ch07,Tight)_Ch06
Area CD9 and not CD41	Area_Object(M02,Ch02,Tight) and not Object(M07,Ch07,Tight)
Aspect ratio CD41	Aspect Ratio_Object(M07,Ch07,Tight)
Circularity CD41	Circularity_Object(M07,Ch07,Tight)
Circularity erode CD41	Circularity_Erode(M07,7)

The axis labels refer to the gating strategy plots in Figs. 1 and 2. The IDEAS feature names are provided with the default channel numbers. See Subheading 3.3 for the specific markers in each channel

the CD41 mask. For example, we can divide the MK population into ploidy classes by DNA content, which is an important parameter of endomitotic progression through the maturing MK compartment. This is achieved by measuring only the DRAQ5 signal within the CD41 mask for each event (as in Fig. 2 #1) (*see* **Notes 28** and **29**). While the ploidy peaks may not be discrete given the relatively small number of MKs sampled by IFC compared to traditional flow cytometry, we circumvent this issue by identifying the 2N and 4N peaks for total DRAQ5 intensity in the original 50,000-event file (Fig. 1a) and use these values to set the 2N and 4N gates. The subsequent ploidy class gates (8N, 16N, 32N) can then be determined based on DRAQ5 intensity doubling. Examples of MKs in each ploidy class are provided in Fig. 3.

4 Notes

1. We utilize PB2 buffer for dissection and washing. PB2 was optimized for the preservation of erythroblast precursors for flow cytometric analyses [11]. Phosphate-buffered saline can be substituted.
2. We routinely use a 40 μm cell strainer to remove large unseparated cell clumps from titurated marrow cell suspensions.

While we retain large polyploid MKs, these cells have been reported to be capable of reaching over 50 μm in diameter in the mouse [12]. Therefore, a larger pore size (100 μm) may be advisable for certain experiments.

3. We use normal rat serum because all of our primary antibodies are produced in rat. Species selection for blocking buffer serum will depend on the species of antibodies used.
4. Anti-CD41 antibodies are commercially available conjugated to a large number of fluors. We prefer to use eFluor450-CD41, which is collected in channel 7, as there is very little signal overlap from other channels into channel 7. Accordingly, for experiments aimed at obtaining many MKs for analysis, a cell classifier can be set to limit event acquisition based on CD41/channel 7 signal intensity even though data acquisition occurs prior to spectral compensation.
5. This IFC gating strategy does not exclude events containing positivity for markers of other hematopoietic lineages (Lin^+), which is a common step in standard flow cytometric analyses. Our strategy was designed to obtain both accurate numbers and ploidy distribution of bone marrow MKs, and we have shown that eliminating Lin^+ events affects both of these end-points [10]. For example, in Fig. 3, the top 8N and 16N MK examples would be eliminated as Lin + cells are present in the same event. Accordingly, we do not recommend Lin depletion with magnetic beads as a strategy to increase MK frequency. For alternate experimental goals that may require a lineage exclusion, such as examining the phenotypic MK progenitor compartment (MKP) [13], we recommend an extensive Lin panel (for example: Ter119, CD3, B220, Gr1, Mac1, Sca1, CD105). The addition of Lin markers works well with the described MK panel when conjugated to PE-Texas Red, which is available as a streptavidin conjugate (to be combined with biotinylated Lin antibodies) and is collected in channel 4.
6. This MK antibody panel allows for the addition of other markers of interest. Specifically, channel 3 is intentionally left open for antibodies conjugated to PE, one of the most commonly available fluors. For example, we previously used this channel to examine surface CXCR4 expression on MKs [14], or it could be used to examine CD150 if interested in examining the MK progenitor population [13].
7. DRAQ5 is a live cell permeable DNA dye [15]. Using DRAQ5 allows for the analysis of primary bone marrow MKs without the fixation and/or membrane permeabilization steps that are required for more commonly used DNA dyes. We find that fixation both skews the distribution of MKs for analysis and complicates the determination of MK frequency [10].

8. For our analyses we utilize a 2-camera ImageStreamX (Amnis EMD Millipore) imaging flow cytometer with INSPIRE acquisition software and IDEAS analysis software (version 6.0). Some modifications (for example channel selection) may be required depending on the model of imaging flow cytometer. For the Amnis FlowCyte with Quantitative Imaging upgrade, lower resolution images will require the adjustment of masks, and the narrower image width compared to the ImageStreamX may limit the viewing of the largest MKs.
9. We utilize CO_2 asphyxiation so we are able to collect blood via the inferior vena cava for parallel platelet analyses. Cervical dislocation and other methods of sacrifice are compatible with femoral marrow collection.
10. We try to flush the femoral marrow core with a single expulsion of 4 ml of PB2 + heparin with the 22-G needle. If more volume is needed, fill the syringe with the 4 ml flushed cell suspension with a 20-G needle, and again pass though the femur with the 22-G needle. Limiting the passage of cells through the smaller 22-G needle prevents excess sheer and cell damage.
11. Be gentle and minimize bubbles when passing the cell suspension through the syringe. Triturate until red marrow particles are no longer visible. We find this generally requires three syringe passes.
12. The staining procedure results in the preferential loss of enucleated red blood cells. Because of this, we have found it necessary to prepare an unmanipulated aliquot of bone marrow stained with only DRAQ5. This sample provides an accurate assessment of the frequency of nucleated cells in the flushed marrow sample which is used in the final calculation of MKs per femur. This step is especially imperative for analyses in which the circulating red blood cells and nucleated cell numbers may be altered in the marrow, such as in the setting of vascular dilation and hemorrhage in the marrow space and hematopoietic cell death following total body irradiation [16].
13. Antibodies from other suppliers may need to be titrated for optimal concentration.
14. The addition of antibodies prepared in PB2 reduces the final concentration of NRS to 12.5 % for staining.
15. For optimal nuclear staining we add DRAQ5 to each sample 5–10 min prior to data collection on the imaging flow cytometer. For experiments with many samples, we stagger so that the addition of DRAQ5 occurs 5–10 min before a given sample is run.
16. In our hands DRAQ5 staining is very sensitive to the number of nucleated cells in a sample. Accordingly, in samples with a

higher frequency of enucleated red blood cells the DRAQ5 concentration may need to be titrated.

17. This MK panel can alternatively be collected entirely in camera 1 channels by recording DRAQ5 in channel 5 and using PE-CD41 in channel 3. The 2-camera channel set-up we describe allows for other markers/fluors to be added to the panel.
18. The unmanipulated DRAQ5-only sample files can be collected in channels 1 (Brightfield) and 5 (DRAQ5), and does not need to undergo spectral compensation.
19. As channel 6 is in use for PE-Cy7, the 785 nm scatter laser must be turned off. While these are typical laser intensity settings, it is important that investigators run the sample that is anticipated to be the brightest first, and set the laser intensity settings to minimize signal saturation (raw max pixel <4000) for that sample.
20. This area lower limit includes red blood cells and excludes smaller debris and beads, allowing for the determination of the frequency of nucleated cells from the total marrow cell suspension.
21. Due to the processing limitations presented by large data files, 50,000 events is the upper limit for each file. Depending on experimental design and goals, we recommend collecting two to five 50,000 event files for each biologic sample. More files may be necessary to obtain sufficient MKs for maturational studies, ploidy analysis, or in experimental conditions in which there are fewer MKs in the bone marrow. A smaller file (10,000 cells) can be collected for the unmanipulated DRAQ5-only file as this is only used to determine the frequency of nucleated cells in the marrow.
22. Be sure to remove all bubbles from the sample before loading into the ImageStreamX in order to avoid the need to return and reload the sample. In our experience, if we have to "Return Sample" and reload it with the "Flush, Lock, and Load" process, we often find fewer MKs in the files collected after the sample is returned by the machine.
23. Single-stained controls for compensation can either be prepared with compensation beads or with marrow cells. As a DNA dye, the DRAQ5 control must be prepared using cells. The DRAQ5-only sample prepared in Subheading 3.2 **step 1** can be used for DRAQ5 compensation. As the DRAQ5 dye is not washed out of the sample, we run the experimental samples followed by the DRAQ5 compensation control, then run bleach prior to collecting other single-stained controls. To collect compensation controls, we set the following cell

classifiers in the appropriate channel for each fluor: Area lower limit = 25 (cells) or 15 (beads), raw max pixel upper limit = 4000.

24. The nucleated cell gate is set based on identifying the 2N ploidy for the entire 50,000-event file on the histogram of DRAQ5 total signal intensity (Fig. 1a). Importantly, many of the events in each file contain cell clumps, which alter the appearance of ploidy peaks. A more typical, clean 2N and 4N distribution can be obtained using the IDEAS "single-cell" building block. The nucleated cell gate is then subsequently used to set the "DNA in CD41" gate for the merged CD41$^+$ file for each sample (Fig. 2, #1). As DRAQ5 signal intensity can vary between samples, it is important to identify the 2N ploidy peak for each sample.
25. For each experiment, it is necessary to prepare the marrow from an age-matched, untreated control mouse and run this sample in parallel with samples from experimental mice. The gates are then set based on this control, and the template applied to the experimental sample files. This control has been termed a "morphometrically relevant biologic control" for IFC analyses [17], and is essential to consistently set gates for the morphometric measurements of the maturational continuum of marrow MKs.
26. It is important to note that this gating strategy is aimed at determining an accurate MK frequency and does not contain a focus gate. It may be necessary to add a focus gate (i.e., gradient RMS feature) for some analyses such as nuclear localization or protein internalization studies.
27. For 7–9-week-old female C57BL/6 mice, IFC-defined MKs comprise 0.3 % of nucleated marrow cells [10]. Using the total femur cells flushed (~4×10^7) and the frequency of nucleated cells in the unmanipulated DRAQ5-only sample (~0.5), we generally calculate 50,000–60,000 MKs/femur.
28. Restricting DRAQ5 intensity measurements to within the CD41 mask allows for accurate DNA content determination even within events in which MKs are associated with other nucleated cells (Fig. 3, top examples of 8N and 16N MKs).
29. The clearest ploidy peak definition is obtained with DNA dyes (e.g., DAPI) following sample fixation and permeabilization. Due to the differential cell loss that accompanies these additional sample manipulations, we find that live cell preparations with DRAQ5 are preferable when interested in determining MK numbers. The determination of ploidy distribution based on DRAQ5 signal intensity doubling as described in the text provides reproducible ploidy measurements that correlate well with published values for C57BL/6 mice [3, 10].

Acknowledgements

The authors gratefully acknowledge the technical support of Dr. Tim Bushnell and the Flow Cytometry Core Facility at the University of Rochester Medical Center. The development of this methodology was supported by funding from the National Institute of Digestive and Kidney Diseases (NIDDK) and the National Institute of Allergy and Infectious Diseases (NIAID) of the National Institutes of Health.

References

1. Odell TT Jr, McDonald TP (1961) Life span of mouse blood platelets. Proc Soc Exp Biol Med 106:107–108
2. Harker LA, Finch CA (1969) Thrombokinetics in man. J Clin Invest 48:963–974
3. Corash L, Levin J, Mok Y, Baker G, McDowell J (1989) Measurement of megakaryocyte frequency and ploidy distribution in unfractionated murine bone marrow. Exp Hematol 17:278–286
4. Metcalf D, MacDonald HR, Odartchenko N, Sordat B (1975) Growth of mouse megakaryocyte colonies in vitro. Proc Natl Acad Sci U S A 72:1744–1748
5. Levine RF, Hazzard KC, Lamberg JD (1982) The significance of megakaryocyte size. Blood 60:1122–1131
6. Geddis AE, Fox NE, Tkachenko E, Kaushansky K (2007) Endomitotic megakaryocytes that form a bipolar spindle exhibit cleavage furrow ingression followed by furrow regression. Cell Cycle 6:455–460. doi:10.4161/cc.6.4.3836
7. Odell TT, Jackson CW (1968) Polyploidy and maturation of rat megakaryocytes. Blood 32:102–110
8. Italiano JE, Lecine P, Shivdasani RA, Hartwig JH (1999) Blood platelets are assembled principally at the ends of proplatelet processes produced by differentiated megakaryocytes. J Cell Biol 147:1299–1312
9. Yu M, Cantor AB (2012) Megakaryopoiesis and thrombopoiesis: an update on cytokines and lineage surface markers. Methods Mol Biol 788:291–303
10. Niswander LM, McGrath KE, Kennedy JC, Palis J (2014) Improved quantitative analysis of primary bone marrow megakaryocytes utilizing imaging flow cytometry. Cytometry A 85A:302–312. doi:10.1002/cyto.a.22438
11. England SJ, McGrath KE, Frame JM, Palis J (2011) Immature erythroblasts with extensive ex vivo self-renewal capacity emerge from the early mammalian fetus. Blood 117:2708–2717,http://dx.doi.org/10.1182/blood-2010-07-299743
12. Ulich TR, del Castillo J, Senaldi G et al (1999) The prolonged hematologic effects of a single injection of PEG-rHuMGDF in normal and thrombocytopenic mice. Exp Hematol 27:117–130
13. Pronk CJH, Rossi DJ, Månsson R et al (2007) Elucidation of the phenotypic, functional, and molecular topography of a myeloerythroid progenitor cell hierarchy. Cell Stem Cell 1:428–442. doi:10.1016/j.stem.2007.07.005
14. Niswander LM, Fegan KH, Kingsley PD, McGrath KE, Palis J (2014) SDF-1 dynamically mediates megakaryocyte niche occupancy and thrombopoiesis at steady-state and following radiation injury. Blood 124:277–286. doi:10.1182/blood-2014-01-547638
15. Martin RM, Leonhardt H, Cardoso MC (2005) DNA labeling in living cells. Cytometry A 67A:45–52
16. Hooper AT, Butler JM, Nolan DJ et al (2009) Engraftment and reconstitution of hematopoiesis is dependent on VEGFR2-mediated regeneration of sinusoidal endothelial cells. Cell Stem Cell 4:263–274. doi:10.1016/j.stem.2009.01.006
17. Filby A, Davies D (2012) Reporting imaging flow cytometry data for publication: why mask the detail? Cytometry A 81:637–642. doi:10.1002/cyto.a.22091

Chapter 20

Sickle Cell Imaging Flow Cytometry Assay (SIFCA)

Kleber Y. Fertrin, Leigh Samsel, Eduard J. van Beers, Laurel Mendelsohn, Gregory J. Kato, and J. Philip McCoy Jr.

Abstract

Hemoglobin S polymerization under hypoxic conditions in sickle cell disorders causes characteristic shape changes to human red blood cells. Previous sickling assays used to investigate the efficacy of novel agents to treat these disorders are laborious and observer dependent. Here, we describe a partially automated, high-throughput sickling assay using imaging flow cytometry.

Key words Sickle cell disease, Sickling assay, Imaging flow cytometry

1 Introduction

Sickle cell disorders comprise a group of inherited blood diseases that share the genetic mutation HBB Glu6Val, causing the production of the abnormal hemoglobin S (HbS). HbS polymerizes inside red blood cells (RBCs) under hypoxic and acidotic conditions, and the formation of such polymers leads to RBC deformation and to the characteristic elongated shape of a crescent or sickle [1].

RBC sickling is not homogeneous, and the percentage of RBCs undergoing sickling in vitro varies largely across different sickle cell disease genotypes, as well as variations in the concentrations of HbF [2]. Sickling assays have been developed to give an objective measure of the percentage of circulating cells in a given blood sample that sickle upon exposure to conditions known to cause HbS shift to the T-state, which is prone to polymerize and cause RBC sickling. Traditionally, sickling assays rely on the subjective determination of the percentage of sickling cells by manually counting 200 RBCs through optical microscopy [3–6]. Therefore, sickling assays based on optical microscopy are cumbersome, time-consuming, and highly observer dependent.

Natasha S. Barteneva and Ivan A. Vorobjev (eds.), *Imaging Flow Cytometry: Methods and Protocols*, Methods in Molecular Biology, vol. 1389, DOI 10.1007/978-1-4939-3302-0_20, © Springer Science+Business Media New York 2016

We have, thus, developed an automated, observer-independent sickling assay by applying imaging flow cytometry, which allows software algorithm-driven classification of 20,000 RBCs per sample, yielding a less subjective and more high-throughput assay. This assay has been shown to sensitively detect the known effects of both patient-dependent and -independent variables, such as HbF concentration and sample pH [7].

The SIFCA procedure involves blood sample collection, deoxygenation, imaging flow cytometry, and image analysis. The automated image analysis requires accurate definition of specific masks, features, and image selection.

2 Materials

Prepare and store all solutions and reagents at room temperature unless otherwise specified. Diligently follow all waste disposal regulations when disposing waste materials.

2.1 Blood Sample Preparation and Deoxygenation Components

1. Vacutainer EDTA tubes (BD Biosciences Inc.).
2. HS-500 Hemox Solution (TCS Scientific Corporation, New Hope, PA): Store at 4 °C.
3. Phosphate-buffered saline (PBS) 1×.
4. Glutaraldehyde 25 % (Sigma-Aldrich, USA): Vials should be stored frozen at −20 °C or less before use.
5. Glovebox (*see* **Note 1**) connected to a vacuum pump and a gas cylinder containing 95 % nitrogen/5 % hydrogen mixture, equipped with an oximeter and a 96-well plate shaker. The apparatus is shown in Fig. 1, with the location of valves and gauges for the gas cylinders and vacuum line indicated.
6. 500 μl, 96-well micro-titer plates with adhesive cover.
7. Table-top centrifuge.
8. 1.5 mL polypropylene tubes.
9. 200 μL and 1 mL pipettes and tips.

2.2 Imaging Flow Cytometry Run and Analysis Components

1. PBS 1×.
2. ImageStream^X or MK II imaging flow cytometer (Amnis Corporation, Seattle, WA, USA) equipped with Amnis INSPIRE and Amnis IDEAS softwares.
3. Amnis Speed Beads suspension.

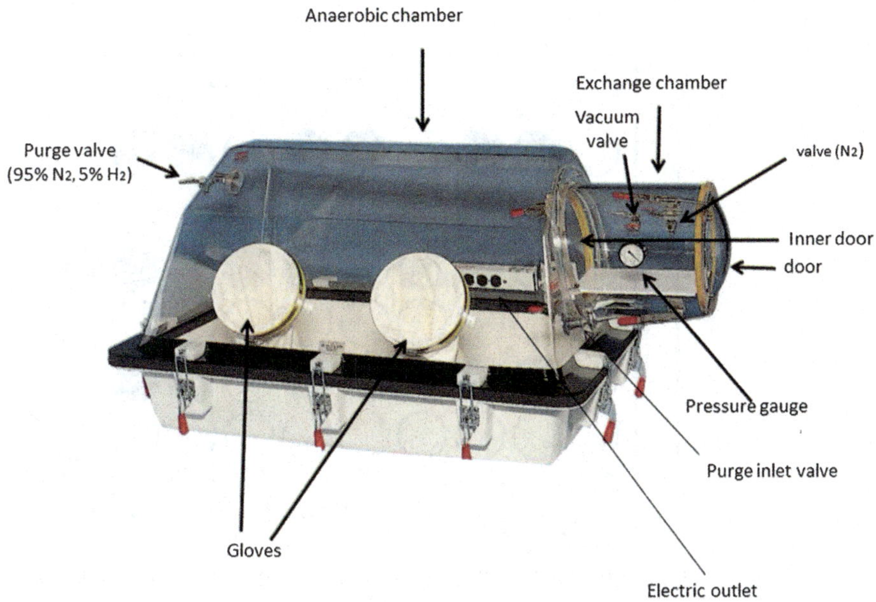

Fig. 1 Glovebox used for sample deoxygenation. The apparatus is shown with the location of valves and gauges for the gas cylinders and vacuum pump indicated

3 Methods

Carry out all procedures at room temperature unless otherwise specified.

3.1 Blood Sample Collection and Preparation

1. Collect 4 mL of peripheral venous blood into Vacutainer EDTA tubes. Sample can be stored at 4 °C for up to 24 h.
2. If sample was stored at 4 °C, place tube on a horizontal shaker at 80–100 rpm until sample is homogenized (approximately 2–5 min).
3. Pipette 10 μL of whole blood into a 1.5 mL polypropylene tube containing 990 μL of Hemox Buffer and mix. This makes a 1 % suspension of blood.
4. Pipette 400 μL duplicates of the 1 % suspension into adjacent wells of a 96-well plate. If more than one sample will be incubated in the same plate, place sample doubles one well apart in both directions, and preferably away from the edges of the plate (Fig. 2).
5. Seal the 96-well plate with an appropriate self-adhesive plastic cover.

3.2 Blood Sample Deoxygenation

1. Before starting, make sure the glovebox is airtight, the gas cylinder contains enough of a 95 % nitrogen/5 % hydrogen mixture, and that the vacuum pump and the oximeter are both

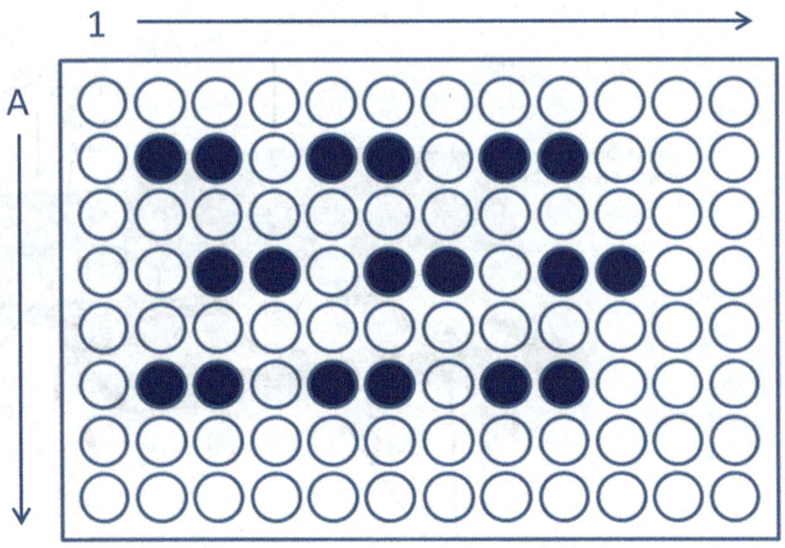

Fig. 2 Schematic representation of sample arrangement in a 96-well plate

operating. Place a 200 μL pipette, 200 μL pipette tips, a new self-adhesive plastic cover, and a proper waste disposal bag for pipette tips and used microplate covers inside the glovebox.

2. Thaw a vial of frozen glutaraldehyde 25 % to room temperature. Total volume of glutaraldehyde should be at least 10 % of the total sample volumes (i.e., 80 μL for two 400 μL samples).
3. Decrease the oxygen concentration inside the glovebox to 2 % by injecting N_2/H_2 mixture and aspirating with the vacuum pump (*see* **Note 1**).
4. Place the 96-well plate inside the glovebox on top of a shaker and remove adhesive cover.
5. Place the vial containing glutaraldehyde 25 % inside the glovebox and leave cap open (*see* **Note 2**).
6. Upon opening the glovebox to place the plate and vial, some oxygen may enter the glovebox, so it will be necessary to readjust the oxygen concentration to 2 %.
7. Incubate the 96-well plate at room temperature and 2 % oxygen for 2 h under constant shaking at 300 rpm.
8. Once the incubation is done, turn off the shaker and pipette 40 uL of glutaraldehyde 25 % into each well containing 400 μL of 1 % blood suspension. Mix between seven and ten times with the pipette to sufficiently mix the glutaraldehyde and incubate for 5–10 min.

9. Reseal the plate with a new cover seal and remove from glovebox.
10. Transfer the contents of each well to a properly identified 1.5 mL polypropylene tube.
11. Wash each sample by adding 1 mL of PBS 1×.
12. Centrifuge at 2000 × *g* for 2 min at 4 °C.
13. Discard supernatant and wash two more times.
14. Resuspend the cell pellet in 100 μL of PBS 1× and store at 4 °C until it is run.

3.3 Imaging Flow Cytometer Sample Run

1. In INSPIRE, for the first run, select the appropriate parameters from the menus to the right: In the Illumination menu, ensure that Brightfield is on and set to 800 for channels 1 and 9, and that the 785 nm laser is set to 0.5; in the Magnification and EDF menu, set the magnification to 60×.
2. Click on "Load"—the equipment will open the sample portal to load the vial containing the sample.
3. Place the open polypropylene tube into the sample portal (*see* **Note 3**).
4. Click on "OK" when the tube is correctly in place.
5. Once the run starts, wait until the images appear to be in focus and centered (about 30–60 s).
6. Click on "New Histogram".
7. Click on the "All" population to select it.
8. Choose "Area_Ch01" as the *X* Axis Feature.
9. This generates a new histogram in the analysis area whose title bar says "All".
10. Click on "Create Line Region" and draw a line that selects all events with an area between 20 and 350. This eliminates the acquisition of small particles and debris.
11. Name this population "R1."
12. At this point, if this is the first sample run, save this layout as a template that can be loaded for future runs.
13. Type the name of the file to which images should be saved. Make sure that the number of events acquired is set to 20,000, and that the events to be acquired come from "R1," not "All," to minimize acquisition of noise. Designate the destination folder into which files will be saved.
14. Click on "Acquire" to start acquisition.
15. Once acquisition is done, click on "Load" to discard the remainder of the sample and proceed to the next tube. The equipment will return the empty tube before allowing loading

of the next sample. Alternatively, click on "Return" to save the remainder of the sample. The equipment will return the tube containing the remaining cell suspension. Click on "Load" again after removing the tube to allow the next sample to be loaded.

3.4 Imaging Flow Cytometry Analysis: Mask Definition

1. Inside the IDEAS software, click on the "Guided Analysis" menu.
2. Click on "Wizards…," click on the "Begin Analysis" wizard, and follow the instructions until the wizard ends to identify single, focused cells (*see* **Notes 4** and **5**).
3. To define new masks, select the "Analysis" menu, then click on "Masks…" to open a dialogue box (*see* **Note 6**).
4. On the left side, a box will show a list of masks already available. To create a new mask, click on "New."
5. Click on "Function" to open a dialogue box called "Define Mask Function".
6. Under "Function," select "Object."
7. Under "Mask," select "M01."
8. Under "Channel," select "Ch01."
9. Under "Image," select "Ch01."
10. The name of the mask will, by default, become the definition of the mask just created.
11. Click on "OK."
12. This returns you to the previous box, where it will show the definition of the mask. It should show "Object(M01, Ch01, Tight)".
13. Click on "OK" to include the new mask. It will show up as the last mask on the list to your left.
14. Click on "Function" to open the dialogue box called "Define Mask Function."
15. De-select "Link Inputs."
16. Under "Function," select "System."
17. Under "Mask," select "Object(M01, Ch01, Tight)."
18. Under "Channel," select "Ch01."
19. Under "Image," select "Ch01."
20. Under "Weight," select "80."
21. Click on "OK."
22. This returns you to the previous box, where it should show the definition of the mask as "System(Object(M01, Ch01, Tight), Ch01, 80)."

23. Click on "OK" to include the new mask.
24. . Click on "Function," and a dialogue box called "Define Mask Function" will open up.
25. De-select "Link Inputs."
26. Under "Function," select "Range."
27. Under "Mask," select "System(Object(M01, Ch01, Tight), Ch01, 80)."
28. Under "Image," select "Ch01."
29. Under "Area," set a Minimum value of 350, and a Maximum value of 5000 (the latter should be default).
30. Under "Aspect Ratio," check that Minimum value is 0, and Maximum value is 1.0 (default).
31. Click on "OK."
32. This returns you to the previous box, where it will show the definition of the mask as "Range(System(Object(M01, Ch01, Tight), Ch01, 80), 350–5000, 0–1)."
33. Click on "OK" to include the new mask.
34. Click on "Function" to open the dialogue box called "Define Mask Function."
35. De-select "Link Inputs."
36. Under "Function," select "Range."
37. Under "Mask," select "Object(M01, Ch01, Tight)."
38. Under "Image," select "Ch01."
39. Under "Area," set a Minimum value of 350, and a Maximum value of 5000 (the latter should be default).
40. Under "Aspect Ratio," check that Minimum value is 0, and Maximum value is 1.0 (default).
41. Click on "OK."
42. This returns you to the previous box, where it should show the definition of the mask as "Range(Object(M01, Ch01, Tight), 350–5000, 0–1)."
43. Click on "OK" to include the new mask, and it should show up as the last mask on the list to your left.

Figure 3 depicts the sequential creation of user defined masks and demonstrates how these tighter masks yield more accurate analysis of single cells and their feature values.

3.5 Features Definition

1. Select the "Analysis" menu, then click on "Features…" to open a dialogue box.
2. On the left side, a box will show a list of features already available. To create a new feature, click on "New."

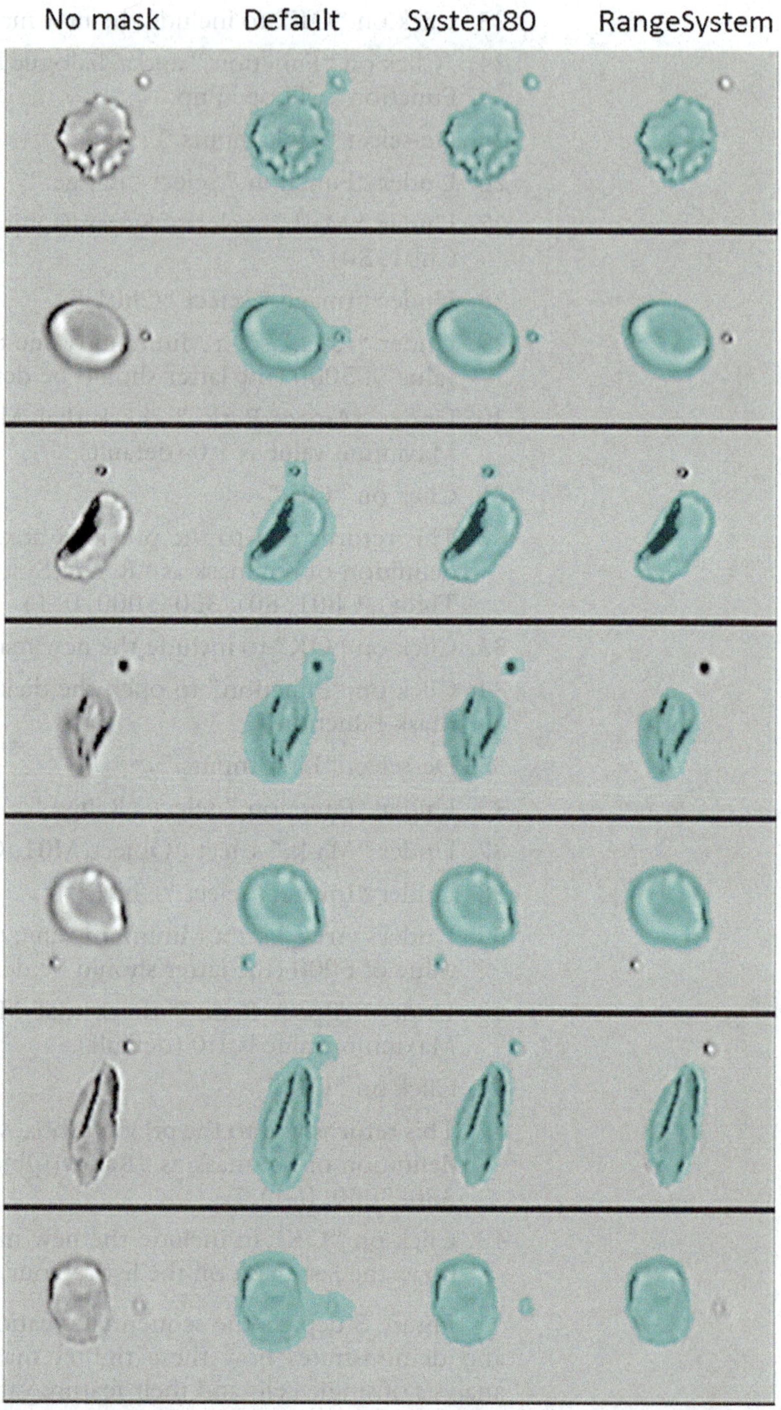

Fig. 3 Bright-field imagery depicting creation and purpose of user-defined masks. The *left column* shows the bright-field image with no mask. The *second column* shows the default mask M01. The *third column*, "System80," shows the user defined System (Object (M01, Ch01, Tight), Ch01, 80) mask. The *right column*, "RangeSystem," shows the user defined Range (System (Object(M01, Ch01, Tight),

3. Select the Feature type "Spot Count" from the pull-down menu next to "single."
4. Select the Object M01 mask from the pull-down menu.
5. Click on "Set Default Name" (typically, it will be Spot Count_ Object M01) and click "OK" (*see* **Note 7**).
6. Repeat **steps 3–5** for all three new masks: System(Object(M01, Ch01, Tight), Ch01, 80), Range(System(Object(M01, Ch01, Tight), Ch01, 80), 350–5000, 0–1), and Range(Object(M01, Ch01, Tight), 350–5000, 0–1).
7. Click on "Close," and the software will calculate the values for each of the features and each cell (*see* **Note 8**).

3.6 Image Selection

1. To define the "Single cells" gate, click on "New Scatterplot."
2. Choose "Area_M01" as the *X* Axis Feature, and "Aspect Ratio_M01" as the *Y* Axis Feature from the pull-down menus.
3. Click on "Create Rectangle Region" or "Create Polygon Region" to draw a region that gates the main cell population, tight enough to exclude particles and debris to the left, and the larger cell aggregates to the right. Name this region "Single Cells." One example using a rectangular region is shown in Fig. 4a (*see* **Note 9**). It is important that cells with low aspect ratio be included in this gate, as cells which are sickled will have lower aspect ratios.
4. To define the "Cells in focus" gate, click on "New Histogram."
5. Click on the "Single Cells" population to select it.
6. Choose "Gradient RMS_M01_Ch01" as the *X* Axis Feature.
7. This will generate a new graph in the analysis area whose title bar says "Single cells."
8. Click on "Selected Bin" in the "Population" pull-down menu.
9. Click on "New Line Region" and draw a line that selects the right portion of the curve containing cells in focus, including

Fig. 3 (continued) Ch01, 80), 350–5000, 0–1) mask. Each image contains a small particle next to a single cell. The default mask includes the small particle and thus elongates the masked area, resulting in skewed feature values, such as aspect ratio and shape ratio. Use of the System (Object (M01, Ch01, Tight), Ch01, 80) mask achieved a tighter fit to the cell and disconnected the small particle mask from the cell mask. Use of the Range(System(Object(M01, Ch01, Tight), Ch01, 80), 350–5000, 0–1) mask allowed for a tight mask of the cell image, but no longer masked the small particle, thus providing a more accurate measurement of the feature values of the cell. Using these masks for generating the spot count and shape ratio features enabled better precision in analysing frames containing only one object as well as more accurate shape ratios

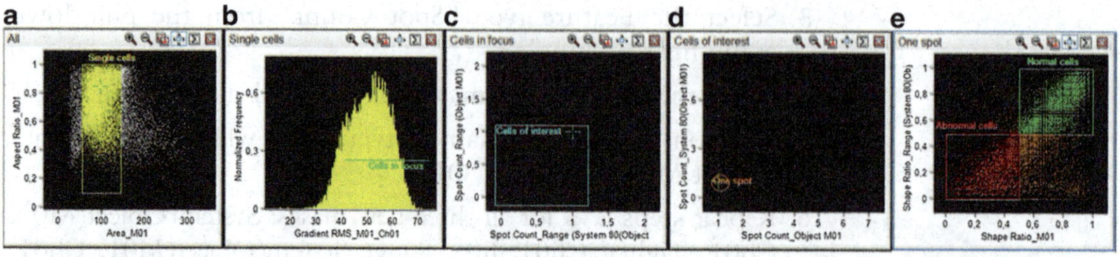

Fig. 4 Example of the graphs obtained in the SIFCA analysis. (**a**) Shows the gate defining single cells. (**b**) Displays the definition of focused images. (**c**) Shows gating out images with more than one cell per image, and (**d**) depicts the refinement of this selection, yielding the "one-spot" population. (**e**) Shows the final analysis gates, defining "normal cells" and "abnormal cells." In (**a**), (**b**), and (**e**), each *dot* represents one individual event. In (**c**) and (**d**), the *dots* represent a population of events containing various numbers of masked events. The 1 spot population was gated and used for further analysis. The use of the spot count feature with multiple masks allowed for the elimination of artifacts from the analysis, such as events having a cell with a particle next to it, thereby giving the masked event a longer shape ratio

the whole right tail. Name this population "Cells in focus." This should look like Fig. 4b.

10. Click on the bins to check the focus of the selected cells to define the cutoff value that selects focused cells. You can manually adjust the value of the *X* Coordinate through the "Regions…" dialogue box under the "Analysis" menu, or by adjusting the left edge of the region (*see* **Note 10**).
11. To define the "Cells of interest" gate, click on "New Scatterplot."
12. Click on the "Cells in focus" population to select it.
13. Choose "Spot Count_ Range (System(Object(M01, Ch01, Tight), Ch01, 80), 350–5000, 0–1)" as the *X* Axis Feature, and "Spot Count_ Range (Object(M01, Ch01, Tight), 350–5000, 0–1)" as the *Y* Axis Feature from the pull-down menus.
14. Click on "New Rectangle Region," and draw a rectangle that gates cells that have an *X* and *Y* value equal or less than 1.
15. Name this population "Cells of interest." This should look like Fig. 4c.
16. To define the "One spot" gate, click on "New Scatterplot."
17. Click on the "Cells of interest" population to select it.
18. Choose "Spot Count_ Object M01" as the *X* Axis Feature, and "Spot Count_ System (Object (M01, Ch01, Tight), Ch01, 80)" as the *Y* Axis Feature from the pull-down menus.

19. Click on "New Oval Region," and draw a circle that gates the single dot containing cells that have an *X* and *Y* value equal to 1.
20. Name this population "One spot." This should look like Fig. 4d. This is the population of cells that will be classified according to shape.

3.7 Final Cell Shape Analysis

1. Inside the Features dialogue box, click on "New."
2. Select the Feature Type "Shape Ratio."
3. Select the "Range(System(Object(M01, Ch01, Tight), Ch01, 80), 350–5000, 0–1)" mask.
4. Click on "Set Default Name" and click on "OK."
5. Repeat the same procedure for the "M01" mask, creating the "Shape Ratio_M01" feature.
6. Click on "New Scatterplot."
7. Click on the "One spot" population to select it.
8. Choose "Shape Ratio_M01" as the *X* Axis Feature, and "Shape Ratio_ Range(System(Object(M01, Ch01, Tight), Ch01, 80), 350–5000, 0–1)" as the *Y* Axis Feature from the pull-down menus.
9. Use the "New Rectangle Region" tool to create two square regions: one top-right region, named "Normal cells", and one bottom-left region, named "Abnormal cells."
10. Select "Regions…" from the "Analysis" menu.
11. Click on "Normal cells."
12. Type in new values for the *X* Coordinates (top 0.5, bottom 1) and *Y* Coordinates (top 1, bottom 0.5).
13. Click on "Abnormal cells."
14. Type in new values for the *X* Coordinates (top 0, bottom 0.5) and *Y* Coordinates (top 0.5, bottom 0) and click on "Close." The resulting plot should look like Fig. 4e.
15. Click on the "$\sum$" icon on the top right corner of this graph. This will show the statistics for this graph, with the "Count" and "%Gated" values for each population (*see* **Notes 11** and **12**).
16. The "% Gated" value will yield what percentage of the cells with an adequate image (i.e., a focused, single cell image) had their shape classified as "normal" or "abnormal". Cells outside these gates are classified as "indeterminate."
17. The SIFCA Abnormal:Normal ratio can be calculated by dividing "Abnormal cells" Count by the "Normal cells" Count (*see* **Notes 13–15**).

4 Notes

1. The glove box (Plas Labs, CAT# Z563013; *see* Fig. 1) contained a gas analyzer (ServoFlex miniMP 5200 by Servomex) to detect concentrations of oxygen, and was equipped with a standard vacuum pump (model number 5KC37NN76X, General Electric). The oxygen concentration inside the glovebox was decreased to 2 % by injecting a mixture of N_2/H_2 and aspirating oxygen with the vacuum pump. We find that determining the optimal rate of gas injection that balances the aspiration by the vacuum pump makes the deoxygenation process less cumbersome and prevents glove damage from excessive pressure. Lowering the oxygen concentration to 2 % with the vacuum pump on and gas cylinder open will result in a final oxygen concentration lower than desired, so, in our experience, reaching 2.5 % is enough to turn off gas and vacuum and wait for the system to equilibrate. Usually, before the system stabilizes, reaction between oxygen inside the glovebox and hydrogen in the gas mixture will yield water vapor and cloud the glovebox while the oxygen concentration continues to drift down.
2. This allows the glutaraldehyde to equilibrate with the hypoxic conditions.
3. Polypropylene tubes with hinged lids should be loaded well open to avoid obstruction. In our experience, many red blood cell samples will be too concentrated to be run as prepared, generating an "insufficient volume" error upon loading the sample. Adding additional 50–100 μL of PBS while mixing the sample may prevent rejection by the equipment when trying to load. Removing any bubbles with the pipette at this point is also crucial to avoid bubbles into the cytometer.
4. Since this assay does not use fluorescent antibodies, there is no need to create a new Compensation Matrix. If prompted about the channels used in the experiment, click on "Skip" until the user exits the wizard, since only bright-field images will be used (Channels 01 and 09).
5. If "View" is set to "All Channels," the user should be able to see all the 12 channel images for each cell. Select "Ch01" from the pull down menu next to "View" to see only one bright-field image of each cell. For a better view of the images with uniformly shaped fields of view: click on "Image Properties." On the "Display Properties" tab, define Display Width and Display Height (lower right corner of the window) as 100, and click on "OK."
6. IDEAS automatically generates some masks, such as M01, M02, and M03. For the SIFCA, you will need to create new

masks. You can name them with simpler names, but make sure not to get them mixed up when creating the features.

7. Although IDEAS also creates several features automatically, you will need to create new features to complete the SIFCA analysis. You may choose to name the features differently, but in this protocol, we have chosen to stick to the default names generated by the software.
8. Every time a new feature is created and the user clicks on "Close," the software will calculate the values for each cell. This will take some seconds, and can take a couple minutes depending on how many features have been created at once, and how many images have been acquired.
9. The *X* and *Y* values that define this gate may change from day to day, and from instrument to instrument depending on fluidics. Every time a region is created, the software will automatically suggest a name (e.g., R1, R2, R3). If the user clicks on "OK" by accident, the name can be changed later by accessing the same dialogue box.
10. The remaining coordinates do not need to be changed, providing the whole right tail of the curve has been selected when defining the line region. In the example shown in Fig. 2, the *X* value used was 41.
11. The user can visualize the separated populations of cells by selecting the "Abnormal" and "Normal" cell populations from the "Population" pull-down menu.
12. If either of these values are not shown by default, right-click on the plot or table to open a dialogue box for "Statistics" and make sure "Count" and "%Gated" are checked.
13. For "% Gated" values to be accurate, make sure the "One spot" population corresponds to a "%Gated" value of 100 %.
14. Alternatively, the SIFCA Abnormal:Normal ratio can be obtained by dividing "%Gated" of "Abnormal cells" by "Normal cells" values.
15. Higher SIFCA Abnormal:Normal ratios are obtained in samples containing a larger number of sickling cells, i.e., samples subjected to lower pH, and samples from patients with lower HbF content.

Acknowledgements

This work was supported by the National Heart, Lung and Blood Division of Intramural Research (1 ZIA HL006013-03 and 1 ZIA HL006149 01). The authors would like to acknowledge James Nichols, RN for blood sample procurement, and Xunde Wang, Ph.D., for assistance with the glovebox figure.

References

1. Barabino GA, Platt MO, Kaul DK (2010) Sickle cell biomechanics. Annu Rev Biomed Eng 12:345–367. doi:10.1146/annurev-bioeng-070909-105339
2. Brittenham GM, Schechter AN, Noguchi CT (1985) Hemoglobin S polymerization: primary determinant of the hemolytic and clinical severity of the sickling syndromes. Blood 65:183–189
3. Charache S, Conley CL (1964) Rate of sickling of red cells during deoxygenation of blood from persons with various sickling disorders. Blood 24:25–48
4. Bergeron MF, Cannon JG, Hall EL, Kutlar A (2004) Erythrocyte sickling during exercise and thermal stress. Clin J Sport Med 14:354–356
5. Horiuchi K, Ohata J, Hirano Y, Asakura T (1990) Morphologic studies of sickle erythrocytes by image analysis. J Lab Clin Med 115:613–620
6. Obata K, Mattiello J, Asakura K, Ohene-Frempong K, Asakura T (2006) Exposure of blood from patients with sickle cell disease to air changes the morphological, oxygen-binding, and sickling properties of sickled erythrocytes. Am J Hematol 81:26–35. doi:10.1002/ajh.20514
7. van Beers EJ, Samsel L, Mendelsohn L, Saiyed R, Fertrin KY, Brantner CA et al (2014) Imaging flow cytometry for automated detection of hypoxia-induced erythrocyte shape change in sickle cell disease. Am J Hematol 89:598–603. doi:10.1002/ajh.23699

Index

Natasha S. Barteneva and Ivan A. Vorobjev (eds.), *Imaging Flow Cytometry: Methods and Protocols*, Methods in Molecular Biology, vol. 1389, DOI 10.1007/978-1-4939-3302-0, © Springer Science+Business Media New York 2016

O

P

Q

R

S

T

U

V

W

CPSIA information can be obtained
at www.ICGtesting.com
Printed in the USA
LVOW05*0901161116
513189LV00002B/5/P

9 781493 933006